DILEMMAS
OF URBAN
ECONOMIC
DEVELOPMENT

URBAN AFFAIRS ANNUAL REVIEWS

A series of reference volumes discussing programs, policies, and current developments in all areas of concern to urban specialists.

SERIES EDITORS

David C. Perry, *State University of New York at Buffalo*
Sallie A. Marston, *University of Arizona*

The **Urban Affairs Annual Reviews** presents original theoretical, normative, and empirical work on urban issues and problems in regularly published volumes. The objective is to encourage critical thinking and effective practice by bringing together interdisciplinary perspectives on a common urban theme. Research that links theoretical, empirical, and policy approaches and that draws on comparative analyses offers the most promise for bridging disciplinary boundaries and contributing to these broad objectives. With the help of an international advisory board, the editors will invite and review proposals for **Urban Affairs Annual Reviews** volumes that incorporate these objectives. The aim is to ensure that the **Urban Affairs Annual Reviews** remains in the forefront of urban research and practice by providing thoughtful, timely analyses of cross-cutting issues for an audience of scholars, students, and practitioners working on urban concerns throughout the world.

INTERNATIONAL EDITORIAL ADVISORY BOARD

RECENT VOLUMES

34   ECONOMIC RESTRUCTURING AND POLITICAL RESPONSE
Edited by Robert A. Beauregard

35   CITIES IN A GLOBAL SOCIETY
Edited by Richard V. Knight and Gary Gappert

36   GOVERNMENT AND HOUSING: Developments in Seven Countries
Edited by Willem van Vliet–– and Jan van Weesep

38   BIG CITY POLITICS IN TRANSITION
Edited by H. V. Savitch and John Clayton Thomas

40   THE PENTAGON AND THE CITIES
Edited by Andrew Kirby

41   MOBILIZING THE COMMUNITY: Local Politics in the Era of the Global City
Edited by Robert Fisher and Joseph Kling

42   GENDER IN URBAN RESEARCH
Edited by Judith A. Garber and Robyne S. Turner

43   BUILDING THE PUBLIC CITY: The Politics, Governance, and Finance of Public Infrastructure
Edited by David C. Perry

44   NORTH AMERICAN CITIES AND THE GLOBAL ECONOMY: Challenges and Opportunities
Edited by Peter Karl Kresl and Gary Gappert

45   REGIONAL POLITICS: America in a Post-City Age
Edited by H. V. Savitch and Ronald K. Vogel

46   AFFORDABLE HOUSING AND URBAN REDEVELOPMENT IN THE UNITED STATES
Edited by Willem van Vliet ––

47   DILEMMAS OF URBAN ECONOMIC DEVELOPMENT: Issues in Theory and Practice
Edited by Richard D. Bingham and Robert Mier

In Memoriam

ROBERT MIER
1942-1995

Rob died on February 5, 1995, of non-Hodgkin's lymphoma.

DILEMMAS OF URBAN ECONOMIC DEVELOPMENT

◆

ISSUES IN THEORY AND PRACTICE

EDITED BY
RICHARD D. BINGHAM
ROBERT MIER

**URBAN
AFFAIRS
ANNUAL
REVIEWS
47**

SAGE Publications
International Educational and Professional Publisher
Thousand Oaks London New Delhi

For information address:

 SAGE Publications, Inc.
2455 Teller Road
Thousand Oaks, California 91320
E-mail: order@sagepub.com

SAGE Publications Ltd.
6 Bonhill Street
London EC2A 4PU
United Kingdom

SAGE Publications India Pvt. Ltd.
M-32 Market
Greater Kailash I
New Delhi 110 048 India

Printed in the United States of America

ISSN 0083-4688

ISBN 0-8039-5919-2 (hardcover)

ISBN 0-8039-5920-6 (paperback)

97 98 99 00 01 02 03 10 9 8 7 6 5 4 3 2 1

Acquiring Editor:	Carrie Mullen/Catherine Rossbach
Editorial Assistant:	Kathleen Derby
Production Editor:	Diana E. Axelsen
Production Assistant:	Karen Wiley
Typesetter/Designer:	Janelle LeMaster
Indexer:	Virgil Diodato
Print Buyer:	Anna Chin

Contents

Preface   xi

1. Is Local Economic Development a Zero-Sum Game?   1
   John P. Blair
   Rishi Kumar

   A Comment on Incentives   21
   Bronwen J. Turner

   One Applied Researcher's Reaction to the
   Zero-Sum Question   24
   Ron Shaffer

   The Zero-Sum-Game Controversy: A Reply   26
   John P. Blair
   Rishi Kumar

2. How Do We Know That "But for the Incentives"
   the Development Would Not Have Occurred?   28
   Joseph Persky
   Daniel Felsenstein
   Wim Wiewel

   Commentary on the Question of Incentives   46
   Valerie B. Jarrett

   Commentary on the Question of Incentives   49
   Nancey Green Leigh

   Response on Incentives   54
   Joseph Persky
   Daniel Felsenstein
   Wim Wiewel

3. How Important Is "Quality of Life" in Location
   Decisions and Local Economic Development?   56
   James A. Segedy

   Commentary on the Significance of "Quality of Life"   74
   Signe M. Rich

Commentary on "Quality of Life" in Location Decisions   76
Brian J. Cushing

Response to Commentaries   80
James A. Segedy

4. Is There Really an Infrastructure/Economic
   Development Link?   82
   David Arsen

   A Comment on Infrastructure and Economic
   Development   99
   Steven C. Deller

   Reply to Steven C. Deller   102
   David Arsen

5. What Is the Role of Public Universities in Regional
   Economic Development?   104
   Michael I. Luger
   Harvey A. Goldstein

   Commentary on the Role of Public Universities   135
   Donald F. Smith, Jr.
   Robert E. Gleeson

   Response to the Commentary   138
   Michael I. Luger
   Harvey A. Goldstein

6. Taxes and Economic Development in the American
   States: Persistent Issues and Notes for a Model   140
   Paul Brace

   Commentary on Taxes and Economic Development   162
   Judith Kossy

   Commentary on Taxes and Economic Development   165
   Brett W. Hawkins

   Reply to Judith Kossy and Brett W. Hawkins   167
   Paul Brace

7. Is Industry Targeting a Viable Economic
   Development Strategy?   171
   Kenneth Voytek
   Larry Ledebur

Commentary on Industry Targeting 195
Donald T. Iannone

Commentary on Industry Targeting 197
Sabina E. Deitrick

Response to Donald T. Iannone and Sabina E. Deitrick 201
Kenneth Voytek
Larry Ledebur

8. Can City Hall Create Private Jobs? 203
Leon Taylor

Commentary on "Can City Hall Create Private Jobs?" 219
William M. Bowen

Reply to William M. Bowen 222
Leon Taylor

9. Do Leadership Structures Matter in Local Economic
Development? 224
Laura A. Reese

Leadership and Economic Development Policy:
A Commentary 241
Robert A. Beauregard

Response to Commentary 243
Laura A. Reese

10. Can Economic Development Programs Be Evaluated? 246
Timothy J. Bartik
Richard D. Bingham

Commentary on "Can Economic Development
Programs Be Evaluated?" 278
Robert Giloth

Commentary: Evaluation Yes, but on Whose Behalf? 284
David Fasenfest

Rejoinder 288
Timothy J. Bartik
Richard D. Bingham

11. Is There a Point in the Cycle of Cities at Which
Economic Development Is No Longer a Viable Strategy?
Or, When Is the Neighborhood Too Far Gone? 291
William W. Goldsmith

Commentary on the Cycle of Cities 320
Peter B. Meyer

Commentary on the Cycle of Cities 324
Carol Waldrop

Reply to Peter Meyer and Carol Waldrop 328
William W. Goldsmith

Index 331

About the Editors 337

About the Contributors 338

Preface

Both the study and the practice of local economic development have come a long way in the past decade. There are some 15,000 economic development organizations in the United States alone, employing between 25,000 and 50,000 practitioners (Boyle, 1990). They are highly educated individuals, more than 90% holding college degrees. A whopping 47% hold advanced degrees, at least in the Midwest (Visser & Wright, 1996).

The economic development practice has two professional associations: the American Economic Development Council and the National Council for Urban Economic Development. Each association has its own serial publication: *Economic Development Review* and *Economic Development Commentary*.

The emphasis on graduate education in economic development has increased. About 27% of midwestern universities offer course work in economic development, about 11% having a formal concentration in the subject at the master's level. Most of these concentrations have emerged in the past decade (Visser & Wright, 1996).

The journal *Economic Development Quarterly* was founded 10 years ago and was "designed to bridge the gap between practitioners, academics and informed citizens in the field of economic development." Its success in uniting practitioners and scholars is undoubtedly due to the high level of educational achievement of the practitioners.

At the same time, the number of books on economic development has increased dramatically. Not long ago, the books on economic development could fit between the bookends on one's desk; today, this is no longer the case. The ED literature today is substantial—and, in general, quite good. Many of the articles in the *Economic Development Quarterly* have had a significant effect on economic development policy and practice (see, e.g., Blair & Premus, 1987; Harrison, 1994; Ledebur & Woodward, 1990; Marlin, 1990; Rubin, 1990; White & Osterman, 1991).

Robert Mier and Joan Fitzgerald (1991) have taken the development of the field one step further, declaring that it is becoming established as a new academic discipline (pp. 268-269). Richard Bingham and Mier (1993) found that more than 50 theories from the various social sciences have been applied to local economic development. Theories, of course, are among the hallmarks of an academic discipline.

All of which is by way of saying that economic development in both practice and study has come into its own. From these roots, *Dilemmas of Urban Economic Development* was born. Given the rapidity with which the field has been growing recently, we thought it would be useful to pause and assess the state of knowledge concerning practical local economic development issues. To do this, we thought, a focus on the field's important questions would be appropriate.

But questions important to whom? We wanted the issues of concern to be important theoretically and practically. We therefore suggested chapter titles to the contributors in the form of questions, such as "Is Industry Targeting a Viable Economic Development Strategy?" and "Can Economic Development Programs Be Evaluated?" We then asked our contributing authors, selected because of known expertise in the relevant area, to assess the state of knowledge regarding the question. We expected that some questions would have clear answers and that the evidence would be so strong as to be unequivocal, as with the question "Can Economic Development Programs Be Evaluated?" Timothy Bartik and Richard Bingham respond, "Of course." They then go on to illustrate how economic development programs "should" be evaluated.

Other questions, however, do not have such clear-cut answers. For example, research findings may be mixed and the evidence on the question noncumulative. In other cases, the questions may have hardly been researched; the problems may be so complex that up to now they have defied answer. In such cases, in which the cumulative research results do not definitively answer the question, we have asked the contributing authors to suggest a methodology that, if implemented, would be likely to answer the question.

One of the reasons there seems to be a closer link in the economic development profession between practice and research is the high educational level of practitioners. They are clearly capable of reading and evaluating academic work and producing high-quality theoretical and empirical work on their own. Therefore, such a book about practice requires practitioner input. Yet practitioner input to economic development books is as difficult to achieve as it is in other areas, and for essentially the same reason: Economic development practitioners are extremely busy people, and writing a book chapter is not among their major functions. So we sought practitioner input by asking for a commentary on each of the chapters; in most cases, we were successful.

We also obtained an academic commentary on each chapter. Our major purpose here was to provide a methodological critique of the material presented.

As we stated before, the goal of this volume is to assess the state of knowledge about major questions in economic development. In this sense, the book is about practice. But it is about theory, too, as most of the chapters are grounded in one or more theories of development.

In the first chapter, John Blair and Rishi Kumar answer the question "Is local economic development a zero-sum game?" by explicitly relying on economic development theory. They use a game-theoretic approach to investigate the question. The critical variables in their model are externalities or spillover benefits, the subsidy or economic development incentive, and profits to the firm. Their models show that local economic development efforts can contribute to the growth of the economy by allowing marginally profitable firms to operate and by tilting locational choices to geographic areas of high spillover benefits. On the other hand, zero- or negative-sum outcomes are also possible: for example, through oversubsidization.

Chapter 2, by Joseph Persky, Daniel Felsenstein, and Wim Wiewel, is closely related to the first chapter in that it too is concerned with incentives. The authors initially critique the issues raised by many of the studies investigating the impacts of incentives. They then report on a new methodology for evaluating the impact of incentives developed at the University of Illinois at Chicago. The model is derived from cost-benefit theory and shows great promise for answering the question "How do we know that 'but for the incentives' the development would not have occurred?"

In Chapter 3, James Segedy draws on traditional location theory (see Blair & Premus, 1993) to assess the impact of "quality of life" on firms' location decisions. Segedy proposes a new economic development paradigm composed of subjective as well as objective parameters.

In Chapter 4, David Arsen addresses the question of the link between infrastructure and economic development. Econometric modeling has provided the basic approach to the question taken by most researchers. Arsen answers the linkage question affirmatively. He then goes on to note that the linkages (and there are several) between infrastructure and development "are complicated and subtle, and many are poorly understood." His chapter does much to improve this understanding.

Chapter 5 investigates the linkages between universities and economic development. Michael Luger and Harvey Goldstein develop and present a conceptual model of the university as a multiproduct organization. This characterization departs from the traditional literature, which measures the regional consequences of higher education institutions mostly in terms of their spending multipliers. Luger and Goldstein conclude that there is a legitimate basis for universities to get involved in economic development activities, but only if the university and its structure, and the region and its economy, are amenable.

In Chapter 6, Paul Brace examines the sticky relationship between taxes and economic development. Utilizing a political economy approach, he searches for the answer to the question "Do taxes have a negative impact on economic development efforts?" As the commentary by Judith Kossy notes, Brace's "model reflects the real complexity of the relationships between taxes, expenditures, and development and the importance of exogenous variables and time lag factors." Brace ultimately concludes that political fears rather than real economic consequences really define the tax/economic development relationship.

Is industry targeting a worthwhile economic development strategy? There is certainly some evidence that it is not (e.g., Wolman & Voytek, 1990). In Chapter 7, Ken Voytek and Larry Ledebur draw on the industrial location theory literature in making the case for industry targeting (Blair & Premus, 1993). They conclude that although it is based on theory, it is more art than science. "Effective industry targeting requires expertise, talent, experience, and knowledge. It is not an endeavor for the uninitiated or the innocent. As an underdeveloped art, it is not a mechanical process. It requires a high degree of intuition."

Leon Taylor examines efforts by city halls to create private-sector jobs in Chapter 8. The chapter is grounded in theories of high-tech development (e.g., Goldstein & Luger, 1993; Markusen, Hall, & Glasmeier, 1986), entrepreneurship (Bates & Dunham, 1993), and human capital (Fitzgerald, 1993). Taylor's answer to the question "Can city hall create private jobs?" is "Sometimes. Possibly always."

In Chapter 9, Laura Reese tackles the question "Do leadership structures matter in local economic development?" She draws on theories of urban politics in her investigation. These include theories of community power, regime, formal structure, and bureaucracy. Reese concludes that governmental structure, administrative arrangements, and leadership do make a difference.

The simple answer to the question posed in Chapter 10—"Can economic development programs be evaluated?"—is "Of course." But as Timothy Bartik and Richard Bingham point out, these programs are not being evaluated. Bartik and Bingham draw on theories of evaluation to illustrate that evaluation in the field of economic development is in its infancy. They then suggest appropriate methodologies for evaluating economic development programs.

In the final chapter, William Goldsmith examines the question of triage. Are places ever "too far gone" to bother with economic development? A Detroit planner recently created a furor by suggesting that parts of Detroit be leveled and turned into farmland. On the other hand, there are hopeful signs of recovery in the South Bronx. Goldsmith argues that impoverished places cannot be abandoned. But his essay is not optimistic.

Rob Mier passed away in February 1995 after a long battle with non-Hodgkin's lymphoma. This book was important to him, and throughout his terrible battle with cancer he edited and commented on most of the chapters. The authors welcomed his input, and the chapters are stronger for them. We can also thank Rob for the unusually insightful commentaries. Rob knew most of the practitioners from his days in Chicago's city hall and secured their contributions.

In one sense, the book is incomplete in that it does not contain the traditional introductory and concluding chapters by the editors. On the other hand, it is unlikely that our introduction would be more appropriate than John Blair and Rishi Kumar's "Is Local Economic Development a Zero-Sum Game?" Nor is there a more fitting conclusion than William Goldsmith's "When Is the Neighborhood Too Far Gone?"

REFERENCES

Bates, T., & Dunham, C. (1993). Asian American success in self-employment. *Economic Development Quarterly, 7,* 199-214.

Bingham, R. D., & Mier, R. (Eds.). (1993). *Theories of local economic development: Perspectives from across the disciplines.* Newbury Park, CA: Sage.

Blair, J. P., & Premus, R. (1987). Major features in industrial location: A review. *Economic Development Quarterly, 1,* 72-85.

Blair, J. P., & Premus, R. (1993). Location theory. In R. D. Bingham & R. Mier (Eds.), *Theories of local economic development: Perspectives from across the disciplines* (pp. 3-26). Newbury Park, CA: Sage.

Boyle, M. R. (1990). An economic development education agenda for the 1990s. *Economic Development Quarterly, 4,* 92-100.

Fitzgerald, J. (1993). Labor force, education, and work. In R. D. Bingham & R. Mier (Eds.), *Theories of local economic development: Perspectives from across the disciplines* (pp. 125-146). Newbury Park, CA: Sage.

Goldstein, H. A., & Luger, M. I. (1993). Theory and practice in high-tech economic development. In R. D. Bingham & R. Mier (Eds.), *Theories of local economic development: Perspectives from across the disciplines* (pp. 147-171). Newbury Park, CA: Sage.

Harrison, B. (1994). The myth of small firms as the predominant job generators. *Economic Development Quarterly, 8,* 3-18.

Ledebur, L. C., & Woodward, D. (1990). Adding a stick to the carrot: Location incentives with clawbacks, recision, and recalibrations. *Economic Development Quarterly, 4,* 221-237.

Markusen, A., Hall, P., & Glasmeier, A. (1986). *High-tech America.* Boston: Allen & Unwin.

Marlin, M. R. (1990). The effectiveness of economic development subsidies. *Economic Development Quarterly, 4,* 15-22.

Mier, R., & Fitzgerald, J. (1991). Managing economic development. *Economic Development Quarterly, 5,* 268-279.

Rubin, H. J. (1990). Working in a turbulent environment: Perspectives of economic development practitioners. *Economic Development Quarterly, 4,* 113-127.

Visser, J. A., & Wright, B. E. (1996). Professional education in economic development: Conflicting expectations for college programs in the Great Lakes region. *Economic Development Quarterly, 10,* 3-20.

White, S. B., & Osterman, J. D. (1991). Is employment growth really coming from small establishments? *Economic Development Quarterly, 5,* 241-257.

Wolman, H., & Voytek, K. P. (1990). State government as consultant for local economic development. *Economic Development Quarterly, 4,* 211-220.

1 Is Local Economic Development a Zero-Sum Game?

JOHN P. BLAIR
RISHI KUMAR

Some observers are concerned that local economic development efforts are a zero-sum game (James, 1984; U.S. Department of Housing and Urban Development, 1988, p. II-2). The argument goes like this: If the size of the national economy is determined by factors beyond the control of local development officials, then one jurisdiction's economy can expand only at the expense of other places.

This chapter shows that local economic development programs can be a positive-sum game in three ways. First, economic development programs can stimulate net new economic activity such as job creation. Second, even if the level of economic development activity were fixed, local economic development programs could affect locational choice so as to increase the benefits from a given leveled economic activity. Spatial redistribution of economic activities can be economically efficient and not necessarily a zero-sum game. Finally, local economic development programs can enhance the business environment by undertaking projects for which the costs of the improvements are less than the benefits to businesses.

Although positive-sum outcomes are possible, so are zero- and negative-sum outcomes. In negative-sum games, resources are wasted or used in inefficient ways. Local economic development efforts may also result in oversubsidization, a zero-sum transfer.

A game-theoretic approach is developed in the first part of this chapter. Key variables affecting the outcomes of local economic development efforts are identified, and the possibilities of various outcomes are explained. Then the model is extended to examine real-world circumstances that could result in zero- or negative-sum outcomes, including oversubsidization. Policy suggestions to reduce undesirable outcomes are also discussed.

■ A Game-Theoretic Perspective

Game-theoretic models typically reduce complex situations to a few elements in order to focus on key relationships. In our local development model, there are three critical variables: (a) externalities, or spillover benefits to local residents; (b) the inducement, subsidy, or economic development incentive; and (c) profits to the firm. By examining the relationships among these three variables, we demonstrate that local economic development efforts may have positive outcomes for individual communities and for society as a whole. However, circumstances can also lead to zero- or negative-sum outcomes and oversubsidization.

Elements of the Game

The external benefits that local residents receive from new economic activity are the primary incentives for local economic development programs. When economic growth occurs in an area, many individuals receive benefits for which they do not pay. Gains to workers due to increased employment opportunities because an establishment located in an area would be an example of an externality or spillover. To the extent that employees are paid above their reservation wage, they receive external benefits.[1] The fact that underemployed or unemployed workers may be willing to pay some amount of money to secure jobs locally indicates the presence of externalities.

Although local workers are usually the primary beneficiaries of external benefits from economic development, property owners may also realize net gains through higher property values, higher rents, or lower vacancy rates. Others may benefit from lower taxes, agglomeration economies, or other factors.

The second element of the local development model is the "inducement." Local jurisdictions offer economic development inducements (incentives, subsidies, or bribes) in an almost endless variety of forms, such as land write-downs, infrastructure improvements, tax abatements, labor force training, and low-interest loans. The inducement need not be a direct cash offer. Our model allows for indirect inducements such as a clean environment or a good-quality educational system, as well as direct cash transfers.[2]

The third element is expected profits to the firms. Our model focuses on business development, although it may be modified to apply to non-profit organizations. The model assumes that a firm (a) will locate in the most profitable jurisdiction and (b) will not operate unless it expects to earn positive profits. The net social benefit from new economic activity is the sum of external benefits received by local residents due to business expansion and profits to the firm.

Any positive profit level will be sufficient incentive for a firm to operate. Profits are defined as what is left after all costs, including compensation for risk, entrepreneurship, capital, and other opportunity costs, have been deducted from revenues (Mansfield, 1991 p. 277). Alternatively, a positive profit may represent payment above the best alternative use of the firm's resources. In our model, expected profits vary for two reasons: (a) the location and (b) the economic development incentive.

The game-theoretic model can be used to illustrate three typical outcomes of economic development efforts: (a) positive-sum outcomes, (b) zero- and negative-sum outcomes, and (c) oversubsidization. In other words, outcomes other than positive sum are possible, depending on the circumstances and the behavior of local economic development officials and the value of the three variables identified in the model.

Positive-Sum Games

Positive-sum games are the most desirable outcomes of local economic development efforts. Positive-sum outcomes occur when local economic development efforts bring about more benefits to society as a whole than would have been the case if economic development programs were absent. Positive-sum outcomes can occur in three ways: (a) An otherwise unprofitable firm is made profitable and therefore provides external benefits in excess of the cost of the inducement, (b) a firm is encouraged to relocate to an area where the citizens value (need) the jobs and other benefits of economic development more than the residents of competitor places, and (c) the economic development incentive provides more benefits to firms than costs to taxpayers.

Case 1P: Otherwise Unprofitable Firm Provides
Sufficient External Benefits

Figure 1.1 illustrates a positive-sum game in which the external benefits could not be attained without an economic development inducement. Two types of payoffs are shown in Figure 1.1 (and the other figures). Payoffs without parentheses are simply potential outcomes *if* the firm were to operate in the city. The payoffs in parentheses indicate the actual payoffs after economic development inducements have influenced the outcome. The inducement flow is depicted by the arrows.

Assume that City A is the only potential location. Before any local economic development efforts, the firm could operate only at a $1 loss. The lack of profitability would prevent the business start-up. The net social benefits from operations at A, $29 ($–1 profit for the firm and $30 for the community), would be lost. (Profits to the firm or stockholders count as social benefits because the firm's owners are part of society.)

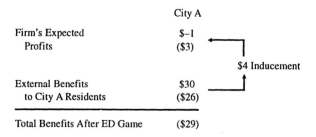

Payoffs in parentheses indicate payoffs after subsidy.
Payoffs not in parentheses are potential payoffs without subsidy.

Figure 1.1. Economic Development Program Stimulates New Activity (Case 1P)

Economic development programs could bring about the socially optimal outcome. Suppose that an inducement of $4 is offered to the firm. If the inducement had to be financed by local residents, their *net* benefits (external benefits less the cost of the subsidy) from the operation at A would fall to $26. The incentive would cause the firm's profits to increase to $3 if it located in A. Hence, due to economic development efforts, the firm would operate, and the external benefits would be captured.[3]

*Case 2P: Location Changed to
High-External-Benefit Area*

The case in which local economic development efforts tilt the locational choice to a community providing greater external benefits is shown in Figure 1.2. Assume that the firm will operate only in one city, so that if the firm locates in A, the profits associated with B are irrelevant and vice versa. (The model could easily be modified to include numerous cities.) Also, the external benefits to residents of the location not selected will be zero.

In the absence of economic development programs, the firm would locate in B. Net social benefits of $2 would result. Unfortunately, the outcome that would result if economic development programs were absent is suboptimal. Total benefits would be greater if City A were the site because of the greater externalities. In this case, net benefits would be $29. If Cities A and B were competing with each other in an attempt to attract the company, officials in A might offer an inducement of $4. Consequently, A would capture the establishment, which otherwise would have located in B. The economic development competition would result

Possible Locations

	City A	City B
Firm's Expected Profits	$-1 ($3)	$2
		$4 Inducement
External Benefits to City A Residents	$30 ($26)	$0
External Benefits to City B Residents	$0	$0
Total Benefits After ED Game	($29)	NA

Payoffs in parentheses indicate payoffs after subsidy.
Payoffs not in parentheses are potential payoffs without subsidy.

Figure 1.2. Locational Choice Altered to More Efficient Location
(Case 2P)

in a location that provided greater social benefits—$29 rather than $2. Hence a positive-sum outcome would occur. Tilting the location of a firm from one city to another is not necessarily a zero-sum game because the location can affect net social benefits from a given level of development activity.[4]

Case 3P: Inducements Have Higher Value Than Cost

A positive-sum outcome can also occur when the local development incentive has greater value to businesses than costs to the community. Suppose that the initial situation is as shown in Figure 1.3. The firm could not initially operate in A profitably. However, suppose that City A initiates a job training program at a cost of $4 and that the program provides $8 of benefits to the firm. The cost/value gap is indicated by the interruption in the transfer flow. Net social benefits would increase to $33. The greater the gap between the community's costs and benefits to the business organizations, the greater the likelihood that a previously unfeasible activity would become profitable.

Many local development efforts can increase benefits relative to costs. Because of the quasi-public-goods characteristics of many economic development projects, a single economic development program may benefit more than one business simultaneously. For example, suppose that a

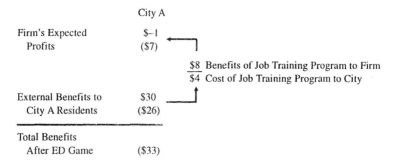

Payoffs in parentheses indicate payoffs after subsidy.
Payoffs not in parentheses are potential payoffs without subsidy.

Figure 1.3. Value Exceeds Cost of Subsidy (Case 3P)

community concentrates its economic development resources on improving the transportation system. This project might serve as an inducement to many businesses. Inducements that have wide impacts often have a greater development bang and will contribute more to national growth than incentive packages that narrowly focus on a single business.

Negative-Sum Games

A negative-sum game occurs when local economic development activities result in lower net social benefits than would have occurred if such activities were absent. There are three principal ways that economic development programs can result in negative-sum outcomes: (a) Inducements could be large enough to support a firm that should not have been subsidized, (b) competitive bidding among communities could result in a suboptimal location, and (c) cost of the economic development incentive could be greater than the value to the firm. These cases are opposite the positive-sum outcomes.

Case 1N: Support for a Nonviable Firm

A nonviable firm is defined as one that could not operate profitably in any city without a subsidy *and* for which the external benefits would be too small to justify a subsidy sufficient to make it profitable. The firm shown in Figure 1.4 illustrates such a business. In the absence of local economic development programs, the firm would not operate because profits could not be made. Suppose, for reasons we will discuss later, that local economic development officials offer a "bribe" sufficient to make the operation profitable for the firm. For instance, an inducement of $20 might be offered, as shown in Figure 1.4. In this case, the firm would operate, but at a cost so high as to outweigh the value of the external benefits. A negative-sum game would result. The residents of City A would be worse off due to the economic development activity ($–17 rather than $0), as would be the nation as a whole ($–13 rather than $0).

Case 2N: A Suboptimal Location

Consider next the case in which a firm is induced to locate in a suboptimal location, as illustrated in Figure 1.5. In the absence of any economic development program, the firm would locate in City A, where both profits and external benefits were greatest. Consequently, City A would be the socially optimal location. However, suppose that economic development officials in City B develop an incentive package of $10 to induce the firm to locate in B. Due to ineptitude or some other reason, economic development officials in location A fail to provide any incentive. Consequently, the firm's profits from the location at B would increase to $20 after accounting for the subsidy, and the firm would locate there. Residents of B would receive only $5 of net external benefits, but they would have received nothing if the firm had located in A. The potential payoff would decline from $41 if the firm had located in A to $25 at B; the outcome would be inefficient. However, the economic development officials in City B would have enhanced the welfare of their residents at the expense of residents of City A. A lesson from Case 2N is that one community may gain at the expense of other places even in zero- or negative-sum game situations. Therefore local economic development officials may at times find it appropriate to engage in zero- or negative-sum competitions.

Case 3N: Costly but Low-Valued Inducements

In addition to administrative costs, many of the traditional incentives offered by localities are more costly to local residents than the value to the firm (Rasmussen, Bendick, & Ledebur, 1984). Suppose that a community provides a new building for a firm, costing the city $1,000,000. However, the firm can purchase a comparable building at a cost of $800,000. The value of the building to the company, and therefore the impact on profits, would be $800,000, whereas the cost would be $1,000,000. Rasmussen and his associates pointed out that in addition to the negative-sum resource waste, there may be unintended beneficiaries of economic development efforts that drive a wedge between community cost and value to a business.[5]

		Possible Locations	
		City A	City B
Firm's Expected	$-16		
Profits		($4) ◄──┐	$-3
		│	$20 Inducement
External Benefits to		$3 ─────┘↑	
City A Residents		($-17)	$0
External Benefits to		$0	
City B residents			$1
Total Benefits		($-13)	NA
After ED Game			

Payoffs in parentheses indicate payoffs after subsidy.
Payoffs not in parentheses are potential payoffs without subsidy.

Figure 1.4. Subsidy Encourages a Nonviable Firm (Case 1N)

A negative-sum game is illustrated in Figure 1.6. City A is the only feasible development location, but the firm may still attempt to secure a subsidy. Suppose that city officials offer development incentives costing $8 to encourage the firm to locate in A but that the company values the incentives at only $3 (thus the interruption in the transfer flow shown in Figure 1.6). As a consequence, net social benefits fall to $26. Net social benefits would have been $31 in the absence of the subsidy. Benefits to residents of City A would fall from $30 in the absence of the subsidy to $22. Profits to the firm would rise to $4.

Oversubsidization and Zero-Sum Games

Oversubsidization is a transfer of wealth from residents to private firms that accomplishes nothing. Oversubsidization occurs when the value of the economic development bribe is greater than the amount necessary to encourage a given level of economic development. Oversubsidization can occur within the context of positive-, zero-, and negative-sum games. However, oversubsidization by itself is a zero-sum game.

Exactly when oversubsidization occurs is a matter of opinion. For instance, the subsidy shown in Figure 1.1 is slightly more than necessary to attract the firm. One could argue that a portion of the subsidy constitutes

Possible Locations

	City A	City B	
Firm's Expected Profits	$11	$10 ($20)	
External Benefits to City A Residents	$30	$0	$10 Inducement
External Benefits to City B Residents	$0	$15 ($5)	
Total Benefits After ED Game	NA	($25)	

Payoffs in parentheses indicate payoffs after subsidy.
Payoffs not in parentheses are potential payoffs without subsidy.

Figure 1.5. Suboptimal Location (Case 2N)

oversubsidization and that this portion is a zero-sum transfer. In practice, it is impossible to gauge exactly the minimum necessary subsidies.

Oversubsidization may harm local residents. For instance, it is possible to provide such a costly economic development incentive that all the external benefits to community residents are captured by the firm (we might coin the term *superoversubsidization* to apply to this case). Suppose economic officials in City A offer $35 to the firm depicted in Figure 1.5. City A residents would end the round of negotiations with negative benefits of $–5, whereas the firm's profits would rise to $46. Notice that the combined benefit would be $41 if the firm located in A with or without the subsidy. The superoversubsidization would harm residents in the subsidizing locality.

Policy Considerations From the Model

The outcomes described above may occur in combinations. For instance, an externality-producing firm may be made profitable (a positive-sum effect) with the use of a subsidy that costs more than the benefits are valued (a negative-sum effect). Whether the outcome is positive or negative depends on which effect is greatest. Economic development officials, therefore, should attempt to minimize negative-sum effects in policy design and implementation. Each element in the game-theoretic model can be important in efforts to minimize negative-sum effects.

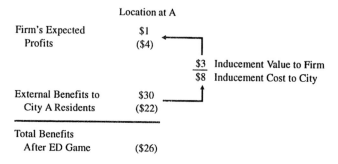

Payoffs in parentheses indicate payoffs after subsidy.
Payoffs not in parentheses are potential payoffs without subsidy.

Figure 1.6.   Subsidy Cost Exceeds Benefits (Case 3N)

A knowledge of expected profits and the factors that influence expected profits can be useful in many ways. Officials should target activities that require modest subsidies to become profitable, while avoiding the provision of unnecessary subsidies. An appreciation of the amount of subsidy necessary to attract a firm will minimize oversubsidization. Economic development practitioners may also find it useful to estimate business profitability in competitor cities to calibrate their incentive offers better during rounds of intergovernmental competition. An understanding of how various types of assistance contribute to firm profits will also help prevent situations in which the subsidy costs exceed business value. For instance, it would make little sense to offer a firm a job training program if 90% of the employees were PhD chemists.

Local officials should also appreciate the magnitude of the externalities attributable to a firm. The cost of the inducement should be less than the external benefits received by community residents. Otherwise, residents will experience a net loss from economic development efforts. Even when the externalities are only roughly estimated, the information can inform officials about the relative importance of various businesses to the community. Such knowledge will help them to avoid "shoot anything that flies" strategies (Rubin, 1988). The wide variations in the subsidies per job created suggest that few communities currently calibrate the cost of subsidies to external benefits (Blair, 1991, p. 247).

When developing incentive packages, officials should avoid offering costly subsidies that have little value to businesses they seek to influence. An improved knowledge of the types of incentives, particularly business value (what the impact of profits will be), can help achieve this end. One approach is to emphasize programs that will benefit a wide variety of existing and potential businesses simultaneously. Such programs might target a particular cluster of businesses, such as advanced-technology firms, if the community was attempting to create agglomeration economies in that sector. Administrative overhead of economic development activity tends to increase costs while providing no benefits to business.

Practicality

The local development model includes three key variables: (a) expected profits, (b) external benefits, and (c) the subsidy cost/value. Even when the key variables are not measurable, the model can be a useful guide to both long- and short-term planning. But the variables in the model can be measured, albeit imprecisely.

Expected profits can be estimated by examining financial data and projections. Businesses routinely provide such information to financial institutions in support of loan requests, and the data are part of the budgeting process. It would be reasonable to allow city officials to examine the information when a firm requests development assistance.

The principal external benefits are new jobs for residents and tax benefits. Established techniques for valuing jobs indicate that the value of a job depends on (a) the likelihood that local residents will get the job, (b) whether the resident is employed or unemployed, and (c) multiplier impacts. Michigan officials carefully evaluated the value of jobs created by the Mazda facility and calibrated their incentive package accordingly (Bachelor, 1991). Baade and Dye (1988) pointed out that different kinds of jobs should be valued differently. The value of a job to a local resident is normally considered to be the difference between the after-tax wage and the individual's opportunity cost. The opportunity cost is usually zero for an unemployed person.

Techniques for estimating fiscal impacts are also established (Burchell & Listokin, 1978) and have been applied to local economic development projects (Rosentraub & Swindell, 1991). Other benefits, such as agglomeration economies and enhanced development prospects, are more difficult to quantify and may best be treated as intangibles. Cost-benefit studies have developed techniques for valuing most economic development outcomes.

The cost of subsidies to the governments is relatively easy to measure. However, the value of in-kind incentives is likely to be more difficult to measure, particularly if the benefits are diffused throughout the community. Nevertheless, the cost of the subsidy should set a lower bound on the business value.

■ Circumstances, Strategies, and Development Policies

This section examines real-world situations that are likely to result in zero- or negative-sum outcomes and oversubsidization. Policies and strategies that may help economic development officials secure positive-sum outcomes will be discussed.

The Laissez-Faire Argument

The laissez-faire argument is that in a well-functioning competitive economy, there would be no need for governmental economic development programs. The external benefits, which are the main motive for economic development programs in our model, would be insignificant or nonexistent in the laissez-faire worldview. Market mechanisms, especially private contracting, would operate quickly to eliminate unemployment and other idle resources.

To understand the logic of the free-market argument, assume that labor is the only resource that can benefit from the spillover effects of economic development. In keeping with the laissez-faire view, also assume that resources are perfectly mobile, wages are flexible, and information is costless. Given these admittedly unrealistic assumptions, unemployment would not be a serious problem. Local unemployment would cause wages to fall to the level at which the unemployed workers were indifferent to the choice between work and leisure (i.e., staying unemployed). Therefore, newly employed workers would not receive positive externalities from newly created jobs (the wage would just equal the opportunity cost of working). The low labor costs would quickly attract labor-intensive firms to the area, reducing unemployment. Also, some workers would leave the area to seek jobs elsewhere, further reducing the unemployment problem and the externalities from economic development. In the absence of positive externalities, there would be no point in providing economic development incentives. Similarly, the compensation of other unemployed resources would fall to reflect a level of compensation equal to other opportunities.

The conditions necessary for laissez-faire strategy to dominate policy decisions are not satisfied in reality. Labor immobility and wage rigidities contribute to unemployment, and the wage paid by most jobs exceeds the reservation wage of the unemployed (Yellen, 1984). Consequently, economic development activity usually generates positive spillover benefits.[6]

Although the conditions of the laissez-faire argument are seldom satisfied, two important policy implications flow from the perspective. First, because different types of businesses will ameliorate different externalities, an analysis of the types of market failures that exist in a community may help local officials decide which types of businesses they should encourage. For example, a city with an unemployment problem may target a different type of firm from that targeted by a city concerned about low wages or an inadequate tax base. Second, direct efforts to correct market failures can provide economic development benefits. For instance, federal labor relocation and welfare portability programs will reduce labor immobility, a market failure. Third, if there is little evidence of market failures in a jurisdiction, the reason to offer large economic development subsidies should be questioned. Further, when low-paying jobs are being provided, local officials should recognize that the external benefits for workers will be less than if high-paying jobs were being considered. The incentive should be adjusted accordingly.[7]

Pitfalls and Strategies in the Real World

Given the many market imperfections in the U.S. economy, there are likely to be numerous opportunities to create positive-sum economic development outcomes. There are also several situations in which zero- or negative-sum games and oversubsidization are likely. This section explores possible pitfalls and strategic principles that may assist economic development planners in avoiding negative-sum outcomes and oversubsidization. However, there is a cautionary note. Our discussion illustrates important pitfalls and potential strategies but is not exhaustive. Further, there are no "cookbook" techniques regarding when to engage in interjurisdictional competition, which businesses to seek, or what incentive packages to offer.

Direct Competition

Some companies deliberately place several communities in direct competition by announcing that a facility will be located in only one of several places. The companies then start negotiating with the competitor cities for economic development incentives. Such situations can easily result in zero- or negative-sum outcomes, particularly when the external benefits in the competing cities are similar. Because only one city will be a "winner," the other competitors will have spent resources to no avail.[8] An interesting twist on the practice of pitting one city against another occurs in the case of competitive plant shutdowns. When International Harvester announced it would close either a plant in Indiana or a plant in Ohio, they were able to extract substantial subsidy offers from both areas before deciding which plant to close. General Motors has employed a similar strategy in its consolidation efforts.

One policy issue is whether cities or states that are placed in direct competition should compete against one another. On the one hand, the cities are forced to compete because if one side refused, that side would lose. On the other hand, the combined welfare of the residents in all competing communities would be better served if the two governments cooperated rather than engaging in a bidding war over a zero-sum outcome. The communities could agree to limit the size of the "bribe." Thus, the transfer of wealth from citizens to the company would be reduced.

Higher-level governments also may help reduce oversubsidization through intergovernmental competition by restricting the use of intergovernmental aid. Frequently, two communities will employ state or federal grants to supplement their economic development incentives. For instance, two communities might both use Community Development Block Grant money to supplement competing incentive packages in head-to-head competition. If the funds were not available to either party for this use, the outcome would probably be the same, except that the transfer of wealth from government to the company would be less.

Intrametropolitan Competition
for Local Customers

Intrametropolitan competition among firms in a local market often is a zero-sum game because the market size is fixed. For instance, assume that the market for department store merchandise is determined by the size of a metropolitan region. If a community subsidized the establishment of a new department store or the construction of additional retail space, the likely consequence would be that the new activity would grow only by taking sales from existing stores. Thus no new external benefits would accrue to the community in return for the subsidy. Similarly, economic development agencies have built or subsidized the construction of office space. Often these new buildings simply attract tenants from other existing buildings without increasing the number of permanent jobs. Thus officials should avoid frequent subsidization of firms that compete in the local market.

Why would economic development officials subsidize companies that compete with existing local business for the same markets? One possibility is that development officials receive more credit when new businesses are formed than when existing businesses remain stable. This explanation suggests that policy should recognize business retention of local service firms as a priority equal to expansion of such firms. Another explanation is that intrametropolitan competition over tax revenues may encourage suburbs to compete with one another. Tax sharing and other forms of intergovernmental competition may be appropriate strategies in this case (Rusk, 1993).

Political Pressures and Insecurities

Insecure or ambitious politicians may pressure local officials to provide subsidies in excess of external benefits and in excess of a firm's minimum requirements. If the oversubsidization does not affect the firm's location, the transfer will simply be a zero-sum game. If a firm moves from its most socially desirable location, a negative-sum outcome will result.

Political officials are often judged by their ability to attract high-profile economic activities. The benefits of such developments accrue in the very short term, whereas the costs are embedded in the form of higher taxes that are spread over many future years (Blair, 1991, p. 524). Many politicians have short-term time horizons; in the extreme, they may look ahead only to their next election. Consequently, a politician may feel the need to provide excessive development incentives to attract economic development activities near election time or if there is a major political need for a success story. Political pressures that could lead to oversubsidization are particularly potent when a plant shutdown is being discussed because the existing workforce is a powerful lobby for doing whatever is necessary to maintain the facility.

Jones and Bachelor (1984) reviewed efforts by Detroit officials to capture the "Poletown" automobile plant, which would replace two older facilities in the Detroit area that were being closed. The authors concluded that the incentives offered to General Motors were excessive. The oversubsidization resulted in a "corporate surplus." The excessive subsidies were the result of General Motor's powerful bargaining position, the weak political position of Mayor Coleman Young, and generous supplemental aid provided by a sympathetic White House. In essence, the mayor needed an "accomplishment" to help ensure a reelection victory. Governmental officials pulled out all the stops in attempts to attract the facility at Poletown.

It is impractical to devise a set of rules that can completely insulate the local economic development process from politics. Civil service structures and citizen review boards can provide some protection. Another solution would be to increase the awareness of this potential problem. Even perceptions that the economic development process was being abused would have some (maybe small) deterrent impact.

Information Asymmetries and Poorly Defined Outcomes

When the circumstances surrounding a transaction are known by one party but not by others, information asymmetry occurs. Although local officials should try to reduce the information problem and clarify the benefits that a new economic activity will bring to a community, they face several significant problems. Private companies tend to withhold critical information from public officials. Therefore, city officials are much less certain regarding the minimum subsidy a business will require to locate in the area (i.e., expected profits). They may also be uncertain regarding the value a firm places on some in-kind subsidies. Local officials may not even understand the nature of the jobs and investment associated with a new business establishment. On the other side of the table, negotiators for businesses are likely to have a reasonable idea of the size of an incentive package they require from a community, what the city can afford, and what it has done for other firms.

The problem of information asymmetry is exacerbated when a business negotiates with several jurisdictions about the same facility. The business enterprise will know each city's offer precisely and have a rough idea of the pressures that economic development officials face and of each community's ability to pay. The community representatives, in contrast, may not even know the other competing cities. Under conditions of information asymmetry and interjurisdiction competition, zero- or negative-sum outcomes become more likely.

Another problem that can result in oversubsidization occurs when a city offers incentives for uncertain future benefits. In a typical case, a firm seeking to receive a subsidy to expand or locate in a community will exaggerate the spillover benefits. Public officials may also exaggerate the scope of derived benefits. Thus communities may "pay" for external benefits that are intimated but never delivered. Increasingly, formal contracts and clawbacks are being used to ensure that businesses live up to their promises. Clawbacks are contractual clauses that require firms to repay part of the development incentive if they do not generate designated levels of economic development activity. However, there are limits to the use of explicit agreements regarding a firm's performance and the use of clawbacks because of the inherently uncertain nature of business outcomes (Peters, 1993).

Entitlement Versus Discretionary Incentives

When governments provide incentives as long as businesses satisfy certain preestablished criteria, oversubsidization is likely. Development incentives will be given to firms without demonstrated need, and the amount given may be out of proportion to need or external benefits.

Enterprise zones in many states and the empowerment zones proposed at the federal level are noticeable examples of programs that automatically entitle businesses to subsidies as long as they satisfy certain criteria. Tax codes in most states also provide incentives intended to stimulate economic development. Entitlement incentives contrast with discretionary programs in which local officials use their judgment and negotiating skills to determine the extent, if any, of economic development subsidies.

Enterprise zones may have a second, related, problem contributing to zero- or negative outcomes. Incentives large enough to encourage new activity will also attract firms that would have operated somewhere else in the metropolitan area (Wilder & Rubin, 1996). Thus, a substantial portion of gains within an enterprise zone will be primarily redistributive within the metropolitan area. If the external benefits of job creation are roughly equal throughout a metropolitan area, the enterprise zone can simply result in the relocation of an establishment and a longer com- mute for employees, but no positive spillover for the region—a zero- or negative-sum outcome. In contrast to entitlement incentives, discretionary policies provide local officials with choice regarding the type or size of the subsidies they wish to provide. Avoiding subsidies when they are unnecessary is the major strength of discretionary programs. The disadvantage is that local officials must make decisions that they may not be capable of making. Can the firm succeed? What are the external benefits? Is the subsidy too large?

Targeting

Targeting industries that do not have cost advantages in the region can have a negative result because if the incentives attract such a firm, it will be in an inefficient location. Community strategies that focus on a cluster of related activities or a particular development path for which the region already has an economic advantage may contribute to positive-sum outcomes. One advantage of focusing development efforts on appropriate related industrial clusters is that economic development officials will gain a better understanding of businesses in that sector. They will, therefore, be in a better position to assess the financial needs of firms and to understand the value of in-kind transfers such as infrastructure and tax abatements. Furthermore, if local officials understand locational needs within the sector, they will be in a better position to assess how well competitor cities meet the industry locational requirements. Thus they can better judge where their community stands in intergovernmental competition. Familiarity with an industrial cluster may also help local officials to estimate better the external benefits they expect to receive from prospective developments.

A further advantage of targeting industrial clusters is that infrastructure development may be better focused and the cost-value problem addressed. For instance, a community might seek to attract a cluster of manufacturing activities that tend to require good transportation, a semiskilled or skilled workforce, waste disposal facilities, or good sites. If economic development officials concentrated their efforts on creating an environment that cluster manufacturing activities seek, they could take advantage of the quasi-public-goods characteristics of some programs. Similarly, if economic development officials targeted high-tech industries, they might develop an amenity-rich environment, sponsor technology transfer programs, and support universities.

■ Conclusions

Local economic development efforts can contribute to overall national development by encouraging otherwise unprofitable firms to operate, tilting a firm's locational choice to areas of high spillover benefits, and providing benefits that outweigh costs. Thus, local economic development efforts can be a positive-sum game. However, zero- or negative-sum outcomes are also possible. Oversubsidization is a common form of a zero-sum game.

The relationship among external benefits to the community, the firm's profits, and the economic development incentives are important determinants of economic development outcomes. These variables should be an important focus in development planning and implementation.

The presence of market failures that create the potential to capture external benefits is a necessary but not sufficient condition for positive-sum economic development efforts. However, there are no formulas for avoiding zero- or negative-sum outcomes. Judgment is necessary.

There are numerous notable pitfalls that economic development officials should recognize, including (a) business efforts to create direct intergovernment competition, (b) intrametropolitan competition over local markets, (c) political pressures, (d) information asymmetry, (e) entitlement subsidies, and (f) poor industrial targeting.

Strategies to ameliorate these potential problems include (a) collusion among local governments and limitations on the use of intergovernmental grants; (b) intrametropolitan competition and a stronger focus on business retention; (c) independent reviews of economic development subsidies; (d) rationalized contractual agreements, including clawbacks; (e) greater reliance on discretionary programs; and (f) targeting of industrial clusters that have nonsubsidy cost advantages in the region.

NOTES

1. The reservation wage is the lowest wage a worker will take in order to work.

2. The form that inducements take can be important because the cost of the incentive may differ from its value to the firm. Therefore, local economic development programs may either create value or waste resources. For instance, an environmental program that cost $100 but provided benefits to a firm of $150 (measured in higher profits) would create wealth.

3. The model assumes that the $4 inducement is not used for some other economic development project.

4. Case 2P represents a theoretical possibility. However, the economic conditions in most areas are similar in that there are many unemployed resources that could benefit from an expansion in economic activity. Consequently, it is likely that most urban areas would receive roughly equal external benefits from a given increase in economic activity. There may be instances in which a firm will bring special agglomeration economies to one particular city, but such instances are probably not common.

5. Consider a tax abatement of $100 annually. To the community, the opportunity cost of the abatement is $100, but the value is less to the firm. Suppose that the firm is in the 30% marginal tax bracket. If the firm pays the $100 tax, its profits will be reduced by only $70 because the tax is deductible. Thus the value of the $100 tax abatement to the firm will be only $70.

6. There are numerous market failures not envisioned in the laissez-faire model. Bartik (1990) suggested that real-world market failure includes unemployment, underemployment, fiscal benefits, agglomeration economies, human capital, research and innovation spillovers, information blockages, and imperfect capital markets. These market failures provide a basis for capturing external benefits and positive-sum local economic development efforts. The types of market failure that exist in a community may help local officials decide which types of businesses they should encourage.

7. Some laissez-faire advocates argue that even if there are obvious market failures, government economic development efforts are likely to make matters worse rather than better because development officials lack the appropriate knowledge (information) and motivation to carry out efficient programs (Wolf, 1979). This criticism applies to poorly managed economic development programs, but it need not be universally valid.

8. Even the city that attracts the new economic activity may lose through superoversubsidization. A potential problem is similar to the winner's curse. Assume that individuals at an auction have differences of opinion regarding the value of an item. Let the opinions regarding value be represented by a normal distribution, and let the median opinion represent the most likely value. The winner of the auction may consequently overpay. The implied pitfall for local economic development officials is that they will spend too much on inducements compared to the external benefits they expect to receive. To avoid the winner's curse, economic development officials should bid less than they estimate the full value of the economic development activity to be.

REFERENCES

Baade, R. A., & Dye, R. F. (1988). Sports stadiums and area development. *Economic Development Quarterly, 2*, 265-275.

Bachelor, L. W. (1991). Michigan Mazda and the factory of the future. *Economic Development Quarterly, 5*, 114-125.

Bartik, T. J. (1990). The market failure approach to regional economic development policy. *Economic Development Quarterly, 4*, 361-370.

Blair, J. P. (1991). *Urban economic development.* New York: Irwin.

Burchell, R. W., & Listokin, D. (1978). *The fiscal impact handbook.* New Brunswick, NJ: Center for Urban Policy Research.

James, F. (1984). Urban economic development: A zero sum game? In R. Bingham & J. Blair (Eds.), *Urban economic development* (pp. 245-267). Beverly Hills, CA: Sage.

Jones, B. D., & Bachelor, L. W. (1984). Local policy discretion and corporate surplus. In R. Bingham & J. Blair (Eds.), *Urban economic development* (pp. 245-267). Beverly Hills, CA: Sage.

Mansfield, E. (1991). *Microeconomics* (7th ed.). New York: Norton.

Peters, A. H. (1993). Clawbacks and the administration of economic development policy in the Midwest. *Economic Development Quarterly, 7,* 328-340.

Rasmussen, D., Bendick, M., Jr., & Ledebur, L. (1984). A methodology for selecting economic development incentives. *Growth and Change, 15,* 18-25.

Rosentraub, M. S., & Swindell, D. (1991). Just say no: The economic and political realities of a small city's investment in minor league baseball. *Economic Development Quarterly, 5,* 152-167.

Rubin, H. J. (1988). Shoot anything that flies, claim anything that falls. *Economic Development Quarterly, 2,* 236-251.

Rusk, D. (1993). *Cities without suburbs.* Washington, DC: Woodrow Wilson Center.

U.S. Department of Housing and Urban Development. (1988). *The president's national urban policy report 1988.* Washington, DC: Author.

Wilder, M. G., & Rubin, B. M. (1996, Autumn). Rhetoric versus reality: A review of studies on state enterprise zone programs. *Journal of the American Planning Association, 62,* 473-491.

Wolf, C., Jr. (1979). A theory of non market failure: Framework for implementation analysis. *Journal of Law and Economics, 22,* 107-139.

Yellen, J. (1984, May). Efficient wage models of unemployment. *American Economic Review, 84,* 117-120.

A COMMENT ON INCENTIVES

Bronwen J. Turner

In the chapter "Is Local Economic Development a Zero-Sum Game?" the authors analyze one aspect of economic development, the increasingly common and controversial use of incentives by communities to induce companies to new locations.

Though agreeing with the basic conclusion that local economic development programs can be a positive-sum game, I question the authors' implication that the use of incentives is the primary method of delivering benefits. In practice, the situation is more complex. If the use of incentives is part of a community's game plan, the community has to be sophisticated in how it uses them in order to benefit. To be truly successful over the long term, a community's economic development program must be multifaceted and well considered, with incentives constituting only one aspect.

■ Use of Incentives

Should a community wish to use incentives, forethought and analysis can help. There are many ways in which it can maximize its gain from this activity while minimizing the risks raised by the authors, such as over-subsidization or erosion of the future tax base.

1. Operate on the basis of an economic development plan for the community that enjoys broad public support. The plan should include an analysis of the competitive advantages of the community and the potential fiscal, social, and environmental implications of business development. Target for recruitment industries that will benefit from those advantages. Make strategic use of incentives in instances in which specific firms or projects will be catalysts for industrywide growth or will spawn areawide revitalization. The plan should outline the parameters within which incentives will be offered. If every business requires an incentive, the community should reassess its plan and business climate to determine if other structural changes are required.

2. Tailor each incentive package to the unique attributes and revenues of a specific project or prospect. Although such a discretionary approach may raise questions of fairness, it also can reduce the risk of providing a company benefits over and above those needed while incurring an unnecessary cost to the community or providing incentives to companies not requiring them. Requesting incentives has become a standard operating procedure for companies, coming more from an attitude of "everybody else does, so why shouldn't we" and communities' willingness to grant them than from a genuine operating-cost requirement. The exception to this approach is the enterprise zone concept or entitlements as described by the authors. In this instance, a community has conducted analysis sufficient to ascertain that granting certain predetermined tax advantages or other concessions in a prescribed area relative to others in the community will result in a net gain to the community overall.

3. *Undertake a multiple-year analysis of the fiscal impacts occurring as a result of the company's operations.* This demonstrates both the size and timing of benefits the community may receive from the company, such as sales tax or property tax revenues, as well as the cost of providing services. Such analysis can assist community officials in determining the nature of the incentive package (i.e., which revenue stream to use), its size (i.e., preferably never give revenues in excess of those required to cover the cost of services), the timing of incentives relative to when costs are incurred, and any clawback provisions. In many instances of multicity or multistate bidding wars, it is extremely difficult for a community to obtain the company financial data necessary to perform a "but-for" test. As a substitute, a fiscal impact analysis can at least help the community ensure that it is not offering more than it will gain.

4. *Invest in the community.* The best components in an incentive package are those resulting in a long-term gain for the community itself, not short-term "breaks" for the company. Examples of these are investments in infrastructure and buildings, particularly in areas in need of revitalization, and improvements in workforce skills. The community gains a double benefit if people excluded from the workforce or incurring public costs (e.g., the unemployed or welfare recipients) are trained and engaged by the company. A related strategy is inducing companies or projects that stimulate other private investment or expansion, such as creating a demand for local suppliers, or that will generate public revenues. With many companies now being "footloose" and having multiple regions in which to locate and operate successfully, a community's best strategy is to create the overall quality of life that the community seeks for itself over the long term and to support the local companies invested in being contributors to this long-term vision.

Despite the widespread use of incentives, the practice raises nagging and unresolved issues of fairness. For instance, why should companies new to a community be compensated in a manner that existing companies that have been contributing to their community through good times and bad may not be? Or even if operating costs are lower and/or profits and benefits are higher in a new location, what about the cost to the original community's economy of losing the business income and job base and the cost to the nation of replicating infrastructure? Being mindful of not gaining at the expense of another community would suggest paying attention to local business retention and expansion and strategic recruitment.

■ A Multifaceted Economic Development Program

A successful economic development program that delivers benefits to a community over time must contain many initiatives in addition to incentives. This point was briefly mentioned by the authors and deserves greater discussion and attention by communities.

The current popularity of incentives may distract communities from the many other economic development program initiatives in which a community should be engaged and that can deliver benefits over a sustained period of time. These include targeting for growth industrial clusters that do not require subsidies (as mentioned briefly by the authors); creating a climate conducive to business growth (improving the regulatory or permitting process improves the operating climate for many or all businesses at what may be little or no cost to the community); investing in the public infrastructure and amenities necessary to redevelop or revitalize commercial and residential areas; providing for education, worker training, and entrepreneurial skill development; providing access to capital; providing technical assistance to emerging companies, especially minority- and woman-owned businesses and those that will supply products and services to local companies; and implementing programs designed to retain and expand existing companies. In an era in which costs of labor, raw materials, and transportation are less important determinants of location by companies, communities must focus on the broad quality-of-life factors—the quality of the environment, housing and cultural opportunities, recreation amenities, community safety, worker "knowledge" skills, the education system, and the business climate. These are the qualities that lead to sustained prosperity.

ONE APPLIED RESEARCHER'S REACTION
TO THE ZERO-SUM QUESTION

Ron Shaffer

The issue of whether local economic development is zero sum is a perennial quandary within academic and political spheres and is even becoming an issue within the business sector as businesses move from the old low-operating-cost paradigm to a quality and value paradigm.

John Blair and Rishi Kumar apply game theory to potential inducement situations to gain insight into potential zero-sum outcomes. The authors place zero-sum outcomes in the context of community and firm, not the more common frame of several communities bidding for a firm. The use of game theory, though helpful, does not really advance some of the more important issues that need to be examined. Further, if the idea of zero sum is cast in the more typical frame of communities competing for a firm, the theory of auctions (Hansen & Lott, 1991; Riley & Samuelson, 1981) is a much more useful methodology.

The prospects of zero sum, often predicted by our conceptual models, are reduced substantially when the assumptions of rational economic actors, free flow of information, resource mobility, equal welfare weights among places and groups, and the comparative statics mode of analysis are discarded. The authors appear to be of similar mind in their commentary on laissez-faire outcomes.

Game theory does not move us forward because it still uses the same basic assumptions, especially comparative statics, as the guide for making a decision. We do not gain insight into why some actions are taken, nor do we gain much insight into the learning that occurs in making decisions regarding incentives.

The analysis presented in the chapter gives little assistance in determining in what form and how external benefits to the community occur as a result of the economic development event (Bartik, 1991; Halstead, Chase, Murdock, & Leistritz, 1984; Pleeter, 1980; Shaffer & Tweeten, 1974; Summers, Evans, Clemente, & Beck, 1976; Weber & Howell, 1982). The inducement decision becomes relatively easy when one can a priori determine the direct and indirect benefits from the event. Game theory gives some insight on how to use those numbers once they are determined but not much insight into determining their magnitude.

The negative cases permit more insight into the dilemma faced by communities regarding potential zero-sum outcomes than do the positive-outcome scenarios. The authors are quite correct that the information asymmetry between the prospective firm and locality prevents local officials from having much confidence in negotiating or estimating the value of the inducement that is possible and necessary. Yet game theory does not reduce that barrier.

The authors have identified the key dilemma for the local practitioner offering inducements to a particular firm in the sentence "In practice, it is impossible to gauge exactly the minimum necessary subsidies." For the local practitioner, inducement programs are often too blunt to have impact at critical junctures of the firm's decision-making process, and the cost of acquiring information to determine the precise type of inducement justified far exceeds the potential benefits received. Thus, the practitioner is pressured to offer some type of inducement package on relatively short notice when the project parameters are still fluid and is also held accountable for the one that got away rather than for the cost of oversubsidization. The practitioner is almost never held accountable for oversubsidization, but annual performance reviews invariably discuss the ones that got away or did not even nibble. Thus no value is placed on the cost of oversubsidization, but an immense cost is assigned to acquiring information to focus the subsidy. In this type of environment, the theory of auctions and winner's curse (Hansen & Lott, 1991; Riley & Samuelson, 1981) give us much more insight than game theory.

The analysis is leveraged on the assumption that because the community is willing to pay the subsidy, the community must perceive the external benefits as exceeding their costs.[1] But empirical analysis suggests that the decision makers offering the inducement and the payers of the inducement are two separate parties, so that incidence is as crucial as magnitude.

Finally, the authors ignore the political reality that we may place different weights on the impacts on selected groups of people and/or places. This means that even though game theory results project positive-, neutral-, or negative-sum outcomes, the crucial decision is the weights.

NOTE

1. If the majority of the businesses changing locations are seeking to minimize their labor costs, the unasked question is how mischievous this assumption is.

REFERENCES

Bartik, T. J. (1991). *Who benefits from state and local economic development policies?* Kalamazoo, MI: W. E. Upjohn Institute.

Halstead, J. M., Chase, R. A., Murdock, S. H., & Leistritz, F. L. (1984). *Socioeconomic impact management: Design and implementation.* Boulder, CO: Westview.

Hansen, R. G., & Lott, J. R., Jr. (1991). The winner's curse and public information in common value auctions: A comment. *American Economic Review, 812*, 347-361.

Pleeter, S. (Ed.). (1980). *Economic impact analysis: Methodology and applications*. Boston: Martinus Nijhoff.

Riley, J. G., & Samuelson, W. F. (1981). Optimal auctions. *American Economic Review, 20*, 381-392.

Shaffer, R., & Tweeten, L. (1974). Measuring net economic changes from rural industrial development: Oklahoma. *Land Economics, 50*, 261-270.

Summers, G. F., Evans, S. D., Clemente, F., & Beck, E. M. (1976). *Industrial invasion of nonmetropolitan America: A quarter century of experience*. New York: Praeger.

Weber, B. A., & Howell, R. E. (Eds.). (1982). *Coping with rapid growth in rural communities*. Boulder, CO: Westview.

THE ZERO-SUM-GAME CONTROVERSY: A REPLY

John P. Blair
Rishi Kumar

We noted that "game-theoretic models typically reduce complex situations to a few elements." Both commentators expressed concerns about real-world complexities that were excluded from our model. For instance, Bonnie Turner highlighted the importance of planning, strategic vision, and community support to an understanding of economic development. In a similar vein, Ron Shaffer observed that our model provides little "insight into why some actions are taken" or "insights into the learning that occurs." Such elaborations on our model are welcome and useful. However, if models were as complicated as reality, they would be long, tedious, unfocused, and seldom used.

The commentaries illustrate a traditional tension between theoretical models that deliberately simplify complex situations and the needs of practitioners to consider almost everything. In some cases, models may be modified to incorporate more complexity, but it may be just as useful to realize that no single model will capture the entire economic development process. We do feel, however, that the game-theoretic model is useful in describing conditions that result in zero- or negative-sum games.

Both commentators were concerned with questions of equity. Turner noted that existing businesses frequently do not receive subsidies as large as new firms do. Unequal subsidies are a concern with any set of business incentives. We believe that some unfairness will beset any discretionary program. In fact, treating businesses equally is probably not a desirable

TABLE 1.1 Communities Compete in a Zero-Cost Manner for a Firm

	Possible Locations		
	City A	*City B*	*City X*
Firm's Expected Profits	$5	$5	$5
External Benefits to City A Residents	$5	$0	$0
External Benefits to City B Residents	$0	$5	$0
External Benefits to City X Residents	$0	$0	$5
Total Benefits After ED Game	$10	$10	$10

economic development goal. It could result in higher costs. The trick is to provide the smallest *necessary* subsidy to each firm..

Shaffer raises a more fundamental equity concern. He suggests that "different weights" should be assigned to the benefits according to who receives them. (This same objection has been raised regarding cost-benefit studies.) Shaffer is correct in the contention that weights could alter the outcome. If benefits flow to high-priority groups, that project could warrant greater business subsidies. The game theory model, however, could easily be modified to accommodate a weighing system. The analyst could simply alter the value of the benefits according to the priority of the recipient group. Most economists agree that alternative weighing systems may be justified under certain circumstances. The implementation problem is that no one can persuasively argue what the alternative weights should be.

Shaffer also states that "the authors place zero-sum outcomes in the context of community and firm, not the more common frame of several communities bidding for a firm." We intended our model to apply to both situations—zero-sum transfers to the firm as well as zero-sum interjurisdictional competition. To clarify how the model can apply to a one-firm, multicity example, consider Table 1.1.

All locations would provide total benefits of $10. If the communities competed in a zero-cost manner, any location would constitute a zero-sum outcome. If resources were expended in the competition, a negative-sum outcome would result. The zero- or negative-sum outcome would hold, regardless of how the benefits were split between the firm and the community.

Finally, although both commentators express concern about aspects of our model, they seem to agree—at least implicitly—that economic development efforts may result in a negative-, zero-, or positive-sum outcome and that oversubsidization may occur. The outcome depends on the size of the externalities, the inducement, and firm profits. These variables are critical to economic development practice.

2

How Do We Know That "But for the Incentives" the Development Would Not Have Occurred?

JOSEPH PERSKY
DANIEL FELSENSTEIN
WIM WIEWEL

Much economic development activity at the state and local level is characterized by a demand to show its real value. Faced with shrinking revenues and growing competition, city governments are increasingly demanding cost-effective programs. On a day-to-day basis, these concerns cannot be reasonably met by traditional evaluation techniques. Public decision makers want serious predictions of benefits and costs, and they want them before the project is begun. Thus, practitioners and analysts are being called on to demonstrate in advance the likely returns to programs and incentives under consideration.

Retrospective evaluations are difficult, but prospective (ex ante) evaluations are even more complicated. Ex post studies must accurately identify and specify a but-for scenario in order to attribute observed outcomes to an incentive or program. They must answer the question of what would have happened in the local economy in the absence of the program. The difference between the observed outcomes and those that would have been expected to occur on the basis of this counterfactual identification is attributed to the workings of the economic development program or incentive. For ex ante evaluations, the analyst must forecast both a but-for scenario and a second scenario incorporating the specific project. Moreover, ex ante work must typically be achieved on a short time line and a small budget if it is to be used in making the decision about funding a project or business.

Under the circumstances, there is great temptation in ex ante studies to accept simplistic and overly optimistic approaches to evaluation. The pitfalls of such superficiality are obvious. As in ex post studies, overstating results may undermine the evaluation and in the end call into question the credibility of the program itself (Barnekov & Hart, 1993; Center for Community Change, 1988). In addition, erroneous analyses may lead to the waste of public resources on undeserving projects or the denial of support for worthy ones.

Although highly variable in their level of rigor, ex post evaluations have generally been more serious than ex ante projections. To be genuinely useful to governmental, community, and private decision makers, ex ante studies must be significantly improved in quality. The absence of a professional consensus with respect to ex ante studies means that techniques are often applied on an inconsistent basis. For the practitioner, the lack of a well-accepted and affordable approach to determining what would happen without the project (the counterfactual case) becomes a daily annoyance as project justifications lack intellectual coherence. Often project proponents' estimates of "created" jobs are accepted at face value, expanded via a multiplier, and treated as if they represent the difference the project makes.

Sharp cost constraints face practitioners who seek to improve on such simple ex ante project evaluations. But progress is possible. This chapter presents a methodology we designed that incorporates a serious counterfactual (no-program) situation in its estimations. The methodology is primarily oriented to projects involving a subsidy or other incentive to a firm to expand or locate a new facility. Although more crude than expensive, case-by-case, full-scale research efforts, this methodology addresses the key issues raised in the ex post literature and makes use of "state-of-the-art" regional models to help establish meaningful counterfactuals. We will first review the key issues raised by the ex post evaluation literature and then present our methodology for ex ante projections.

■ Key Modeling Issues Raised
by the Ex Post Literature

Bartik and Bingham, in Chapter 10 of this book, provide a useful review of methodologies used in the ex post evaluation literature. (For important earlier reviews, see Bartels, Nicols, & van Duijn, 1982; Foley, 1992; Richardson, 1988.) More specifically, they consider process evaluations, before-and-after evaluations, comparison-group evaluations, randomized-control-group evaluations, and evaluations based on economic modeling. Obviously, neither process evaluations nor before-and-after evaluations can be used in an ex ante setting. In principle, surveys, comparison groups, and randomized control groups can be used to make ex ante forecasts. For example, businesspeople might be questioned concerning their likely investment plans with and without a hypothetical program, and the results could be used to forecast the effect of the program. Or a small pilot effort with a control group might be analyzed from an inferential perspective to generate expected behavior with and without a program. Although these techniques are common in ex ante cost-benefit analysis in such areas as environmental policy and health policy, they have not played a major role in ex ante program evaluations in the economic development field. They are generally far too costly and time-consuming when the merits of a specific project are being considered in advance.

Thus, the vast majority of ex ante studies take the form of economic modeling. Not surprisingly, the level of sophistication embodied in these models can differ dramatically. At the most elementary level, the impact of a subsidized activity—say, a new branch plant—is equated with the expected employment of the plant times a simple economic base multiplier. Here the assumption is that "but for" the subsidy, there would be no employment change. At the other extreme, an evaluation may make use of an elaborate regional model to forecast the world with and without the subsidized activity. Many off-the-shelf regional input-output and forecasting models (such as the Regional Economics Models Inc. [REMI] model) routinely generate a full-scale regional control (baseline) model that replicates the economy of the area being studied (Treyz, Rickman, & Shao, 1992). This is then compared with an alternative simulation based on specific demand-side changes exogenously dictated, such as the opening or closure of a new manufacturing plant. The difference between the alternative and the control simulations represents the effect of the changes.

Regardless of the apparent sophistication of the model used to evaluate a proposed project or program, the practitioner/analyst does well to be sensitive to a number of important modeling issues raised by the ex post literature and too often overlooked in ex ante evaluations. Key among these are the following:

1. Does the model seriously assess whether public expenditures on the program/project represent a vital input to the firms involved or simply serve as "deadweight" spending?
2. Does the model have a clear logic for determining the number of retained jobs as well as the number of new jobs?
3. Does the model consider the possibility that a program might displace or replace existing activity by creating "unfair competition" between assisted and nonassisted firms (the business displacement issue)?
4. Does the model allow for the possibility that workers taking project-related jobs might have found employment elsewhere in the absence of the project (the labor opportunity cost issue)?

5. Does the model allow for the possibility that programs or projects suppress or inhibit the development of new economic activity (the deterrence issue)?

6. Does the model make clear what might be expected from the resources dedicated to the program in the absence of the project/program (the program opportunity cost issue)?

Each of these basic questions deserves attention.

Deadweight Spending

Perhaps the most upsetting finding of the ex post survey literature is the high proportion of business respondents who assert that their economic behavior would have been the same in the absence of public subsidies (Bovaird, 1992; Foley, 1992; Storey, 1990). Thus, any model used to determine the difference a project might make must consider, however imperfectly, the possibility that much of the private activity related to the program might actually have been undertaken in the absence of the program.

One question is whether, in the absence of a subsidy or incentive, a firm would have been able to raise the money at market rates. But the firm might demand a higher level of return than market rate borrowing would allow. This is a much harder counterfactual situation to establish. It may also suggest that subsidizing healthy firms is an efficient strategy for expanding local economic activity.

New Versus Retained Employment

New and retained jobs each need to be analyzed differently to determine whether they are really due to the program or subsidy. In the case of new employment, the plausible counterfactual situation might be the firm's inability to increase market share without the subsidy and thus to create new employment. In the case of retained employment, the counterfactual situation may mean that a firm would leave the area in the absence of the program. Thus new and retained employment need different specifications of the but-for condition.

In the ex post literature, the survey approach has been used to define this distinction between program-created and program-retained employment. Willis and Saunders (1988), for example, analyze the employment effectiveness of the Mid Wales Development Agency in this way. They use a two-step survey of firm managers to elicit, first, the amount of employment attributable to the agency and, second, the amount of new versus retained employment among the total employment attributed to the agency. At each stage, a different counterfactual situation is specified.

Ex ante claims for retained jobs being associated with a program/ project must be scrutinized carefully. At a minimum, the logic of job retention claims should be made clear. Does it seem plausible that the firm would actually leave in the absence of a subsidy? As we will discuss in the section on our model, this is more likely to be the case if, for example, the firm leases rather than owns its space, or if it has attractive offers from other places.

The Business Displacement Issue

To determine the number of jobs created and retained by a project, we must be able to identify any employment displacement resulting from that project. *Displacement* here is defined as the reduction in output and/or employment in nonassisted firms resulting from the competition generated by assisted firms. The greater the degree of displacement, the less effective the program. It can be argued that in highly competitive markets, subsidizing one group against another simply results in no net employment or income growth. This view assumes that economic development is a zero-sum game, a contention that is no longer universally accepted even at the national level (see Bartik, 1991). In most cases, the degree of displacement must be determined empirically.

An ex post study by Loh (1993) of grants, loans, and business incentives operating in Ohio suggests that displacement can be quite extensive. The reviewed programs, according to Loh, did not increase the number of employed people. However, they did significantly raise incomes.

A number of ex post surveys have addressed the displacement issue. In these, respondents are requested to estimate the extent to which the assistance they have received has affected the output of other firms. This has been done for evaluations of self-employment financial assistance for the unemployed (Elias & Whitfield, 1987), the impact of a development agency (Willis & Saunders, 1988), enterprise zones (Department of the Environment, 1989), regional enterprise grants (Department of Trade and Industry, 1990), and enterprise board financial assistance (Monck, 1991). Evidence shows that the respondents' perception of the displacement effects varies widely, ranging from 10% (Willis & Saunders, 1988) to 50% (Department of the Environment, 1989).

A number of difficulties are associated with these surveys (Elias & Whitfield, 1987). First, the degree of displacement caused by the program runs along a continuum. Unlike the simplest "but-for" situations, which involve a straightforward discrete choice assessment of whether the development would have occurred but for the program, displacement can be measured continuously. Second, displacement is not geographically constrained. It can occur beyond the market confines of the firms and places being assisted, and respondents generally have little knowledge of these effects. Finally, the assessment is elicited from those who are responsible for causing the displacement and not from those experiencing it. The latter are very difficult to identify, and the ability of the former to evaluate these effects must be subject to considerable bias.

In view of these difficulties, other assessment approaches have been attempted. The first involves deriving displacement effects on the basis of the product produced by the assisted firm (Persky, Felsenstein, & Wiewel, 1993; Willis & Saunders, 1988). If an assisted firm is a net exporter (out of the region of analysis), then the likelihood of its displacing local firms is lower than if it serves local markets. Input-output models calibrated for a small area routinely provide this information at the industry level in the form of the regional purchase coefficient (RPC). This is an index of local self-sufficiency, representing the amount of local demand that is supplied locally in each industry.

The location of a firm within the area of analysis can also have a significant effect on where its customers come from and whether in its absence local customers would be served by firms from outside or inside the region. RPCs and other applied measures of export activity are an average that may not accurately describe the specific firm (e.g., major retail and service firms at the periphery may draw from outside the area, whereas firms located in neighborhood business districts are more likely to serve a strictly local clientele). Thus the displacement effect should be estimated taking these issues into account as much as possible. A market survey would be the ideal method. In its absence, a simple distance decay function can be used to model displacement impacts, on the assumption that the more centrally located the business, the more likely it is to displace local business. These approaches are operationalized for ex ante evaluations later in this chapter.

The Labor Opportunity Cost Issue

Many workers finding employment in project-related jobs might, in the absence of the project, have found other jobs either in the community or elsewhere. As individuals, these workers recognize that their net gains only equal their new wages less the wages in their alternative job. From the perspective of the community, calculating net benefits requires consideration of the chain reaction through the labor market resulting from the operation of the program (so-called "vacuum effects"). For example, if a training program results in a worker's changing jobs, then two possibilities exist for the position he or she vacates. If it is filled by a previously unemployed local resident, no offsetting loss has occurred, and the output/income produced by the newly trained worker can be fully credited to the training program. If the vacated position is not filled by someone from the community, then the loss of output/income that this imposes must be deducted from training program benefits.

Labor opportunity costs are likely to be high when new jobs in a community are taken by in-migrants or potential out-migrants. These individuals would be likely to find work elsewhere in the absence of the project. However, those distant, now vacant, jobs will do little if any good to community residents. Under these circumstances, only the net gain to the individuals involved and not their entire income should be counted as project benefits.

Bartik (1991) estimated that fully 75% of metropolitan employment growth can be expected to fall into the hands of in-migrants. These data strongly suggest that labor opportunity costs be taken seriously in the evaluation of economic development projects.

Another form of labor opportunity cost consists of transfer payments. When the funding for such payments originates outside the region of interest, their loss represents a cost to the project.

The Deterrence Issue

This issue has received only scant attention in the literature. *Deterrence* refers to the process whereby the program inhibits the development of economic activity that might have matured in its absence. Aside from displacing local economic activity, the program may deter other economic activity from setting up in the local area or from being attracted from outside. The mechanism of deterrence is similar to that of displacement. The assisted firm or place reduces the availability and increases the price of labor and other factors and thus either displaces or deters other economic activity in the local area.

This effect has been measured ex post through the use of shift-share analysis (Harris, Lloyd, McGuire, & Newlands, 1987). This approach allows for the factoring out of national and industrial influences and the isolation of the specific program or economic development event as represented by the differential shift. In this instance, the deterrence effect of the North Sea oil boom on the economy of Aberdeen, Scotland, was examined. It was shown that although the oil industry created much local employment, it also inhibited the appearance of fast-growing, non-oil-related activities in the local economy.

The Program Opportunity Cost Issue

The correct identification of the counterfactual situation also calls for an assessment of the alternative uses of the resources of the program. The opportunity cost of a program relates to the extent to which the place or firm is better off from this use of resources, as compared with the next best alternative. For example, in estimating the local income impact associated with a small-business assistance program, all extra earnings may not be attributable to the program if most of these benefits would have been attained through operating a different program at the same cost. There is, therefore, a high opportunity cost associated with the program, and this has to be discounted from any calculation of total program net benefit. On the other hand, if the presence of the program causes previously underutilized resources (such as land and labor) to be fully utilized, then the opportunity cost associated with the program is low. Incentives such as tax abatements, and not just direct financial assistance, will also incur an opportunity cost. The key issue in this context is the extent to which new taxes would appear in the absence of the program (i.e., the "but-for" situation).

The methodological problem here is attaching shadow prices to (or evaluating the opportunity costs of) different economic development programs (loans, grants, tax abatements, training, and so on). In the case in which the impact of a program is measured in terms of value added, its net benefit will be value added minus the opportunity cost of the inputs (Hamilton, Whittlesey, Robison, & Ellis, 1991). If the program competes for limited funds with city or regional functions such as social services, then a high opportunity cost will have to be subtracted from any benefits. In general, programs based on standard governmental transfers have a low opportunity cost. In contrast, competitive funds for economic development might have a higher cost.

The actual magnitude of these opportunity costs needs to be determined on a case-by-case basis. Although the opportunity cost of a loan, for example, is generally taken as the difference between the market rate and that of the program (Hamilton et al., 1991), in some instances the market rate may not be appropriate. This can be the case with loan programs for small businesses in which the assistance is composed of an explicit subsidy (the direct cost of the program) and an implicit subsidy (the opportunity cost of the loan; Howland, 1990). Because small firms often have no access to private-capital markets, the cost of diverting funds to them that otherwise would have been more profitably employed elsewhere in the private market is not a relevant issue. A further approach has been to adopt an opportunity cost in conjunction with a survey strategy to determine the extent to which the program was truly additional to the capital needs of the firm (Mount Auburn Associates, 1987).

Generic or precise determinations of program opportunity costs are therefore hard to make, and a case-by-case approach needs to be adopted.

However, to improve the rigor of this approach, sensitivity analysis can be used across a range of values to determine how much final impacts would be affected by setting shadow prices at different levels.

■ A New Methodology for Ex Ante Assessment

Clearly, the ex post literature sheds light on many of the problems involved in specifying appropriate counterfactuals. Quasi-experimental techniques avoid many of these problems by using a control group. However, such an approach is not easily adopted to ex ante evaluations. Rather, as suggested above, ex ante work most often involves modeling. These models, however, should be founded on the knowledge acquired in ex post studies of all sorts. Although ex ante practice in economic development varies widely, it is only a slight exaggeration to state that current standards of formal analysis are both vague and naive. This does not mean that practitioners do not appreciate many of the subtleties discussed above. From their real-world experience, practitioners have considerable understanding of the problems raised in establishing a counterfactual scenario. For just that reason, they often base their decision making on their qualitative understanding and not on any quantitative evaluations.

The dangers of the qualitative approach are threefold: Such an approach is difficult to justify, open to charges of inconsistency, and vulnerable to political manipulation. Supplementing insight with formal analysis helps to provide a consistent framework within which decisions can be made and understood. In trying to meet the needs of the city of Chicago's Department of Planning and Development, we have developed a cost-benefit approach to economic development projects (Persky et al., 1993). The resulting methodology, as embodied in the University of Illinois-Chicago Urban Economic Development (UICUED) Cost-Benefit Spreadsheet, cannot answer definitively all the issues raised in the last section. However, it does force practitioners and analysts to consider these issues systematically. In doing so, it contributes, we hope, to the quality of real-world decisions.

As in all cost-benefit frameworks, the UICUED spreadsheet takes seriously the establishment of the counterfactual situation. Project proponents often point to large numbers of new or retained jobs and the related income as direct project benefits. The basic logic of the spreadsheet holds that such project-related employment and income figures must be adjusted downward because some and perhaps all of the employment would have occurred in the absence of the project.

The practitioner/analyst using the spreadsheet begins by answering a number of questions related to the project being evaluated. The first three of these directly address the deadweight spending issue. The central purpose here is to establish the likely behavior of the firm in the absence of a subsidy. To be considered for further analysis, the project must get a positive response to at least one of these questions:

1. Is gap financing required? (i.e., has a participating financial institution denied the loan application?)
2. Are significant incentives being offered by other municipalities?
3. Is Chicago a high-cost location for this type of firm? Under "costs," the analyst is urged to consider all of the following: wages, utilities and other input costs, transportation and distribution costs, and taxes.

These questions leave wide discretion to the practitioner/analyst. Yet at the same time they force consideration of the relevant issues at hand.

Table 2.1 shows key answers of a practitioner in the hypothetical case of a loan to a small-business printing company, Associated Paper Forms. This company anticipates expanding into a new line that requires $750,000 of new capital. A participating bank has indicated a willingness to provide $375,000 if the city will finance the rest on a low-interest loan. Clearly, the firm requires gap financing; hence this question is answered in the affirmative. The investment will not be undertaken in the absence of the loan.

Next, the UICUED spreadsheet focuses on the question of job retention. To win credit for retaining jobs (i.e., to establish a counterfactual that includes job losses in this firm in the absence of the project), the project must meet one of the following conditions (in addition to a yes answer to Question 2 or 3 above): The firm leases its plant and has a demonstrable space constraint, the firm leases its plant and faces substantial neighborhood costs (security, inadequate public services, and/or congestion), the firm has been bought by a corporation that now possesses duplicate facilities, or the firm faces technological obsolescence. Each of these questions builds on the literature on enterprise shutdowns (LeRoy, 1988; Ranney, 1988). In this way, the spreadsheet encourages the practitioner/ analyst to treat claims of job retention carefully.

In the case shown in Table 2.1, Associated Forms fails to qualify under any of these questions. Hence, no claim for retained jobs is appropriate for this loan.

As suggested in the last section, the spreadsheet approaches the displacement problem by concentrating on the industry group directly involved in the project. The displacement effect is based on estimates of regional purchase coefficients included in the city of Chicago's REMI model. Once the industry group is entered, the spreadsheet classifies as jobs that would have existed even without the subsidy the proportion of all jobs corresponding to the regional purchase coefficient. A small allow-

TABLE 2.1 The Center for Urban Economic Development Cost-Benefit Evaluation Spreadsheet on Associated Paper Forms: Basic Input Data

GAP FINANCING?	(YES = 1, NO = 0)	1
COMPETING INCENTIVES?	(YES = 1, NO = 0)	0
HIGH CHICAGO COSTS? Consider: a. Wages b. Taxes c. Utilities d. Input Costs e. Distribution Costs	(YES = 1, NO = 0)	0
DEMONSTRABLE SPACE CONSTRAINT?	(YES = 1, NO = 0)	0
HIGH NEIGHBORHOOD COSTS? Consider: a. Local Service Deficiencies b. Security c. Local Transport Problems	(YES = 1, NO = 0)	0
FIRM LEASES BUILDING OR BUILDING FULLY DEPRECIATED OR RESIDENTIAL ENCROACHMENT?	(YES = 1, NO = 0)	0
DUPLICATE CAPACITY?	(YES = 1, NO = 0)	0
POTENTIAL TECHNOLOGICAL GAP?	(YES = 1, NO = 0)	0
MEETS FEDERAL RETAINED GUIDELINE?	(YES = 1, NO = 0)	0
SIC CODE		26
PERIPHERAL LOCATION?	(YES = 1, NO = 0)	0

ance in this process is made for local serving firms located close to the city boundary, where presumably they draw considerable business from suburban residents who would have otherwise shopped outside the city. The scaling of this adjustment is still somewhat crude. But in principle it can be refined through research on neighborhood/suburban shopping and purchasing patterns.

Associated Forms, a printing firm, is in Standard Industry Classifications (SIC) code 26 and is situated in a nonperipheral location. These answers, as seen in Table 2.1, then set the expected local and export product shares and the suburban competition factor as reported in Table 2.2. If the analyst has case-specific information on the firm's customer locations, these data can be used to override the model's estimates.

Table 2.2 also reports employment multipliers and income multipliers based on the REMI model of Cook County. For each two-digit industry, a standard REMI simulation of modest size is included in the spreadsheet

TABLE 2.2   The Center for Urban Economic Development Cost-Benefit
Evaluation Spreadsheet on Associated Paper Forms:
Calculated Parameters

	Calculated Values
LOCAL/PRODUCT SHARES (from REMI)	0.103
EXPORT/PRODUCT SHARES (from REMI)	0.897
SUBURBAN COMPETITION FACTOR	0.1
GROWTH THRESHOLD FACTOR	1
NEW JOB FACTOR	0.907
RETAINED JOB FACTOR	0
AVERAGE JOB MULTIPLIER (from REMI)	2.233
INCOME FACTORS FROM REMI (per direct job) All $ values in 000's of 1994 $s	
a. Change in Total Labor and Property Income	$86,683
b. Change in Social Insurance Contribution	$6,433
c. Change in Transfer Payments	($5,519)

data bank. Here, too, outside information can be used in place of that
included in the spreadsheet itself. For example, if the scale of the project
is large, an auxiliary run of the REMI model might be useful to pick up
deterrence effects as the project drives up wage rates and other costs.
These would undoubtedly reduce multipliers throughout.

Table 2.3 shows the actual spreadsheet. The projected job expansion
of 32 workers is discounted to 29 because the local portion of expected
sales will compete with that of other area firms. This figure is then
subjected to the REMI multiplier to give 62 jobs, which in turn is cut back
to 38 to account for the share of jobs taken by suburban residents. The 38
Chicago-resident jobs becomes the basis for the actual dollar calculations
of disposable income that follow.

The labor opportunity cost question is also addressed in the spreadsheet
model. Our formulation emphasizes that many of the workers who will
gain or retain Chicago employment as a result of the project would have
found jobs elsewhere in the country in its absence. Considerable evidence
from the migration literature (Cushing, 1993; Herzog, Schlottman, &
Johnson, 1986) suggests that high rates of mobility are much more
common for skilled and educated workers than for semiskilled and low-
skilled workers. As a result, we assume that 75% of the top quintile's
earnings (i.e., occupations identified by the REMI model as having wages
in the top quintile) would have existed in the absence of the project, 50%
of the next two quintiles, and 0% of the last two. In addition, all lost
transfer payments are considered as labor opportunity costs.

TABLE 2.3   The Center for Urban Economic Development Cost-Benefit
Evaluation Spreadsheet on Associated Paper Forms:
Basic Spreadsheet

All $ values in 000's of 1994 $s
SIC Code: 26

BENEFITS	Project Fiscal Impact	Project Present Value	1994 Year 0	1995 Year 1
Projected Job Expansion		32.0	----	32.0
Current Employment		96.0	—	96.0
New and/or Retained Direct Employment		29.0	---	29.0
Indirect Employment		32.5	—	32.5
Total Employment		61.6	—	61.6
Total Chicago Resident Employment		37.7	---	37.7
Chicago Resident Disposable Income		5,567	---	962
Chicago Resident Disposable Income		5,356	--	926
(Net of Resident-Paid Property and Local Taxes)				
Income Group 5		566	--	98
Income Group 4		285	—	49
Income Group 3 (net of transfers)		1,530	—	264
Income Group 2 (net of transfers)		2,609	---	451
Income Group 1 (net of transfers)		366	---.	63
Opportunity-Adjusted Chicago		3,952	----	683
Resident Disposable Income				
Income Group 5 (25%)		141	--	24
Income Group 4 (25%)		71	----	12
Income Group 3 (50%)		765	—	132
Income Group 2 (100%)		2,609	.---	451
Income Group 1 (100%)		366	---	63
Tax Revenue Generated				
All Chicago-Based Governments	342	342	—	55
Property Tax	228	228	—	37
Other Taxes	114	114	—	18
City Government Budget Only	168	168	---	27
Property Tax	66	66	—	11
Other Taxes	102	102	—	16
Loan Repayment (Planned)	359	359	—	63
Loan Repayment (Expected)	287	287	—	50
Total All Chicago-Based Governments' Benefits	629		—	105
Total City-Government-Budget-Only Benefits	455			77
Total Benefits		4,581	—	788

TABLE 2.3 Continued

All $ values in 000's of 1994 $s

COSTS	Project Fiscal Impact	Project Present Value	1994 Year 0	1995 Year 1
Funding Sources:				
Note: Do Not Count as a Cost Any Funding That Would Not Come to Chicago in the Absence of the Project				
Loan Sources:				
Federal CDBG	375	375	375	0
State CSBG (Zero Opp. Cost)	0	0	0	0
Other State Sources	0	0	0	0
City of Chicago Loans	0	0	0	0
Grant Sources:				
Federal and State Grants	0	0	0	0
City of Chicago Funds				
Department Funds	0	0	0	0
Bond Issue	0	0	0	0
TIF Funds	0	0	0	0
Value of All Local Tax Forgiveness	0	0	0	0
Total Project Loans (Opp. Cost)	375	375	375	0
Total Project Grants	0	0	0	0
Total City Administrative Costs	10	10	10	0
Cost Foregone (would have been spent even in absence of project)	0	0	0	0
Total City Government Budget Costs	385		385	0
Total Costs	385			

The spreadsheet in Table 2.3 shows the resulting discounting by income class for the Associated Forms case. The original present value of disposable income of $5.6 million is reduced to $3.8 million because of the labor opportunity cost adjustments.

The above approach to the labor opportunity cost question should not be confused with distributional weights. The basic argument here is that income groups are differentiated to allow a more meaningful estimation of the "but-for" condition. Of course, in the process of dealing with these adjustments of the baseline, the spreadsheet must disaggregate disposable income attributed to the project by earned income levels. As a result, the spreadsheet does generate a distribution of both direct and indirect benefits across five wage groups.

These distributional data provide important information for programmatic analysis. Unfortunately, many public officials and private developers, eager to get projects built, too easily lose sight of the fact that many public programs have been justified and approved on the basis of their supposed distributional impacts (Barnekov & Hart, 1993). Distributional effects as well as efficiency should enter into public decision making. Too few evaluations, whether ex post or ex ante, provide these crucial data.

The spreadsheet also contains an explicit calculation of fiscal benefits. These include new tax revenues, as estimated from simple regression equations between various taxes and disposable income. Expected loan repayments are also counted as project benefits. All of these entries for Associated Forms are shown in Table 2.3.

Developing this multipurpose spreadsheet encouraged the Chicago Department of Planning and Development to consider the opportunity cost of their funds. Members of the department were interviewed and quizzed on their understanding of what funds would be used for in the absence of allocation to projects. Because costs on the spreadsheet are entered under "Type of Source," in principle we can set different shadow prices for each source of funds. As these discussions unfolded, there was only one program that staff felt should be treated with a shadow price sufficiently different from its full price. This program involved external funds from the state that were notoriously difficult to qualify for. As a result, the city could make use of only a fraction of the funds it was entitled to. The shadow price of these funds was set at zero because funds went unused every year.

Table 2.3 also shows the cost data developed for Associated Forms. This is a fairly simple case. The only costs are the $375,000 loan out of federal Community Development Block Grant (CDBG) funds and the administrative costs. A major advantage of the UICUED spreadsheet is its ability to consider a range of programs in a consistent framework. Thus, grants, loans, and tax abatements can all be included if used on a specific project. Across projects and funding types, the spreadsheet gives directly comparable calculations. This allows the instrument to be used to analyze the implications of alternative subsidy packages.

The UICUED spreadsheet goes on to calculate the key summary parameters of the project (Table 2.4). In the case of Associated Forms, these look quite good: net present value of benefits of $4.2 million and an internal rate of return of 42.5%. The return is so high because the bulk of the loan is expected to be repaid and a large portion of wages is included in benefits. Even the fiscal impact is positive, with the current value of city taxes outweighing the costs of the project.

The UICUED spreadsheet also estimates a neighborhood impact index designed to reflect the more subtle nonmonetary externalities of local development. Discussing the construction of this index is beyond the

TABLE 2.4 Summary of the Impact to Chicago of the Associated Forms Project

Parameter	Amount
Benefits-Costs (Total)	$4,196
Benefits-Costs (All Chicago-Based Governments)	$244
Benefits-Costs (City Government Budget Only)	$70
Capital Cost per Direct Job	$13.26
Capital Cost per Total Job	$6.25
P.V. Benefit per Direct Job	$157.75
P.V. Benefit per Total Job	$74.38
Internal Rate of Return	42.49%
Fiscal Internal Rate of Return	7.27%
City-Government-Only Internal Rate of Return	2.42%

scope of this chapter, but it is designed to force serious dialogue between financially oriented staff and planning- r neighborhood-oriented staff in the city's department of planning and development.

Again, it is important to emphasize that the approach embodied in the UICUED spreadsheet does not provide definitive answers to all relevant questions. Many blanks must be filled in by the practitioner/analyst. For example, the spreadsheet does not itself hazard a guess as to whether Chicago is a high-cost location for the firm in question. It leaves this up to the judgment of the user. At root, what this approach accomplishes is to incorporate the knowledge of the practitioner into the ex ante analysis in a systematic fashion. This creates a desirable partnership between academics and practitioners. The purpose of any evaluation methodology is to help organize available information. In their accumulated experience, practitioners possess much of the most valuable information on economic development. Effective evaluation requires that this often informal information be used to the fullest extent possible.

■ Conclusions

Although far from standardized, ex post evaluations of economic development programs have yielded considerable insights. A range of methodologies, including surveys, quasi-experiments, trend analysis, and modeling, have been used to construct counterfactuals and hence to allow measurement of program effects.

Although practitioners have real interests in the results of evaluations, they more often confront the task of estimating project effects in advance. Most projects are evaluated in an ad hoc manner, and many formal evaluations remain perfunctory. For many practitioners, the game is to "shoot anything that flies, claim anything that falls" (Rubin, 1988). But such efforts discourage comparability and make difficult any broader consideration of programmatic effectiveness.

The basic message here centers on the need to establish well-defined ex ante evaluation procedures that guarantee some level of reproducibility. Although the methodology of the UICUED cost-benefit spreadsheet can undoubtedly be improved on, it does begin in a manageable way to address the key questions raised in the ex post evaluation literature: deadweight spending, job retention, displacement, labor opportunity costs, deterrence, and program opportunity costs. Most important, the UICUED system emphasizes the organized use of practitioners' insights, rather than their replacement with a lifeless mechanism.

REFERENCES

Barnekov, T., & Hart, D. (1993). The changing nature of U.S. urban policy evaluation: The case of the Urban Development Action Grant. *Urban Studies, 30*, 1469-1483.

Bartels, C. P. A., Nicols, W. R., & van Duijn, J.J. (1982). Estimating the impact of regional policy: A review of applied research methods. *Regional Science and Urban Economics, 12*, 3-41.

Bartik, T. J. (1991). *Who benefits from state and local economic development policies?* Kalamazoo, MI: W. E. Upjohn Institute for Employment Research.

Bovaird, T. (1992). Local economic development and the city. *Urban Studies, 29*, 343-368.

Center for Community Change. (1988). *Bright promises, questionable results: An examination of how well three government subsidy programs created jobs*. Washington, DC: Center for Community Change.

Cushing, B. J. (1993). The effect of social welfare systems on metropolitan migration in the U.S., by income group, gender and family structure. *Urban Studies, 30*, 325-338.

Department of the Environment. (1989). *An evaluation of the enterprise zone experiment*. London: H.M. Stationery Office.

Department of Trade and Industry. (1990). *Evaluation of Regional Enterprise Grants*. London: H.M. Stationery Office.

Elias, P., & Whitfield, K. (1987). *The economic impact of the enterprise allowance scheme: Theory and measurement of displacement effects*. Warwick, UK: University of Warwick, Institute for Employment Research.

Foley, P. (1992). Local economic policy and job creation: A review of evaluation studies. *Urban Studies, 29*, 557-598.

Hamilton, J. R., Whittlesey, N. K., Robison, M. H., & Ellis, J. (1991). Economic impacts, value added and benefits in regional project analysis. *American Journal of Agricultural Economics, 73*, 334-344.

Harris, A. H., Lloyd, M. G., McGuire, A. J., & Newlands, D. A. (1987). Incoming industry and structural change: Oil and the Aberdeen economy. *Scottish Journal of Political Economy, 34*(1), 69-90.

Herzog, H. W., Schlottman, A. M., & Johnson, D. L. (1986). High technology jobs and worker mobility. *Journal of Regional Science, 26*, 445-459.

Howland, M. (1990). Measuring capital subsidy and job creation: The case of UDAG grants. *Journal of the American Planning Association, 56,* 54-63.

LeRoy, G. (1988). *Early warning manual against plant closings.* Chicago: Midwest Center for Labor Research.

Loh, E. S. (1993). The effects of jobs-targeted development incentive programs. *Growth and Change, 24,* 365-383.

Monck, S. (1991). Job creation and job displacement: The impact of enterprise board investment on firms in the UK. *Regional Studies, 25,* 355-362.

Mount Auburn Associates. (1987). *Factors affecting the performance of U.S. Economic Development Administration sponsored revolving loan funds.* Springfield, VA: National Technical Information Service.

Persky, J., Felsenstein, D., & Wiewel, W. (1993). *A methodology for measuring the benefits and costs of economic development programs.* Chicago: University of Illinois at Chicago, Center for Urban Economic Development.

Ranney, D. (1988). Manufacturing job loss and early warning indicators. *Journal of Planning Literature, 3,* 22-35.

Richardson, H. (1988). A review of techniques for regional policy analysis. In B. Higgins & D. J. Savoie (Eds.), *Regional economic development: Essays in honour of Francois Perroux* (pp. 142-168). Boston: Unwin Hyman.

Rubin, H. (1988). Shoot anything that flies, claim anything that falls: Conversations with economic development practitioners. *Economic Development Quarterly, 2,* 236-251.

Storey, D. J. (1990). Evaluation of policies and measures to create local employment. *Urban Studies, 27,* 669-684.

Treyz, G. I., Rickman, D. S., & Shao, G. (1992). The REMI economic-demographic forecasting and simulation model. *International Regional Science Review, 12,* 221-254.

Willis, K. G., & Saunders, C. M. (1988). The impact of a development agency on employment: Resurrection discounted? *Applied Economics, 20,* 81-96.

COMMENTARY ON THE
QUESTION OF INCENTIVES

Valerie B. Jarrett

In the city of Chicago, the primary source of economic development funding is federal money from the Department of Housing and Urban Development (HUD). Due to this fact, the first priority of any proposed project is that it meet HUD's very restrictive guidelines. The bias of these federal dollars is directed toward community development rather than traditional economic development activities. Consequently, the public benefit of a project carries considerable weight with HUD. For example, it is much easier for a grocery store to meet HUD guidelines than for a manufacturer who wishes to relocate in an attempt to reduce costs.

In addition to determining HUD eligibility, Department of Planning and Development (DPD) staff also examine several other aspects of any economic development funding request. An extensive financial analysis is undertaken that includes both a credit analysis of the company and an examination of the business owner's credit history. If a company appears to be in relatively good financial condition, staff will attempt to determine the appropriate level of government assistance. The contribution the company makes to the Chicago economy and its impact on the neighborhood and its residents—which includes jobs, planning and transportation issues, and minority ownership—are all factors used in determining an appropriate level of government subsidy.

A few years ago, it became apparent that the DPD's method of determining economic costs and benefits associated with firms requesting assistance was largely ad hoc. This was true in terms of both overall contribution to the city's economy and contribution to city government coffers. In an era of increasingly constrained financial resources, demand for economic development assistance began to outweigh the supply of available dollars by ever-increasing amounts. As a result, the department was faced with even more difficult funding decisions. The need for a consistent method of assessing the economic impact of alternative proposals became crucial.

The previous chapter discusses at length the details of the cost-benefit model commissioned by the city. From the department's perspective, it was essential that the city have an economic model that was user friendly and could be operated on package software. Although the city leases a

Regional Economic Models Inc. (REMI) model for the city of Chicago, in its present form the REMI model does not suffice. The city uses the REMI model only for technically sophisticated analyses, such as the proposed Third Chicago Area Airport, the downtown circulator, and casino gambling in Chicago. The level of knowledge it takes to operate the model correctly ruled it out: Staff making economic development assistance decisions are finance experts, not economists. However, the REMI model does contain information on the Chicago economy that is crucial for estimating the economic costs and benefits of any proposed economic development project. By using the expertise of staff at the Center for Urban Economic Development (CUED) at the University of Illinois at Chicago (UIC), it was possible to extract certain basic economic information from the REMI model and incorporate it into a simple spreadsheet that could be manipulated by project staff.

To operationalize city economic development policy positions, the UIC staff conducted interviews and focus groups with staff involved in every aspect of economic development decision making. On the basis of these conversations, the model was designed to incorporate the additional assistance that the city was offering to businesses located or locating in any of the city's five enterprise zones. In addition, this process helped identify a number of important factors to consider, including owning versus leasing of a building.

Two points deliberated extensively by DPD and CUED staff were how to treat benefits accruing to individuals not living in the city and how to treat benefits accruing to individuals who moved into the city as a result of government assistance to businesses. In the first case, CUED staff wanted to count all benefits, whereas in the second case they felt that benefits should be omitted. DPD felt that it was more appropriate to count only benefits accruing to city residents, regardless of their tenure. This case illustrates how strict theory must often be adapted when dealing with certain real-world situations. Although, to the outsider, the city's position may seem myopic, it should be understood that Chicago's government exists to serve its residents. As such, the primary concern is the impact of development decisions on all residents, regardless of how long they have lived in Chicago.

The DPD has had the CUED model at its disposal for about 1 year. During this time, it has been used on a few controversial projects. but it has not been incorporated as part of our routine economic development analysis. There are several reasons for this. First, there is no real cost associated with not conducting a cost-benefit analysis. In contrast, failure to comply with HUD's rigid requirements will result in the city's repaying funds to Washington. In light of this, DPD staff often tend to focus on the concrete objective of ensuring that every aspect of the analysis and documentation required by HUD is completed, rather than the more conceptual goal of good economic analysis. Second, the cost-benefit model was housed in the DPD's policy, research and planning division. This division has little involvement in project funding decisions, thus making its involvement in project analysis difficult. Although staff in the various program divisions were trained in how to use the CUED model, their main priority remains in conducting the analysis required by HUD. Finally, external organizations such as community groups or the press tend not to be overly concerned with measures of economic impact or cost-benefit analysis except in the most controversial cases. Most often these are projects that tend to be larger and more sophisticated than the typical economic development proposal. In such instances, the DPD will frequently conduct a REMI analysis. However, there have been several cases when an analysis needed to be conducted very quickly, and in these instances the CUED model was used.

The CUED model has its limitations. Theoretically, the tax and expenditure components are fairly weak. This is an area on which we hope to conduct further work because it is one of the most important aspects of an economic development deal from the perspective of city government. "Will the city break even?" is a question asked of any deal—particularly by the city's budget office. In addition, the DPD has certain policies and/or preferences that are not based on pure economic criteria. Issues to be considered include minority and local ownership, visual impact on the community, potential transportation concerns, and tax delinquency. The DPD worked with the CUED in an attempt to think through issues such as these that should be taken into consideration by the finance officer. On the basis of these findings, a checklist was made to ensure that all analysts would consider the same noneconomic factors when making a recommendation. The CUED did attempt to construct a neighborhood index that quantifies these factors. However, at present this index is still too early in the development stage to use.

In conclusion, although the CUED's cost-benefit model has not become an integral part of every proposed economic development project, it has, in the last year, served the department well in the analysis of a number of relatively complex projects. If HUD reduces the quantity of documentation and analysis required for use of federal economic development funds, the use of the model on all proposals will become much more feasible. Finally, as previously noted, the model was never intended to be the sole tool used in the DPD's decision-making process. The city must consider many noneconomic factors before providing government subsidy for private developers. Good economics, though necessary, is not a sufficient condition.

COMMENTARY ON THE
QUESTION OF INCENTIVES

Nancey Green Leigh

Very little evaluation of economic development projects is done either before or after a project is funded and implemented. In principle, it would seem only logical to place a higher priority and devote greater attention and effort to the before, or ex ante, evaluation activity. After all, is it not better to decide beforehand whether a potential investment in an economic development project is going to yield worthwhile results than to discover after the fact that the expenditure was a waste? The authors of this chapter seek to improve the possibilities for, and methodologies of, ex ante evaluations, and their efforts deserve a warm reception from the economic development field.

The current lack of ex ante activity has meant that many economic development practitioners and public-project decision makers have passively accepted the claims of project proposers of the number of jobs that their projects will create. Persky, Felsenstein, and Wiewel suggest that this passive acceptance of the claims of those seeking public-investment dollars derives from the fact that there currently does not exist a well-accepted, affordable approach to determining what would happen in the absence of the proposed project. In many cases, they are correct: Public-investment decision makers do want to be able to improve their ability to make project funding decisions on the basis of a more accurate knowledge of what the outcomes will be. We should also acknowledge, however, that such funding decisions are often made on the basis of politics rather than economics. Given such realities, we might consider whether there is anything about the ex ante evaluation methodology proposed here that would—because of its ease of use or potentially wide acceptability—serve to encourage even the more politically motivated decision makers to undertake ex ante evaluation activity.

We encounter a catch-22 when we seek to expand and improve ex ante evaluation activity; that is, extensive and high-quality post ante evaluations are required to design effective ex ante evaluations. Hence, to start at the beginning, one must examine the outcomes of the past. As they pursued their goal of creating an affordable ex ante methodology, the authors sought to account for the key issues that have arisen in the conduct of ex post evaluation activity. Their approach to making the ex ante methodology affordable ruled out the use of surveys, comparison groups, and randomized control groups because these are too costly and time-consuming. Instead, it relies on the use of what they call "state-of-the-art" regional models of the relationship between sectoral investments, job creation, and public and private income generation. Here, as shall be made more clear later, lies another catch-22: The "state-of-the-art" regional modeling on which the "affordable" ex ante methodology is built is, itself, not so inexpensive. Nor is it particularly accessible or feasible for use in smaller local economies. The authors developed their ex ante evaluation methodology for the city of Chicago—one of the largest urban economies in the nation.

The chapter presents a valuable review of the key issues (e.g., labor and program opportunity costs, business displacement, "deadweight" spending) raised in the ex post evaluation literature that should be taken into consideration in any ex ante activity. Underlying each of these issues is the need to establish an appropriate "counterfactual" case—that is, what would happen without the project. A deterrent to the use of a more quantitative approach by greater numbers of economic development practitioners, according to Persky et al., is that practitioners are well aware of the difficulties encountered in devising counterfactual scenarios. Hence they tend to rely on qualitative factors in their decision making, and in so doing, they leave themselves open to justification difficulties and political manipulation.

Cost-benefit analysis is the underlying rationale of the economic model presented in this chapter. An attempt is made in the model to force practitioners to consider systematically a number of the quantitative issues raised from the ex post evaluation literature. Persky et al. have devised a spreadsheet approach (the UICUED Cost-Benefit Spreadsheet) that is relatively user friendly as a vehicle for collecting and evaluating proposed project information. Underlying the spreadsheet's design is the logic that the data collected and questions answered will enable the practitioner/ analyst to adjust downwardly the claimed job and income creation made by the proponents of a project. In an attempt to address what a firm's behavior would probably be if the subsidy were not received, three questions in the model focus on the need for gap financing, other municipalities' willingness to offer incentives, and whether Chicago is a high-cost location. The next information entered into the spreadsheet seeks to establish whether the proposed project will lead to the retention of jobs.

Dealt with next in the spreadsheet is the issue that I see as the most problematic in the authors' methodology. Persky et al. seek to account for the "displacement" problem—defined as "the reduction in output and/or employment in nonassisted firms resulting from the competition generated by assisted firms"—of a proposed project. The post hoc literature suggests that numerous difficulties are encountered in trying to generate accurate measures of the displacement effect: What needs to be measured is a continuous rather than discrete variable, displacement can occur beyond the firm and geographic market boundaries of the project, and the data on displacement typically do not come from those who experience the displacement (identification of whom might be difficult if this was attempted) but rather from those creating the displacement.

The technique the authors employ appears to get around these issues in that it does not require much in the way of specific human input that may be inaccurate either because the right humans (i.e., the unknown victims of displacement) are not found to question or because self-serving humans are not relied on as primary sources (i.e., estimates are not being given by the seekers of support). Persky et al. employ the regional purchase coefficient (RPC) of an input-output model (the REMI version) estimated for the Chicago economy as an index of "local self-sufficiency, representing the amount of local demand that is supplied locally in each industry." The displacement effect of a subsidy to a proposed firm is based on the proportion of jobs corresponding to the RPC of the industry in which the firm is found.

A range of problems are associated with the model's reliance on RPCs, as those familiar with input-output (I-O) modeling will be well aware. For example, because the model uses RPCs that are computed at the two-digit, more aggregated industry level, inaccuracy can result because the two-digit level in effect averages what can be a wide variation in the capital-to-labor ratio of subsectors within an industry. A highly capital-intensive production process at the five-digit level will not displace the same amount of labor as would be suggested by the two-digit RPC. The reverse could be true as well.

On another level, the technical coefficients of regional I-O models are derived from the national I-O model (U.S. Department of Commerce, 1994), which is problematic for at least two reasons. To begin with, the pace at which the national coefficients have been updated (the 1987 version was published in the spring of 1994) cannot accurately reflect some of the dramatic shifts we have seen in capital-to-labor ratios over the last decade as American industry has restructured (i.e., automated) to withstand international competition. Further, the technique assumes that a region's industry structure will be the same as the national industry structure, which we know is not true. The distribution of industry across space is not uniform: Some industries are found only in certain regions, and the economic structure of industries can vary on the basis of the difference in availability and price of inputs in different locations.

In a recent review of the use of regional models such as REMI, Jensen (1990) raised the question "Are we ethically able to apply tables with (perhaps significant) built-in errors, to economic analysis, particularly for those who are unable to appreciate the possible significance of these errors?" (p. 20). This is an important question to consider for economic development practitioners and for the economic development consultants and researchers who sell their services to the practitioners.

Jensen (1990) has described two types of groups that have tended to adopt the use of regional I-O models in economic analysis. On the one hand are specialist consultants in I-O analysis who develop their models "with a high level of expertise through continuing research programs" (p. 13). It is the second group of generalist consultants—for whom I-O modeling is only one of a range of tools in an economic consultant's bag —that he sees as problematic. Lacking sufficient skills in I-O methodology, this group tends to provide black-box answers to the economic development impact questions of its clients.

A city like Chicago, with a complex, large economy and presumably sophisticated economic development staff, would perhaps be less likely to be sold on black-box solutions than a smaller city and its decision makers. In this sense, the integration of regional product coefficients from the REMI model within the ex ante evaluation tool for public-investment decisions seems plausible. The size and complexity of Chicago's economy are also such that the magnitude of some of the well-known problems associated with I-O analysis is smaller. For example, determining how much of local production is really consumed locally is subject to a smaller margin of error for a large, complex economy, and this in turn would presumably lead to smaller margins of error in estimating displacement effects.

In essence, then, Persky et al.'s ex ante regional model evaluation tool would seem more appropriate for larger than for smaller economies. However, the authors do not attach any cost figures to their model for the city of Chicago, although this is another significant criterion by which to judge the likelihood of its adaptation by smaller urban economies and their development departments. Although the cost of the computing power needed to run regional models has certainly decreased in recent years, the price of the models and the technical expertise to refine a packaged model for a region has not.

Although, as we noted earlier, the modeling technique that Persky et al. employ appears to avoid inaccuracies such as those produced when the right humans are not found to question or when self-serving humans are relied on as primary sources for the estimates of displacement impacts, we must still consider whether the different types of inaccuracies associated with their model are any "better" or more desirable. We are back to the catch-22 here, for the only way to decide if their inaccuracies are preferable to those from human input is through ex post evaluation.

In the meantime, despite the issues raised above, this reviewer would like to see other (appropriately sized) cities adopt Persky, Felsenstein, and Wiewel's regional model. First, it helps to spread the notion of conducting ex ante evaluation activity with its underlying commonsense logic that it is better to consider systematically beforehand whether a potential investment in an economic development project may yield worthwhile results.

Second, despite the likelihood of some significant sources of error in the modeling results, the modeling of a regional economy has pedagogical value. It helps to illuminate an economy's structure—the interlinkages between different industries and between households and industries. It forces the systematic consideration of the impacts of investment decisions and, in this case of a metropolitan economy, seeks to differentiate those impacts for different neighborhoods and for city versus suburb. A feature particularly noteworthy of Persky, Felsenstein, and Wiewel's model is their effort to differentiate benefits across income or wage groups, such that explicit consideration can be given to the issue of income inequality in the process of economic development.

In concluding, let us return to our earlier question of whether there is anything about the ex ante evaluation methodology proposed here that would encourage its adoption by more politically motivated decision makers. The nature of the decision process in this methodology, which relies relatively little on direct human input to generate the multiplier impacts of a proposed investment, may help to diffuse claims of politically manipulated data, thereby giving it wider acceptability. Further, once the model has been developed, the evaluation process is relatively user friendly. These two features may encourage its adoption and, in so doing, strengthen the notion of what ex ante evaluation should be: an automatic step in the practice of economic development.

REFERENCES

Jensen, R. C. (1990). Construction and use of regional input-output models: Progress and prospects. *International Regional Science Review, 13*, 9-25.

U.S. Department of Commerce, Bureau of Economic Analysis. (1994). *Benchmark input-output accounts of the United States*. Washington, DC: Author.

RESPONSE ON INCENTIVES

Joseph Persky
Daniel Felsenstein
Wim Wiewel

We are grateful for both commentators' generally favorable observations and realize that in offering a response to their criticisms we risk the possibility of an infinite regress, this being an evaluation of the evaluators' evaluation of the evaluators' methodology for evaluating. Still, let us press on and share a few thoughts.

We certainly agree with Nancey Green Leigh on the importance of ex post evaluation. It comes as no surprise that after the fact we can see more clearly than before the fact. Long-run policy making gains greatly from drawing on ex post analyses. But immediate project decision making obviously requires ex ante studies on a case-by-case basis. There is no debate about which is better or worse because their uses are so different.

Apart from the difficulties inherent in predicting the future, ex ante studies carried out on a regular basis must be reasonably priced. In this case, the UICUED spreadsheet builds around the existing REMI model of the Chicago area. However, other much cheaper county-level models are currently available. For our limited purposes, they would serve nearly as well. Generating the necessary multipliers should not be a major barrier to others' using similar methodologies.

More fundamentally, Leigh questions the adequacy of using regional purchase coefficients (RPCs) as indicators of economic displacement. At the theoretical level, the use of RPCs in this context is grounded in the logic of the economic base model. Surely new firms catering to local markets are much more likely to displace other regional firms than are export-oriented companies.

At the empirical level, we agree that current measurements of RPCs are coarse. However, they are certainly robust enough to give useful information on the matters at hand. The issue is not the accuracy of the third decimal point, but the achievement of reasonable and solid quantitative estimates.

Turning to Valerie Jarrett's comments, we agree that the fiscal accounting in the spreadsheet is primitive. The city of Chicago needs to develop a serious model relating economic activity to various tax streams, a task well beyond the scope of our project. Once such a model was built, its output could be easily incorporated into the UICUED spreadsheet.

The UICUED spreadsheet focuses on the economic benefits of a development project. Yet we do recognize the existence and importance of nonmonetary benefits and costs. Although not discussed at length in the present chapter, the issue of neighborhood spillover effects receives considerable attention in the spreadsheet. A weighting system devised in conjunction with the staff of the Chicago Department of Planning and Development attempts to quantify, if not monetize, a number of these externalities. Of course, this is an area rich in research possibilities.

Understandably, Jarrett, as Commissioner of Planning and Development for the City of Chicago, must be responsive to a range of political pressures. For the moment, she is certainly correct that few community or business groups are likely to demand serious cost-benefit analysis of city projects. However, over time, these groups are developing increasing sophistication. At least one such group in Chicago has already used REMI-generated simulations to present its case.

We are the first to admit that the UICUED spreadsheet can be improved in many ways. But it still represents a real advance in ex ante project evaluation. Its insistence on demonstrating the "but-for" condition, its commitment to reckoning displacement effects, and its emphasis on the distribution of benefits and opportunity costs across income groups encourage the systematic evaluation of both equity and efficiency issues.

3

How Important Is "Quality of Life" in Location Decisions and Local Economic Development?

JAMES A. SEGEDY

■ The Changing Paradigms of Location Decision Making

The traditional slogan of firms' location decision makers has always been "location, location, location." But in reality, location strategy has always included a broad range of needs and concerns. Thomas Lyons and Roger Hamlin (1991, p. 23) identified 11 of what they described as "traditional" features of productive political economy:

1. Land
2. Access to markets and materials
3. Labor (quantity and skills)
4. Capital
5. Energy
6. Finance
7. Management
8. Taxes
9. Regulatory climate
10. Research
11. Quality of life

Of these features, only one, "access to markets and materials," could be directly described as "location," yet obviously, many could be construed

AUTHOR'S NOTE: I would like to thank Kevin K. Brown for his assistance in preparing this document and researching the information used in it.

as features of location. Economic development practice must focus on meeting many if not all of these needs to attract business to a community.

But catering to these needs is not always enough. Local officials have often gone to great lengths to offer full arrays of infrastructure development programs, tax giveaways, or land price write-downs to attract a major facility, only to have firms leave when the incentive programs end. There are also many examples of communities investing in the development of industrial parks that now sit empty. All of this suggests that location decisions are neither simple to explain nor certain in their outcome. At best, present-day firm location policy is based on a shifting paradigm.

Although there is no doubt that bottom-line hard costs are, and will continue to be, critical concerns of businesses and business location decision makers, communities are also beginning to view economic development more in terms of long-term economic "vitality" or "maturity" than in terms of expansion. "Growth," according to the Organization for Economic Cooperation and Development (OECD; 1982), "is not an end in itself, but rather an instrument for creating better conditions of life" (p. 7). Edward Jeep, Regional Director of the Chicago Office of the Economic Development Administration (EDA), described this recognition of a need for a shift in priorities:

> The base problem is not *what* is being done in local economic development. It is not *how* it is being done. Those aspects of development are faring well. It is the problem of *why* it is being done. *Why* and *for whom*. It is the problem of the public purposes of development. . . . We began to lose touch with our larger human purposes as the center of our enterprise, the *why* of it, and the *for whom* of it. (p. 238)

In short, over the past few years, "quality of life," the 11th and last feature of Lyons and Hamlin's list of firm needs, appears to be moving up the scale of importance for local officials (Gottlieb, 1992). At the same time, others are finding that "quality-of-life" issues are also becoming more important to firms considering facility location (Gottlieb, 1992; Myers, 1987a)—a potentially important shift in economic development practice. Because literature on the topic is scant, this chapter should be viewed as an early report on the "quality-of-life" variable and its emerging importance to economic development practice.

To begin with, the introduction of quality-of-life issues demonstrates that current economic development thinking is more broad based than the "smokestack-chasing" approach to business attraction. The evolving view of economic development as total-community development also parallels the shift toward more proactive and socially sensitive development strategies. Such a change in location decision criteria and economic development practice is, however, probably less a factor of changing economic paradigms than a reflection of evolving technologies, the changing nature of the economy and businesses in general, and a new awareness of social and human factors. Thomas Daniels, in his study *Big Business and Small Towns* (1993), described the confluence of these three factors in his discussion of the corporate trend toward decentralization. The late 20th-century revolution in firm structure has been inspired by a range of technological and production changes such as telecommuting, "rightsizing" of resident corporate personnel, outsourcing, and the "spinning off of [corporate] divisions into separate, geographically dispersed companies" (p. 3). All of this has provided opportunity for thousands of communities, many small and rural, to get into the corporate location process at last through the marketing of their amenities and the quality of life that such amenities provide. David Heenan (1991) found not only that these decentralized firms relocate into smaller, more amenity-rich communities but that they do so at lower cost (in terms of land, labor, and taxes) than that for comparable metropolitan sites. The small town, reported Daniels, provides "traditional values of the importance of family and community, offering the potential to balance work, family, and community. By contrast, the sense of community and comfortable environment are often lacking in hectic, congested metropolitan areas" (p. 4).

Firm decisions based on quality of life in small towns are not without problems. There is often a paucity of skilled labor and capital in rural regions, and research-intensive and high-tech firms still have greater difficulties in making the transition. However, growing numbers of corporations have successfully "forged partnerships with their communities, so that even though they may resemble company towns, the dominant company has a real stake in the community" (Daniels, 1993, p. 5).

One way of putting quality of life into the economic development process is benchmarking. In Oregon, a strategy of firm attraction is grounded in environmental, public-service, and job quality standards (Oregon Progress Board, 1993; see Figure 3.1). This form of strategy is especially important because the businesses traditionally the most sensitive to quality-of-life issues are those that are more "footloose," or less bound by traditional determinants of location decision making. The foundation of our traditional manufacturing base and capital, in terms of both plant and equipment, has become very mobile. In such a fluid environment, placing issues of community renewal and quality in the privileged position of benchmarks works to the benefit of the firm and the community. Neither community loyalty nor political boundaries have much meaning any longer; therefore, if the community is not an attractive place to live, the company will simply relocate. And if firm mobility is governed

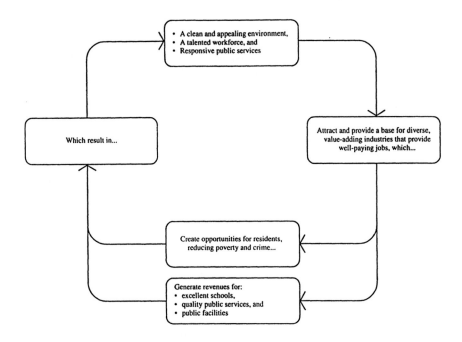

Figure 3.1. The Quality of Life/Economic Development Loop
SOURCE: Adapted from Oregon Progress Board (1993, p. 9).

by quality-of-life benchmarks to begin with, the loss of the firm will also be buffered by the benchmarks.

The classic variables of land, labor, and capital, as well as utility rates, taxes, and the elusive "business climate and entrepreneurial environment" (Gottlieb, 1992, p. 8), are now essentially available universally. As the focus of economic activity shifts more toward the service sector, we are beginning to see other transformations. As suggested above, the decentralized industrial base is finding a home in smaller communities that can offer relatively inexpensive labor and land costs. Heenan (1991) argued that quality-of-life amenities found in small towns offer many advantages over the more traditional urban, suburban, and even "Edge City" locations. Fueled by recent advances in telecommunication and product delivery services, the profit motive can be satisfied virtually anywhere (Daniels, 1993; Heenan, 1991). One need only look to mail-order giants such as Lands' End and L.L. Bean, as well as the growing number of computer software distributors that have thrived in small towns and on the small-town identity (Daniels, 1993). The new variable has become "psychic income" (Foster, 1977; Gottlieb, 1992; Oakey & Cooper, 1989). This translates into a higher value placed on more intangible variables, such as quality of life and amenities. It can even be argued that these factors directly influence a firm's economic "bottom line." Phrases such as "half the pay for a view of the bay" are indications that workers, at both white- and blue-collar levels, are often willing to trade a lower pay scale for a higher quality of life or community amenities (Alonso, 1972; Gottlieb, 1992; Webber, 1984).

A growing body of evidence supports the importance of quality of life and local amenities as critical location determinants. Thirty years ago, "quality of life" did not appear in the top 10 location factors listed by any study (Heenan, 1991). One early study made brief reference to "community appearance" (Yaseen, 1960). More recently, this has changed (see Table 3.1). Some experts insist that quality of life, though not necessarily mentioned by name, has never dropped out of the top three as a reason that people want to invest capital, expand existing businesses, or foster job creation (Brown, 1994; McNulty, 1986; McNulty, Jacobson, & Penne, 1985). Others suggest that the "the aggregate phrase 'quality of life' frequently rank(s) . . . in the top half of all factors" (Gottlieb, 1992, p. 6). Table 3.1 shows the relative ranking of quality of life and amenities as location factors in a firm's choice of a region or state.

Readers of *Site Selection* magazine have recognized quality of life as the most influential location decision-making factor since 1988 (Heenan, 1991). This evidences a change, especially in the popular press. The more academic and analytic literature is a bit more cautious and critical. For example, there is evidence that these quality-of-life variables correlate more with residential satisfaction, as evidenced by white flight reactions, than with a firm's decision, based on its corporate strategy and bottom line. This is further confounded by anecdotal evidence that the "pull" factors associated with individual taste and sense of place are countered by the homogenization of place in the suburban landscape (Gottlieb, 1992). This homogenization of place has left few alternatives to firm and individual relocations seeking the uniqueness of character and community identity that can still be found in many smaller, nonmetropolitan communities. Urbanized areas and large suburban communities have responded with promotional campaigns that feature the cosmopolitan big-city amenities and the diversified quality of life that can be found only at the scale of the larger city and region. The "on again, off again" gentrification movement is clearly part of such practices.

TABLE 3.1 Relative Ranking of Quality-of-Life Factors in a Firm's Choice of a Region or a State

	Greenhut & Colberg (1962)	Foster (1977)	Vaughn (1977)[a]	U.S. Congress (1979)	Schmenner (1982)	Rees (1984)	Festervand et al. (1988)	Lyne (1988)	Hart et al. (1989)	CRUEUE (1990)[b]	Haigh (1990)	IMPULSE (1991)
Quality of life			8/16		3/12	7/10			3/37		4/19	
Cost of living							9/13	11/12				4/21
Cost of housing							3/13	7/12		23/25		15/21
Proximity of housing		2/16							16/37	22/25		16/21
Cultural amenities		7/16		7/26			3/13	9/12	20/37		15/19	14/21
Recreational amenities		8/16					4/13	8/12				
Climate	13/16	15/16	9/16				7/13	12/12	18/37	19/25		3/21
Good schools		1/16		6/26			1/13	5/12				
Government services	11/16			5/26			5/13	3/12		6/25		
Easy commute		10/16										
CEO preference	15/16			16/26								
Environmental quality		5/16					4/13		6/37			7/21
Public safety		4/16		2/26			6/13		12/37			5/21

SOURCE: Adapted from Gottlieb (1993, p. 13).
a. Cited in Eisinger (1988).
b. Center for Real Estate and Urban Economics.

■ Quality of Life: What Factors Are Included?
Which Ones Are Important?

Even a cursory review of the literature suggests that there is little agreement and even less evidence on exactly which quality-of-life factors actually influence economic development and which do not (Andrews & Withey, 1976; Kumcu & Vann, 1991). There have been numerous attempts to quantify and qualify these factors (Andrews & Withey, 1976; Dalkey, 1972; Kumcu & Vann, 1991), but to date, no study demonstrates a clear, scientific cause-and-effect relationship between quality-of-life factors and location decision making. Though "it's not . . . difficult to find widespread agreement among development and corporate real estate professionals as to their importance in many of today's site-selection decisions," there is no consensus about what are the key quality-of-life factors (Venable, 1991, p. 740). The most one can report is that factors gaining some favor among economic and community development organizations seem to be those that relate to community-level and perceptual community attributes (Dalkey, 1972). Table 3.2 provides a summary of these factors, as suggested by the readers of *Money* magazine (Smith & Nance-Nash, 1993).

The de facto importance of these factors is reflected in local chamber of commerce promotional materials, news releases, presentations, and various other promotional activities (Kumcu & Vann, 1991) and reinforced by findings from the recently completed Indiana Total Quality of Life Initiative pilot project, involving 50 Indiana communities (Segedy & Truex, 1994). In that study, community and economic development factors were organized into 12 quality-of-life factors. Each community was invited to define further the nature and scope of these factors and to assign them a relative importance. As seen in Figure 3.2, educational opportunity ranked highest in priority, with economic vitality ranking near the middle. Fusi's (1991) Geo-Life Survey investigated a similar set of criteria from the perspective of economic development and site selection professionals (Figure 3.3).

■ Quality of Life: Reading and
Responding to a Changing Culture

The apparent importance of some of these quality-of-life factors, discussed in the previous section, has not alleviated confusion among community and economic development professionals over their application to location decisions. Most economic development programs continue to place primary (if not exclusive) emphasis on infrastructure, workforce development, and monetary incentives as means to attract and

TABLE 3.2 Alphabetical List of Quality-of-Life Attraction Factors

■ Affordable car insurance	■ Low property taxes
■ Affordable medical care	■ Low risk of natural disasters
■ Clean air	■ Low risk of tax increase
■ Clean water	■ Low sales tax
■ Close to big airport	■ Low unemployment
■ Close to colleges/universities	■ Many hospitals
■ Close to relatives	■ Museums nearby
■ Close to skiing area	■ Near a big city
■ Diversity of local firms	■ Near amusement parks
■ Far from nuclear reactors	■ Near lakes or ocean
■ Good public transportation	■ Near national forests and parks
■ Good schools	■ Near places of worship
■ High civic involvement	■ New business potential
■ High marks from ecologists	■ Plentiful doctors
■ Housing appreciation	■ Proximity to major league sports
■ Inexpensive living	■ Proximity to minor league sports
■ Lack of hazardous wastes	■ Recent job growth
■ Local symphony orchestra	■ Short commutes
■ Low crime rate	■ Strong state government
■ Low housing prices	■ Sunny weather
■ Low income taxes	■ Zoos or aquariums

SOURCE: Smith and Nance-Nash (1993).

retain businesses. Yet as concerns arise as to the narrowness of this approach—that the organizations promoting such policies and public investment have forgotten their duty to their constituents, who are more concerned with community amenities than with economic expansion—quality of life "has emerged as an increasingly visible political concern to elected officials and planners" (Furuseth & Walcott, 1990, p. 76). The cost of this new politics has been billions of dollars poured into the enhancement of community quality of life by local government and businesses, with little or no focus and with equally little indication as to programmatic effectiveness (Fusi, 1989).

In the midst of this shift, professionals must learn to "read" and respond to the new corporate and governmental cultural movements represented in these new standards of quality of life. For example, current trends suggest that more and more corporate personnel are evidencing a reluctance to relocate (Browning, 1980; Malecki, 1984). This will undoubtedly further the demand to develop more local quality-of-life amenities because retaining such workers depends heavily on what the community can offer in terms of amenities. In short, "a good quality of life attracts good employees" (Marlin, 1988, p. 39).

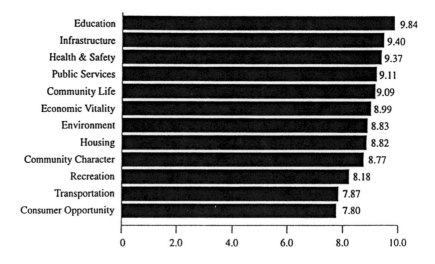

Education 9.84
Infrastructure 9.40
Health & Safety 9.37
Public Services 9.11
Community Life 9.09
Economic Vitality 8.99
Environment 8.83
Housing 8.82
Community Character 8.77
Recreation 8.18
Transportation 7.87
Consumer Opportunity 7.80

0   2.0   4.0   6.0   8.0   10.0

Figure 3.2. The Relative Importance of Community Quality-of-Life Factors
SOURCE: Adapted from Segedy and Truex (1994).

Another indication of changing cultural forces that reflect a concern for quality of life is the increasing adoption by American businesses of total quality management (TQM) and quality-of-work-life initiatives. These programs recognize the value of empowerment in improving product quality and worker productivity. Bruce McClendon (1992) and others have expanded the concept to the public sector and governmental functions. The parallels between the new emphasis on community quality of life and the TQM movement arise from the same basic concept: Whether in the home, the workplace, or the community, the straightest path to productivity and economic prosperity is quality of life. We are seeing, therefore, not only a change in corporate decision making and political/ governmental processes but a parallel change in economic development players. Each of these is becoming aggressively involved with economic development policy and community marketing. "Economic development advances multilaterally as a result of the collective marketing effort of multiple, public and private, development agents within the community complicating the process of public policy making" (Vann & Kumcu, 1995, p. 1). It is in this context that the community's resources are mobilized

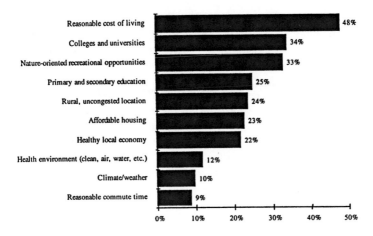

Figure 3.3. Importance of Quality-of-Life Factors in Attracting New Corporate Facilities
SOURCE: Fusi (1991, p. 735).
NOTE: Percentage totals are greater than 100% because more than one option could be chosen.

and allocated to achieve desired economic development outcomes and an optimum quality of life for the society (Samli, 1992).

Not only is the retention of businesses and personnel important in maintaining a strong and diverse economic and tax base; it serves also to establish a community identity. This identity serves not only as an indicator of community and economic stability but as a quality-of-life factor attractive to new business and residential location. If quality of life can be demonstrated as a strong "pull factor" in attracting firms, the converse may also be true. Myers (1987b) and Taylor (1987) suggested, in two separate studies, that the perception of negative factors such as high crime rates and poor educational opportunities can lead to skilled-workforce migration. This is particularly true for industries that communities find most desirable: high-tech manufacturing (65%), research and development (63.4%), and education (62.4%). According to Myers (1987b), "Analysis of the data also suggests that skilled persons who perceive a decline are more likely to plan departure . . . in the next five years" (p. 274). Taylor (1987) further noted the growing realization of corporate executives that "the quality of [their employees'] lives has a direct impact on [their] bottom line through absenteeism, loyalty, productivity and health-care costs" (p. 3).

Beyond the articulation and development of community amenities, as factors of attraction in this new economic development paradigm, are the contributions of community involvement and civic initiative. This new development model featuring quality of life must maintain a healthy, supportive balance between private-sector growth and development and public-sector resources and services (culture, infrastructure, parks, schools). John Marlin (1988) argued that "civic participation, as the infrastructure of public enterprise, plays the same role in ensuring good government that an entrepreneurial infrastructure plays in ensuring economic growth" (p. 42). Civic initiative is essential to the economic development professional's new "reading" of community development, helping to ensure governmental success, enterprise development, entrepreneurial activity, economic vitality, and, ultimately, an enhanced quality of life. It has even been suggested that a "well-developed community spirit" may be the greatest factor in the attraction and retention of businesses and community residents (Glaser & Bardo, 1991).

■ The Emergence of Quality of Life in Economic Development

Quality of life is an elusive commodity. Although there is overall agreement as to its general social importance, there is almost an equal amount of disagreement as to what its essential economic development features are. At the level of social consequence, Maslow (1943), in an outlined hierarchy of needs, suggested the importance of self-actualization in the life of our communities. This was further promoted in 1964 by the introduction of President Lyndon Johnson's Great Society:

> The task of the Great Society is to ensure our people the environment, the capacities and the social structures which will give them a meaningful chance to pursue their individual happiness. Thus the Great Society is concerned not with how much, but with how good—not with the quantity of goods, but with the quality of our lives. (Johnson, quoted in Campbell, 1981, p. 4)

The importance of a holistic view of community life was also furthered by being recognized in the National Environmental Protection Act standards of 1971. Recognition of a variety of social, cultural, and environmental indicators does not challenge the dominance of economic indicators such as gross national product and consumer price indices; rather, it validates their importance in the growth and change of American communities. Beyond this, however, Szalai and Andrews (1980) suggested that the true origins of the quality-of-life concept lay in popular guides to communities from the late 1950s. These initial studies were highly impressionistic descriptions of communities' quality of life in terms of "liveability" and scales of relative ranking. Taken as a whole, they showed that the conditions and amenities of life in one community could be described very differently from the conditions in another community. As such, the relevance and validity of community-based comparative studies of subjective indicators as sources of "universal" targets or benchmark factors of quality of life were, and remain, highly suspect.

Although the relationship of community quality of life to residential satisfaction became relatively well established in the literature, studies such as Robin Widgery's (1982) analysis of citizen satisfaction with community vitality and quality of life showed that such subjective measures may need to be read differently by those interested in broader issues of community (Widgery, 1982). The study demonstrated the importance of aesthetics, media, recreational opportunities, neighbors, sense of security, civic leadership, housing quality, family/friends, public services, educational system, ethnic and cultural diversity, transportation, medical services, and neighborhood and regional context, as well as traditional economic vitality, to the overall well-being of the individual in the community. Although it can be argued that the well-being of the individual citizen is important, Dowell Myers (1988) found that the traditional psychological perception centered on individual-community satisfaction can divert attention away from communitywide planning and development activities. In his study of Austin, Texas, Myers discovered in an assessment of individual satisfaction and well-being that there was no consensus among his respondents on the meaning of community quality of life. At the same time, he found strong consensus on the importance of quality of life in directing public policy and community development.

Further confusing the issue of importance of quality of life in economic development policy is the uniqueness of each community and its perceived value to its citizens. For a resident of New York City, a population density of less than 100 persons per acre stimulates feelings of isolation; for a rancher living on the Great Plains, another house within a mile induces feelings of claustrophobia. In communities with large Amish populations, aggregate educational attainment levels are often quite low because the Amish generally do not value or receive formal education beyond the eighth grade. In other communities, few would argue the importance of educational opportunity to a community's well-being and economic development potential, but in the Amish community, this specific quality-of-life benchmark is not relevant to the community. In the words of one local planning director, "Residents don't need a bunch of statistics to tell them about quality of life in their town" (Landis & Sawicki, 1988, p. 336).

Perceptions of community quality-of-life factors, therefore, are highly contextual or subjective, and the task of employing them as part of an

TABLE 3.3   The Four Pitfalls of Local Economic Development

1. A Failure of Desire:
   Failure to develop a vision that is shared within the larger community, and a failure to achieve consensus on a core of values that is connected with that vision that can be embodied in the local development enterprise

2. A Failure of Patience:
   A misunderstanding of leadership: a failure to understand that development leadership is a facilitation of learning

3. A Failure of Perspective:
   Failure to see process, not events, as the substance

4. A Failure of Nerve:
   Failure to go to the roots of a problem and address it there

SOURCE: Jeep (1993, pp. 239-242).

economic development strategy is to communicate an accurate profile of community life to potential outside firms and employees. One way to portray this might be to employ "a hedonic model that incorporates variations in both wages and housing expenditures" (Blomquist, Berger, & Hoehn, 1988, p. 105) as well as a variety of satisfaction studies and livability indices. The perceptions of the individual may ultimately be a more important ingredient in the location decision than normative livability rankings. The extent to which community leaders and location decision makers recognize this might well add greatly to the success of their economic development policy.

■ Making It Work: The Application of Quality of Life to Local Economic Development Practice

Gottlieb (1992) has identified three problems in producing an objective articulation of the relationship between local quality of life and economic development:

> 1) Local environmental conditions have a relatively small effect on human happiness; 2) the relative importance of different amenities at a place is subjective, and can never be estimated by a researcher creating an index of "objective" data; and 3) quality of life indices rarely incorporate the value of diversity—as opposed to quantity—of amenities. (p. 20)

These are indeed valid concerns, and, as this chapter suggests, community and economic development practices are not purely scientific or objective pursuits. In his discussion of the pitfalls of local economic development (Table 3.3), Jeep (1993) argued that a return to public purpose cannot be facilely dismissed as a freighting of the market system with debilitating and artificial constraints. Public and private goods cannot be seen to be incompatible. But, something vital to both private and public purposes has been increasingly missing from the scene—an understanding and consideration of the more substantive relationships between economic structures and the well-being of community and people. (p. 238)

From this, one can infer that a community must be responsive to these issues and redirect its economic development policies and activities toward the enhancement of local quality of life. A community needs to move beyond conventional economic development mind-sets to a more sensitive form of development that considers local needs and goals as a whole.

The new economic and community development paradigm should be based on eight fundamental goals:

1. Enhance the quality of life of the local community.
2. Promote localization of economies.
3. Promote value-added processes.
4. Develop existing community identity.
5. Develop community identity.
6. Promote social equity.
7. Empower citizens to become involved in community life.
8. Find linkages and encourage community diversity.

This integration of quality-of-life issues into community and economic development practice suggests several fundamental characteristics key to its success: (a) It must be grassroots based, (b) it must be tied to the local context, (c) it must involve the broadest base of community support possible, (d) it must be action oriented, and (e) it must be evaluated and adjusted regularly. In many ways, this represents a radical shift from traditional economic and community development practice. Economic development policies are best developed locally, where they can be associated with the unique resources the community has at hand. State and federal policy should therefore be redirected toward (a) enhancing awareness and understanding of the role of quality of life in the development of local communities; (b) strengthening leadership, communication, and consensus building in the local community; and (c) providing more diverse technical support and fiscal resources for local economic and community development programs. Such ideas represent a redirection of local economic development efforts away from the boardroom and to-

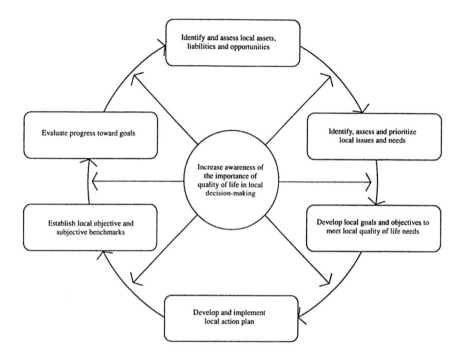

Figure 3.4. Local Quality of Life/Economic Development Process
SOURCE: Adapted from Brown (1994, p. 76).

ward the needs of the community. The new model for local economic development (see Figure 3.4) is an ongoing, citizen-involved process that must include both subjective and objective benchmarking elements in its implementation. It represents an economic and community development paradigm that relies less on techniques and recognizes

> human aspirations, the insights, and the energies that are requisite to the success of authentic development. . . . We have evolved a de facto development theory that is so adverse to human value commitments that it prevents us from evoking or even recognizing the very human energies that are essential to succeeding in our enterprise, that enterprise in which we are otherwise so technically competent, and to which we are genuinely committed. (Jeep, 1993, p. 242)

■ Conclusion

The role of quality of life in location decision making and in local economic and community development is not new, but the reliance on subjective as well as objective parameters is only now becoming recognized as a key determinant in the process. Amenities, aesthetics, quality and diversity of educational and cultural opportunities, equity and citizen empowerment, and perceptions of safety and environmental quality— these are all factors that need to be added to land, labor, and capital in the new economic development paradigm. As the results of a survey of nearly 1,300 businesses, conducted by the Joint Economic Committee of the U.S. Congress (U.S. Congress, 1979), suggested,

> A city's quality of life is more important than business related factors. . . . The results of this survey suggest that individual programs and policies which respond to a particular business need will probably be of limited success in encouraging firms to expand or attracting new firms if they are not a part of a comprehensive effort to upgrade the quality of life in the city. (p. 3)

The bottom line has changed.

REFERENCES

Alonso, W. (1972). Location theory. In M. Edel & J. Rothenberg (Eds.), *Readings in urban economics*. New York: Macmillan.

Andrews, F. M., & Withey, S. B. (1976). *Social indicators of well-being: Americans' perceptions of life quality*. New York: Plenum.

Blomquist, G. C., Berger, M. C., & Hoehn, J. P. (1988). New estimates of quality of life in urban areas. *American Economic Review, 78*, 89-108.

Brown, K. K. (1994). *Assessing, benchmarking, and evaluating community quality of life: The building blocks of successful local economic development*. Unpublished undergraduate thesis, Ball State University.

Browning, E. J. (1980). *How to select a business site: The executive's location guide*. New York: McGraw-Hill.

Campbell, A. (1981). *The sense of well-being in America: Recent patterns and trends*. New York: McGraw-Hill.

Center for Real Estate and Urban Economics. (1990). *Housing prices, other real estate factors and the location choice of firms* (Quarterly Report).

Dalkey, N. C. (1972). *Studies in quality of life: Delphi and decision-making*. Lexington, MA: D. C. Heath.

Daniels, T. L. (1993, October). *Big business and small towns: Threats and opportunities in industrial deconcentration*. Paper presented at the annual meeting of the Association of Collegiate Schools of Planning Conference, Philadelphia.

Eisinger, P. (1988). *The rise of the entrepreneurial state*. Madison: University of Wisconsin Press.

Festervand, T., Lumpkin, J., & Tosh, D. (1988, Summer). Quality of life in the industrial site location decision. *Journal of Real Estate Development, 4*, 19-27.

Foster, R. (1977). Economic and quality of life factors in industrial location decisions. *Social Indicators Research, 4*, 247-265.

Furuseth, O. J., & Walcott, W. A. (1990). Defining quality of life in North Carolina. *Social Science Journal, 27*, 75-93.

Fusi, D. S. (1989). Major quality of life improvements impact communities' wide-ranging spectrum of lifestyles. *Site Selection, 34*, 924-930.

Fusi, D. S. (1991). Education continues to score high as a factor in quality of life location equation. *Site Selection, 36*, 732-738.

Glaser, M. A., & Bardo, J. W. (1991). The impact of quality of life on recruitment and retention of key personnel. *American Review of Public Administration, 21*, 57-72.

Gottlieb, P. D. (1992, October). *Amenities as an economic development tool: Is there enough evidence?* Paper presented at the annual meeting of the Association of Collegiate Schools of Planning Conference, Columbus, OH.

Greenhut, M., & Colberg, M. (1962). Factors in the location of Florida industry. In J. Karaska & D. Bramhall (Eds.), *Locational analysis for manufacturing*. Cambridge, MA: MIT Press.

Haigh, R. (1990). Selecting a U.S. plant location: Management decision process in foreign companies. *Columbia Journal of World Business, 25*, 22-31.

Hart, S., Denison, D., & Henderson, D. (1989). A contingency approach to firm location: The influence of industrial sector and level of technology. *Policy Studies Journal, 17*, 599-623.

Heenan, D. A. (1991). *The new corporate frontier: The big move to small town, U.S.A.* New York: McGraw-Hill.

IMPULSE Research Corporation. (1991). *Survey of major U.S. corporations on relocation plans for the 1990's*.

Jeep, E. G. (1993). The four pitfalls of local economic development. *Economic Development Quarterly, 7*, 237-242.

Kumcu, E., & Vann, J. W. (1991). Public empowerment in managing local economic development: Achieving a desired quality of life profile. *Journal of Business Research, 23*, 51-65.

Landis, J., & Sawicki, D. (1988). A planner's guide to the *Places Rated Almanac*. *Journal of the American Planning Association, 54*, 336-346.

Lyne, J. (1988). Quality of life factors dominate many facility location decisions. *Site Selection, 33*, 868-869.

Lyons, T. S., & Hamlin, R. E. (1991). *Creating an economic development action plan: A guide for development professionals*. New York: Praeger.

Malecki, E. J. (1984). High technology and local economic development. *Journal of the American Planning Association, 54*, 262-269.

Marlin, J. T. (1988). *Cities of opportunity: Finding the best place to work, live and prosper in the 1990's and beyond*. New York: Master Media.

Maslow, A. H. (1943). A theory of human motivation. *Psychological Review, 50*, 370-396.

McClendon, B. W. (1992). *Customer service in local government: Challenges for planners and managers*. Chicago: APA Planners Press.

McNulty, R. H. (1986). Pollyanna, or is the glass half full? *Annals of the American Academy of Political and Social Science, 488*, 148-156.

McNulty, R. H., Jacobson, D. R., & Penne, R. L. (1985). *The economics of amenity: A policy guide to urban economic development*. Washington, DC: Partners for Livable Places.

Myers, D. (1987a). Community-relevant measurement of quality of life: A focus on local trends. *Urban Affairs Quarterly, 23,* 108-125.

Myers, D. (1987b). Internal monitoring of quality of life for economic development. *Economic Development Quarterly, 1,* 268-278.

Myers, D. (1988). Building knowledge about quality of life for urban planning. *Journal of the American Planning Association, 54,* 347-358.

Oakey, R., & Cooper, S. (1989). High technology industry, agglomeration and the potential for peripherally sited small firms. *Regional Studies, 23,* 347-360.

Oregon Progress Board. (1993). *Oregon benchmarks: Standards for measuring statewide progress and government performance.* Salem, OR: Author.

Organisation for Economic Cooperation and Development. (1982). *OECD list of social indicators.* Paris: Author.

Rees, J. (1984). High technology location and regional development: The theoretical base. In Office of Technology Assessment (Ed.), *Technology, innovation, and regional economic development.* Washington, DC: Government Printing Office.

Samli, A. C. (1992). *Social responsibility in marketing.* Westport, CT: Quorum.

Schmenner, R. (1982). *Making business location decisions.* Englewood Cliffs, NJ: Prentice Hall.

Segedy, J. A., & Truex, S. I. (1994). *Indiana total quality of life 1993 pilot report.* Muncie, IN: Ball State University, Department of Urban Planning.

Smith, M. T., & Nance-Nash, S. (1993). The best place to live now. *Money, 22,* 124-142.

Szalai, A., & Andrews, F. M. (1980). *The quality of life: Comparative studies.* Beverly Hills, CA: Sage.

Taylor, H. (1987). Evaluating our quality of life. *Site Selection, 156,* 1-4.

U.S. Congress, Joint Economic Committee, Subcommittee on Fiscal and Intergovernmental Policy. (1979). *Central city businesses: Plans and problems.* Washington, DC: Government Printing Office.

Vann, J. W., & Kumcu, E. (1995). Achieving efficiency and distributive justice in macromarketing programs for economic development. *Journal of Macromarketing, 15*(2), 5-22.

Venable, T. (1991). Recreation, cultural attraction top list of communities' quality of life improvements. *Site Selection, 36,* 740-742.

Webber, M. J. (1984). *Industrial location.* Beverly Hills, CA: Sage.

Widgery, R. N. (1982). Satisfaction with the quality of urban life: A predictive model. *Journal of Community Psychology, 10,* 37-48.

Yaseen, L. C. (1960). *Plant location.* New York: American Research Council.

COMMENTARY ON THE SIGNIFICANCE
OF "QUALITY OF LIFE"

Signe M. Rich

This chapter describes the emerging importance of quality of life to corporate decision makers. In the highly competitive contest between cities for corporate relocations, traditional economic incentives are proving less of a factor than are more intangible qualities of community life. Although difficult to prove statistically, this thesis seems to be experientially valid. For example, at a recent reception for corporate executives seeking to relocate out of California to Albuquerque, there was little interest in economic incentives offered by the city. Instead, discussion focused on the quality of civic involvement and generated a great deal of interest in whether the community supported a new performing arts center.

This chapter correctly identifies several reasons for this trend: decentralization of corporate decision making, spinning off of clusters of related companies, and ascendancy of smaller businesses. Segedy points out that many emerging technology- and research-based industries are able to locate anywhere and that many variables, such as capital, labor, taxes, and labor, are now available universally. As more communities offer the same incentives, those incentives lose their power, and there is less emphasis on deal making and more on those unique features that distinguish a community from its competitors.

This trend is good news for communities, as competition based on economic breaks for business can ultimately be self-defeating. Tax abatement and cheap labor result in less money available for infrastructure, transportation systems, cultural amenities, education, and social equity, creating a downward cycle for future investment and lowering the quality of life for existing residents.

Residents are becoming more aware of these impacts. As citizens come to understand the potential synergistic relationships between economic development and quality of life, they are taking the lead in redefining economic development as a broader and more energizing concept of economic vitality. At a 1994 Shared Vision Town Hall on economic development in Albuquerque, consensus emerged that *economic vitality* meant more than economic development and the recruitment or creation of new businesses. Rather, it concerned a sense of the well-being of the community, as exemplified by such quality-of-life attributes as culture, environment, education, safety, transportation, and equal opportunity.

What does this new paradigm mean for the practitioner of economic development? The economic development profession has traditionally managed the basic tools of capital, infrastructure, tax incentives, and development of a skilled workforce. It is easier to offer an industrial revenue bond or source of equity capital for a growing business than it is to influence civic dynamics and create community. It is also more difficult to measure results of community building.

Planning departments generally concern themselves with land use and regulatory processes rather than with developing a civic culture in the larger community. (It should be noted that communities that know how to treat businesses respectfully with fair and timely development processes are also at a competitive advantage, resulting in the current emphasis on customer satisfaction through total quality management.)

Segedy's chapter implies that small towns are the new competitors, as out-migration that previously moved from center cities to suburbs is now moving from suburbs to small towns. Such growth at the edge has historically proven detrimental to those areas left behind. This argues for the importance of economic strategies at the regional level that encourage all communities to know who they are and build on their unique strengths.

Economic development professionals must learn to mobilize community resources as part of their task of creating a healthy investment climate. In aligning themselves primarily with the business sector, practitioners run the risk of cutting themselves off from an essential constituency that understands and has the potential for leading the new paradigm. The current movement toward economic development at the neighborhood level can be viewed as an attempt to overcome polarization between business and civic interests by applying business incentives to community-based organizations, enabling them to undertake commercial projects. Processes for enhancing civic effectiveness and quality of life can and should be added to the repertoire of economic development agencies.

The Albuquerque town hall forum identified the following components of a program to enhance economic vitality:

- Balancing resource stewardship, cultural development, and economic development
- Increasing the wealth of the community and encouraging its investment back into the community
- Developing local human resources
- Encouraging the attraction, development, and retention of responsible and responsive business
- Cultivating attributes of place and culture that make the community unique
- Ensuring that benefits and opportunities accrue to all citizens

- Enhancing opportunities for achieving employment of a diverse and multicultural population
- Ensuring fair access to capital and technological resources
- Identifying a unified common direction for the community as a whole
- Developing the ability to work together and think regionally

Many of these components could be formulated as measurable time-bound goals. Such indices would help refine a community's understanding of the dynamics of economic vitality and ability to assess the effectiveness of various strategies. For example, a sense of community could be measured by opportunities people have for dialogue and working together. Technologies for citizen participation must be developed for economic development professionals to operate effectively within the new economic development paradigm.

At the juncture of people and economics is a sense of quality of life —culture, art, music, neighborhoods, historic districts, excellence in design of public places, inclusion, and social equity. These building blocks of communities are also assets for economic investment.

Economic development as an emerging, dynamic profession attuned to the realities of a changing world should involve building relationships and partnerships with the community. This means involving people at the beginning of the process. Because economics are the lifeblood of communities, affecting jobs, income, and community life, economic development has potential for uniting a community behind a common theme in which everyone has a stake. The survival of each community depends on its ability to function with inclusiveness, to bring people in to assume a role in decision making. Economic development professionals have a unique opportunity to help communities build the human infrastructure and big-picture strategies needed to compete effectively for jobs and wealth.

COMMENTARY ON "QUALITY OF LIFE" IN LOCATION DECISIONS

Brian J. Cushing

In the foregoing chapter, James Segedy illustrates the potential importance of "quality of life" for business location decisions and economic development policies. Its actual importance, however, is still unknown.

Reading Professor Segedy's chapter raises some important issues that scholars and practitioners must address to understand the effect of quality-of-life considerations on location decisions.

Before we make any progress on this topic, there has to be some uniformity in the use of the term *quality of life*. Many scholars employ a very broad definition that includes traditional economic factors (e.g., income opportunities, employment opportunities, and cost of living) along with amenity factors (e.g., climate, quality of schools, and cultural and recreational opportunities)—all things considered, how satisfying is a particular location? Others restrict *quality of life* to amenities and proceed to discuss the relative importance of economic opportunities and amenities. The broader term is the bottom line for households. To address the policy issue regarding the relative usefulness of traditional economic development policies versus policies with more emphasis on locational amenities, the more restrictive definition is more informative. The differing definitions of *quality of life* make it impossible to draw any strong conclusions from the literature.

As Segedy illustrates, there is some evidence that despite the definitional problem, amenities attract business firms. But we do not have much idea of exactly how quality of life affects business location decisions. A firm may choose a location on the basis of the owner's attraction to local amenities, with labor being attracted by the employment opportunities offered by the firm. Alternatively, the firm's location decision may be purely financial: The local amenities attract labor, resulting in a relatively large labor supply and low labor costs for the firm. The former case might require an individualized approach not amenable to a general economic development policy. The latter case would call for a general policy ultimately aimed at very traditional economic forces: providing an adequate labor pool and other resources required for efficient, low-cost production. One of the means for providing a labor pool might be the development of attractive local amenities.

How could research scholars and policy specialists proceed to understand better the significance of quality of life in location decisions? We already have much information on the labor supply effects of amenities (or the broader definition of *quality of life*). In the literature on compensating differentials, Rosen (1979), Roback (1982), and others have firmly established that households accept lower wages and higher rents in exchange for better amenities. In the migration literature, Graves (1983) and many others have demonstrated that amenities are an important determinant of household location decisions (although there is still much debate regarding the relative importance of economic opportunity and amenities). Besides increasing our understanding of the labor supply effects of amenities, this literature has provided a wide variety of measures for amenities that would be useful in studying business location decisions.

The major impediment to studying directly the link between amenities and business location decisions is the lack of good data. For households, we have innumerable data sources, including many microdata samples that provide detailed information on individual households. For firms, not only do we lack microdata samples, but the aggregate firm data are much more highly aggregated than household data.

Three complementary research approaches would be beneficial for studying the significance of quality of life in business location decisions. First is a metropolitan-area- or county-level simultaneous equations model that includes equations for employment growth, migration, and wages. Greenwood's (1981) model of growth and migration is a prototype that researchers could modify to include directly wages, greater disaggregation by industry and/or occupation, and a set of variables to capture locational amenities. Though rather general, the model has the advantage of using a broad-based national sample that could be replicated for different time periods. The systems estimation approach would shed light on the issue of whether amenities directly affect business location decisions or indirectly affect them through labor supply and wages.

There is a substantial economics literature on intraurban firm location decisions and the effect of federal, state, and local economic development policies. The studies in this literature generally focus on a single metropolitan area and employ some detailed data on firms. Most of the research has considered the manufacturing industry, but more recent research, such as that by Ihlanfeldt and Raper (1990), has considered other sectors. Because these models are able to abstract from amenities such as climate, they are useful for learning more about the influence of local taxes, public service levels, and other differential characteristics of central cities and suburbs.

Community surveys are the basis for the third type of research. Professor Segedy makes several references to this type of research, including work by Myers (1987a, 1987b, 1988). Segedy correctly points out that scholars may never be able to measure adequately some community characteristics, such as community spirit, for use in empirical research. Surveys are also the only way to find out how often businesses make location decisions for very personal reasons that economic development policy cannot influence: for example, Alonso's (1972) instance of the manufacturing firm that located in Worcester, Massachusetts, because the owner's mother-in-law lived in Worcester and his wife insisted on residing in the same city. Community survey studies may confirm the findings of the broader studies or indicate that the other types of studies miss some crucial factors.

In the end, the important policy question is whether economic development policy should incorporate quality-of-life considerations, amenities in particular, and if so, how much emphasis policy should place on quality of life relative to traditional economic incentives such as tax breaks. The answer to this is likely to depend on the economic situation of the community and the trade-off that residents are willing to make to attract additional employment. Bartik (1991) surveyed the research on effectiveness of economic development policy and concluded that traditional economic incentives make a difference. Even urban enterprise zones, the polar opposite of quality-of-life development policy, are effective in some situations (see papers in Green, 1991). Improving quality of life to retain current employment is a moot point if there is little employment to retain. Likewise, quality-of-life policies may not be sensible if an area sees itself as having little chance of competing with surrounding communities in terms of amenities. The research approach suggested above is likely to show that both traditional economic incentives and amenities influence business location decisions. More important, it may inform us regarding the relative importance of these two factors, the mechanism through which amenities affect location decisions, and how their relative importance varies by type of firm and location. These are critical issues for formulating economic development policies.

REFERENCES

Alonso, W. (1972). Location theory. In M. Edel & J. Rothenberg (Eds.), *Readings in urban economics*. New York: Macmillan.

Bartik, T. (1991). *Who benefits from state and local economic development policies?* Kalamazoo, MI: W. E. Upjohn Institute.

Graves, P. (1983). Migration with a composite amenity: The role of rents. *Journal of Regional Science, 23*, 541-546.

Green, R. (Ed.). (1991). *Enterprise zones: New directions in economic development.* Newbury Park, CA: Sage.

Greenwood, M. (1981). *Migration and economic growth in the United States.* New York: Academic Press.

Ihlanfeldt, K., & Raper, M. (1990). The intrametropolitan location of new office firms. *Land Economics, 66*, 182-198.

Myers, D. (1987a). Community-relevant measurement of quality of life: A focus on local trends. *Urban Affairs Quarterly, 23*, 108-125.

Myers, D. (1987b). Internal monitoring of quality of life for economic development. *Economic Development Quarterly, 1*, 268-278.

Myers, D. (1988). Building knowledge about quality of life for urban planning. *Journal of the American Planning Association, 54*, 347-358.

Roback, J. (1982). Wages, rents, and the quality of life. *Journal of Political Economy, 90*, 1257-1278.

Rosen, S. (1979). Wage-based indexes of urban quality of life. In P. Mieszkowski & M. Straszheim (Eds.), *Current issues in urban economics*. Baltimore: Johns Hopkins University Press.

RESPONSE TO COMMENTARIES

James A. Segedy

I found both commentaries to be insightful, very helpful, and quite different from each other. The comments were general in their discussion of the issues, rather than specifically addressed to issues and points raised in my chapter. To that end, my responses will be reflective of the spirit of ongoing discussion.

Professor Cushing suggests that there needs to be a greater uniformity in the definition of *quality of life*. In many ways, he is absolutely correct. From an empirical point of view, the lack of easily defined parameters or measurable indicators is indeed a challenge to traditional scientific inquiry. He also points out that the broader variables associated with the concept of "quality of life" are more aligned with households and individuals than with corporations. I might suggest that that is exactly my point. Location (and some may argue corporate) decision making is becoming more "human," individual, and subjective and less "detached." There must be a realization that it is individuals who make decisions, not corporations.

As I suggested in the early part of my chapter, this is also a reflection of the changing nature of corporate structure to a more decentralized, more highly mobile system. At the same time, the community "playing field" has become more level in terms of the traditional location factors (land and capital). This leaves quality of life and workforce as the differentiating factors. I believe that a combination of factors is driving this transformation: both an enhancement in the value placed on quality of life by the individual and a lessening in importance of the traditional economic advantages of geography. The point is well taken that if communities are not "equal" in terms of economic advantage, either by circumstance or by public policy (enterprise zones, "small town triage," etc.), the importance of amenities will be much less of a factor (I suppose that one could therefore apply Maslow's hierarchy to economic development). It is here that economic development policy must redirect its focus. Do we focus our attention on leveling the "playing field" and letting the unique quality of life and amenities associated with the individual community be the determinant, or do we apply the principles of the "Buffalo Commons"?

I also believe that the importance of quality of life in economic development practice parallels the recognition of existing business expansion and retention and economic localization as critical factors in the local community's economic vitality. These factors, again, favor the more individual/household-scale decision-making process, which places a higher value on quality of life, amenities, and so forth.

I agree that the more analytical approaches need to be pursued, but I also believe that public policy needs to recognize the less empirical sides of the decision-making process. It is here that I disagree with Cushing, not in principle or method, but in application. I do not believe that we should, or even can, differentiate between traditional economic incentives and amenities. Public policy should reflect the importance of both.

Signe Rich's comments seem to reinforce the points made in my chapter. I am very heartened that the points and examples presented in the comments provide supportive evidence for the role of amenities, culture, involvement, diversity, equity, and identity (the building blocks of life quality) in economic development policy and planning. The point that it is much easier to offer a revenue bond than to build a sense of community or identity is well taken. I could not agree more. It is perhaps here that the goal of my chapter lies. Economic development policy and practice should not look toward the "quick fix," but instead look toward the goal of quality of life, and our indicators and benchmarks must reflect this.

4

Is There Really an Infrastructure/Economic Development Link?

DAVID ARSEN

Recent years have witnessed a remarkable swell of interest in public infrastructure spending as a strategy to promote economic development. Though specialists in regional and local economic development have long recognized infrastructure investment as a possible, maybe even enlightened, growth policy, the genesis of this new attention lies elsewhere in two largely unrelated developments. On the one hand, startling research by David Aschauer (1989a, 1989b) and Alicia Munnell (1990a, 1990b) suggested that a secular decline in public-capital stock growth was an important contributor to the nation's mysterious and troubling productivity growth slowdown. Then Bill Clinton, drawing on ideas developed by Robert Reich (1991), promoted public investment as a centerpiece of his economic growth program during the 1992 presidential campaign. It is hardly surprising that the response of policy-conscious economists has been a vast increase in empirical research on the effects of public infrastructure on private-sector economic performance.

This chapter surveys and evaluates available infrastructure research for the lessons it provides for economic development practitioners primarily at the state and local level. Recent econometric research greatly augments the very limited information available from case studies of infrastructure's development impacts. Yet the reader should be warned at the outset that concrete insights of use to practitioners are still very slim. The empirical research has not converged to consensus on fundamental issues. Moreover, even if the technical points now in dispute are resolved, the type of empirical tests typically conducted cannot in any case provide much help to policy makers charged with making actual investment decisions.

The answer to the assigned question posed in this chapter's title is an unambiguous "yes." Indeed, there are several links. The connections between infrastructure and economic development are complicated and subtle, and many are poorly understood. Most of these linkages are not uniform over time or across locations. The most useful insights are to be found in posing a range of more refined questions that recognize the significance of a variety of conditioning factors pertaining to specific features of infrastructure projects and the economic base of an area. The chapter offers a number of suggestions for practitioners who seek to promote economic development through infrastructure investment.

■ Defining and Measuring Infrastructure

Most research on the economic effects of infrastructure has focused on a subset of the public-capital stock, generally called *core infrastructure.* Roughly one half of total public capital, or two thirds of nonmilitary public capital, consists of core infrastructure, hereafter designated simply as *infrastructure.* Infrastructure includes roads, bridges, water supply facilities, sewers, airports, mass transit, and electric and gas plants. This definition excludes substantial portions of publicly owned capital such as military capital, school buildings and equipment, prisons, city halls, and fire stations. The vast majority of infrastructure capital, roughly 90%, is owned by state and local governments, although much of this was financed with federal intergovernmental grants. Infrastructure is a stock of capital generated by flows of public investment. There is wide agreement that public investment may stimulate demand and consequently generate a short-run expansion of employment and income, but this chapter's focus is on the supply-side economic effects of the stock of infrastructure capital.

The economic significance of infrastructure lies not so much in the stock of physical capital per se as in the flow of services it provides to households and firms. Conceptually, it is the amount of services that affects decisions, represented, for instance, as a variable in an individual's utility function or a firm's production function. Broadly speaking, infrastructure services reflect both the quality and quantity of public infrastructure. Beyond that, however, the specific level of infrastructure services is a function of several variables. One, certainly, is public investment spending. But there are many others. Infrastructure services will depend on the design (materials and technology) of public capital, as well as the efficiency and creativity of supporting administrative systems. Utilization rates or the degree of congestion will affect service levels. Alternatively, services will depend on complementarities between an area's infrastructure and its particular economic base. It turns out that this distinction between public-capital stock and the services it generates is very important for evaluating infrastructure research and policy.

Whatever the theoretical appeal of infrastructure services, it is a concept that does not readily lend itself to empirical operationalization. To measure it, one needs detailed information on the physical characteristics —capacity, quality, and utilization—of specific infrastructure projects and networks. The data requirements are daunting, if not completely overwhelming, except for well-defined small-sample or case studies. Essentially, these data requirements make incorporating direct measures of infrastructure services in regression analysis unfeasible. This is a serious limitation. For econometrics provides the most accessible methodology for establishing correlations and relationships between measures of economic development and infrastructure while controlling for other factors.

Confronted with the choice between using an imprecise measure of infrastructure that is compatible with econometric techniques and devising more accurate infrastructure measures that are not suitable for econometrics, most empirical analysts take the former option. With few exceptions, the literature does not attempt to measure infrastructure services at all, but rather assumes (usually implicitly) that services can be approximated by measures of infrastructure stock. The usual approach is to measure the infrastructure stock in monetary terms derived from the value of past public investment adjusted for depreciation (e.g., Costa, Ellson, & Martin, 1987; Holtz-Eakin, 1993b; Munnell, 1990a). Given annual investment flows, public-capital stocks can be estimated using the perpetual inventory method:

$$K_t = (1 - d)K_{t-1} + I_t \qquad (1)$$

where K_t is the end-of-year capital stock in year t, d is the rate of depreciation, and I_t is real investment in year t. Implementation of this procedure requires two additional pieces of information: an estimate of a benchmark public capital-stock, K, on which investment flows accumulate and an estimate of the depreciation rate. In cross-section studies, an area's capital stock is typically standardized by population size.

The great advantage of perpetual inventory accounting is that it generates a measure of infrastructure stocks that can be consistently applied over time and across areas, and within which any combination of specific types of infrastructure can be estimated and aggregated into a single variable. The Bureau of Economic Analysis (BEA) estimates benchmark aggregate (nationwide) public-capital stocks in selected years for both the federal and the state-local government sectors. The BEA, however, does not disaggregate its national-public capital stock estimates by geographical area. So, to obtain benchmark capital stocks, researchers pursuing regional questions must devise estimating procedures to apportion the aggregate state-local capital stock figures across states or regions. Estimates of benchmark state infrastructure stocks are sensitive to the particular assumptions adopted in the apportioning procedure (Holtz-Eakin, 1993b). Because public capital is rarely sold once it is constructed, there are no direct measures of public-capital depreciation. Accordingly, the BEA estimates of depreciation rates entail more guesswork for public than private capital.

Though such capital stock estimates may be as good as one can do for large-scale empirical research, they are obviously very crude proxies for the infrastructure services that actually affect economic development. Two infrastructure networks that embody exactly the same dollar amount of investment may yield vastly different flows of productive services, or the same services may be generated by very different investment expenditures. The latter situation is particularly relevant. Indeed, there is certainly far less interstate variation in infrastructure services than in infrastructure expenditures.

The use of per capita public-capital stock estimates to proxy infrastructure service levels produces some highly dubious implications in practice. In Holtz-Eakin's (1993b) careful estimates, we might attribute the fact that Alaska's (per capita) capital stock is over four times the national average to anomalies associated with that state's low population density (a distortion that is relevant, though less significant, for all jurisdictions). What, then, are we to make of the estimates showing that Nebraska's per capita stock is double Indiana's? Is it really plausible that the level of infrastructure *services* as they are perceived by a typical firm is twice as high in Nebraska? How about Eberts's (1990) estimates indicating that the per capita public-capital stock (services?) in metropolitan New York City exceeds San Diego's by 200%? Rather than offering a reliable proxy for variations in infrastructure services, it would seem just as plausible to assume that the services of interest to business are essentially uniform across states so that variations in expenditures largely reflect institutional, historical, demographic, and climactic factors.

It is easy to recognize why this is so. Some areas show lower measured infrastructure stock simply because certain functions, notably water supply, have been assigned to the private sector. By standard measure, high public-investment spending due to an area's high labor or land costs, or even graft and corruption in the public works department, would be reflected as an increase in the measured infrastructure stock. The capital costs associated with providing a given flow of water services vary substantially by region. Old and young urban areas vary dramatically in the share of infrastructure spending that must be devoted to the replacement of obsolete capital versus the expansion of services (Eberts, 1991). And it is far more expensive to run a quarter-mile of water pipe through Manhattan than through a newly developing rural area. Finally, of particular relevance to this chapter, standard measures of infrastructure offer no way to distinguish between two jurisdictions with equal spending, where one jurisdiction directs the funds to infrastructure used strictly by households, whereas the other concentrates them on industrial park infrastructure highly valued by firms.

The difficulty of constructing measures of infrastructure that capture the stock's relevant features for economic development is doubtless a major obstacle to research in this area. More satisfactory direct measures of some aspects of public-capital services (e.g., traffic flow) are feasible for detailed case studies. It is tempting to suggest that if interjurisdictional variations in infrastructure services could be measured as readily as tax rates, we would know a great deal more than we do about the infrastructure/development link. Although that is probably so, one should not presume that the explanations for observed correlations would be entirely evident. For apart from measurement issues, the conceptual linkages between infrastructure and economic development are also complicated.

■ Conceptual Linkages Between Infrastructure and Development

There is little question that infrastructure investment is associated with economic development. There is a good deal less consensus about the direction of causation between these two variables. Economic development tends to call forth infrastructure improvements both by expanding demand for infrastructure services and by increasing fiscal resources to finance public investment. Though few would dispute this sort of linkage, most research on the infrastructure/development linkage, as well as the focus here, concerns the reverse causation: the potential for infrastructure investment policy to initiate or promote economic development.

Despite widespread agreement that public infrastructure can be an important factor in facilitating economic growth, the precise nature of the linkage is by no means clear. Past empirical research on infrastructure has not devoted much attention to detailing the theoretical conception of infrastructure's impact on development before searching for possible statistical correlations. The linkage, like so many aspects of internal firm decision making, is treated as a black box. Although there has been speculation on an assortment of specific linkage mechanisms, essentially one finds two primary ways in which the linkage is conceptualized.

First, is the view of public infrastructure as a business location incentive. Here, infrastructure matters insofar as the quantity and quality of the capital stock in an area attracts new business investment or induces existing firms to stay and expand. Interjurisdictional competition for business location is implicit in this view. Naturally, this aspect of the infrastructure/development link has been of primary interest to regional and local economic development specialists.

If infrastructure is to serve as a site advantage for business location, then it must either provide distinctive and attractive services or, more likely, offer the standard public-capital services that firms can obtain at any number of alternative suitable sites at a particularly low cost to the firm. The degree to which firms require the services of different types of infrastructure will vary across industries. But for any given firm, expenditures for transportation, water, sewerage, and so forth represent costs of doing business. These infrastructure services might even be thought of as "inputs" to the firm's production. So the relevance of this first conception of the infrastructure/development link turns on the extent to which a location lowers business costs by subsidizing firms' utilization of these services. Obviously, the level of this subsidy can only be gauged relative to the taxes and fees a firm faces in the jurisdiction.

It is striking, therefore, that only limited progress has been made in matching business taxes and fees with infrastructure in the analysis of this location incentive conception. Even a community with elaborate and unique public capital will not attract firms unless the tax price is right, including necessarily a price that does not exceed the cost of obtaining equivalent services through private sources. Ultimately, given competition from other jurisdictions, the effectiveness of infrastructure as a location incentive depends on the extent to which resident taxpayers can be persuaded (coerced?) to subsidize business infrastructure services, unless these costs can be exported to nonresidents. Of course, the empirical task is easier described than done. Eberts (1991) and Munnell (1990a) addressed this concern by including quite aggregate (metropolitan area, state) measures of tax levels as explanatory variables.

The second conception sees public infrastructure as an input that modifies firms' production processes such that productivity and output increase. This is the conception of the infrastructure/development link found in most of the recent econometric research. Infrastructure interacts with and improves the efficiency of private factors of production, increasing labor and total factor productivity. Probable corollaries of this enhanced efficiency are lowered unit costs and higher firm profits.

By drawing attention to the potentially productivity-enhancing aspects of public capital, this research has clearly helped to redress a glaring blind spot in conventional macroeconomic analyses of productivity growth. In principle, this aspect of the infrastructure/development link should reveal itself in aggregate time-series data for the nation as a whole and is entirely independent of interstate competition for business. Researchers who pursue the productivity-enhancing conception of infrastructure turn to interstate analyses, not because of interest in infrastructure's role as a location inducement, but rather to overcome statistical problems associated with aggregate time-series analyses.

In practice, these two conceptions may not be entirely distinct. A given infrastructure network may simultaneously affect firm choices in both ways. On the other hand, an area's public capital may fail to improve productivity and still stimulate development through the subsidy effect of the infrastructure-tax package. In addition, the variables representing economic development in the two conceptions are not strictly equivalent. When an area attracts new business investment, economic development—by most definitions—is furthered. Eberts (1991), for example, used business openings as a dependent variable. But public infrastructure could increase productivity without promoting local economic development. Suppose firms operate in product markets characterized by excess capacity and imperfect competition. In that case, the firm's response to an area's new productivity-enhancing public infrastructure could be to maintain output while cutting employment and private investment, a move that is inconsistent with usual notions of local economic development.

The difficulty of conceptualizing infrastructure/development links is not, however, principally due to these alternative mechanisms through which public capital affects firm decisions. Rather, the feasibility of precise and unqualified statements about these links is seriously limited by the fact that the economic consequences of infrastructure are contingent on both the level and nature of existing economic development and the extent of the existing infrastructure stocks. There is no uniform or stable relationship between infrastructure investment and associated economic development benefits. This complexity arises, in large measure, from (a) "network" characteristics of infrastructure stock, (b) the "lumpiness" of infrastructure investment, (c) complementarities among infrastructure networks, and (d) variations in the need for infrastructure to offset negative externalities of private development.

Most infrastructure capital (roads, sewers, and water distribution) is part of interlocking networks of public investment. Hulten and Schwab (1991a) stressed that the economic benefits of any one component of the network depend on the size and configuration of the entire network of capital. The value of connecting Point A with Point B depends on whether the link also puts A in touch with C and D. The economic benefits of building a road, canal, or railroad between two isolated points across an expansive undeveloped area will be far less than those generated by an equivalent investment connecting the same area to established networks in developed areas. Likewise, there will be differences in marginal and average benefits. Although the initial roads into an undeveloped area may open it for growth, it is doubtful, unless congestion sets in, that successive waves of highway spending will generate benefits proportional to the initial stock.

Infrastructure investment is lumpy. In general, the economic benefits of infrastructure are premised on public capital's reaching a certain minimum feasible scale, before which benefits may be nonexistent (Hulten & Schwab, 1991a). It does not pay to build a bridge halfway across a river. There also may be important economies of scale in construction. It costs less to build one eight-lane bridge than two four-lane bridges.

Complementarities among various infrastructure networks are also critical. A new airport may promote growth, but only if it is suitably connected to ground transportation networks. Road repair spending to promote redevelopment in a central-city industrial area may be pointless without equivalent repairs to water and sewer networks. The economic benefits of investment in fixed-rail mass transit will depend on the congestion and current capacity of highways connecting the same points. Accordingly, efforts to establish correlations between a particular type of infrastructure investment and development without accounting for such interactions among types of infrastructure may actually be highly conditional and misleading.

Finally, much of public-capital investment is driven by imperatives created by private-sector development, particularly the need to mitigate negative externalities. For example, given New York City's high population density, commerce would be crippled if streets and highways had to carry all ground transportation. Massive investment in a subway system is essential in New York merely to establish a level of transportation services for the movement of people and goods, yet this level may be inferior to those in other cities that spend virtually nothing on mass transit. Though it may be unfeasible to measure, the share of infrastructure expenditures devoted to offsetting externalities versus extending productive services to business varies across jurisdictions. This variation inhibits efforts to establish unbiased correlations between infrastructure investment and development.

For all of these reasons, economic development is not a linear function of infrastructure investment. A variety of infrastructure/development links exist, but these links are highly contingent. Yet in the regression-based empirical literature to which we now turn, the infrastructure/development link is implicitly assumed to conform to just this sort of simple linear relationship.

■ Evaluating the Empirical Evidence

The economic impact of infrastructure is one of those interesting topics of empirical research in which the more people examine the issue, the more they disagree. Rather than departing from a technical review of the econometric model specifications and *t*-statistics, it may be useful to begin with one reader's observations on the context for the controversy.

Several "early" studies (e.g., Costa et al., 1987; Duffy-Deno & Eberts, 1991; Eberts, 1986, 1991) suggested significant positive effects of infrastructure on measures of state or metropolitan-area development in cross-sectional regressions. These studies located themselves in regional development research. The results could be interpreted as supporting public investment as an economic development strategy, especially for state and local governments. All of this might have been assimilated into the regional development literature were it not for its connections to what soon proved to be much more controversial research.

David Aschauer (1989a) made much stronger claims regarding infrastructure. He also changed the organizing question that came to dominate subsequent research. Now the question was not simply whether infrastructure could promote development but rather whether there was too little public capital. Aschauer claimed there was a shortfall substantial enough to justify a significant expansion of government investment expenditures. For he claimed not only that infrastructure was an important determinant of economic performance but also that a decline in infrastructure stock relative to GNP was a primary cause of the disappointing downward shift in the nation's long-run growth trajectory after the early 1970s (Aschauer, 1991).

Moreover, Aschauer obtained his results with standard state-of-the-art macroeconomic time-series empirical methods. Unlike the previous regional/metropolitan studies, Aschauer's results, if accepted, required the conclusion that public capital was productive, not just a possible inducement to business location. Aschauer's timing could not have been better. In short order, Bill Clinton and his advisors were advocating public investment to revive the economy. Suddenly the topic received attention from a number of prominent economists and think tanks on the left and right. And although the debate within the journals has been strictly nonideological, the partisan stakes are evident to any observer.

In the modeling proposed by Aschauer and others, infrastructure is linked to private output as an input in an aggregate production function whose services enhance the productivity of both labor and capital. By taking infrastructure out of the total factor productivity term A, the aggregate production function can be expressed as $Q = A*F(K, L, G)$, where K is the private capital stock, L is labor inputs, and G is public infrastructure stock. By assuming a Cobb-Douglas form and being expressed in logs, this equation becomes

$$\ln Q = \ln A* + a\ln K + b\ln L + c\ln G \qquad (2)$$

Aschauer's (1989a) original aggregate time-series estimates of (2), as well as Alicia Munnell's (1990b) reestimates, indicated very large impacts of public capital on private output. Indeed, these estimates imply an output elasticity with respect to increased public infrastructure in the range of 0.3 to 0.4, roughly double the estimated elasticity associated with increased private capital. A number of critics (Aaron, 1990; Gramlich, 1994; Hulten & Schwab, 1991a; Tatom, 1991; and even Munnell, 1992) found this effect implausibly large. Aschauer (1993) offered a response to critics.

Munnell (1990a) sought additional evidence by examining the relationship between public capital and measures of economic activity at the state level. Part of this study estimated production functions for states, and another part explored the role of infrastructure in state employment growth. Hence, Munnell connected the aggregate production function research with the regional development studies. Munnell's estimates of the output elasticity with respect to public capital are much smaller, and arguably more plausible, at the state level, roughly half the estimate at the national level. Together with the regional studies noted above and Garcia-Mila and McGuire (1992), this research suggests that public capital has a positive impact on state- or metropolitan-level output, investment, and employment growth.

Aside from doubts about the size of the estimated infrastructure coefficients, critics have focused on three concerns about the econometric evidence. First, they argue that the direction of causation runs from economic growth to increased infrastructure, not the other way around. Second, they maintain that common long-run trends in output and infrastructure data produce spurious correlation in times-series estimates. Third, they argue that unobserved, state-specific effects correlated with public capital actually account for the estimated infrastructure-economic activity correlations in the state-level studies.

Several authors (Eisner, 1991; Hulten & Schwab, 1991a, 1991b, 1993; Tatom, 1993) raise a legitimate concern about direction of causation that applies to both national- and state-level regressions. From this perspective, observed infrastructure/economic activity correlations probably reflect demand-side rather than supply-side effects. Where this issue has been addressed explicitly (e.g., Duffy-Deno & Eberts, 1991), the evidence suggests that causation runs in both directions. This should raise caution about the interpretation of single-equation models in which infrastructure is taken as exogenous, but it still suggests a positive economic effect of public capital.

The charge of spurious correlation in aggregate regressions is also plausible. Infrastructure and productivity both grew rapidly in the 1950s and 1960s, then slowed in the early 1970s, possibly for entirely unrelated reasons. One solution to this problem is to "detrend" the data by taking first differences: that is, to recast the analysis in terms of whether changes in infrastructure can explain changes in output. When Hulten and Schwab (1991a) and Tatom (1991) did this, they found little or no impact of infrastructure on output. This, however, is not a very satisfactory test. The objective of the production studies is to relate the stock of infrastructure to aggregate supply, but when infrastructure increases, aggregate demand is what changes in the short run. First-differencing destroys the long-term, supply-side relationship that is what one wants to estimate (Munnell, 1992).

One advantage of a pooled-time-series, cross-section methodology is that it offers greater variation in the data, lessening this "common trend" problem. The cross-section studies, on the other hand, have been criticized on the grounds that the infrastructure variables actually capture the effects of omitted, "intrinsic," state-specific productivity determinants, such as location, climate, and mineral endowment (Holtz-Eakin, 1993a). When Holtz-Eakin (1994) and Evans and Karras (1994) introduced statistical controls for such effects, infrastructure was found to have little or no effect on output. Though interesting, such evidence would be more compelling if we had some clue, any direct evidence, that might help identify the omitted state-specific effects. After all, if it really is something such as location or climate, then these are effects that can be represented explicitly in the regression. An alternative interpretation of these contrary results might focus on state-specific errors in the extensive assumptions involved in simply generating the state-level measures of infrastructure stock and/or state-specific errors in the relationship between estimated infrastructure stock and actual infrastructure services.

Some researchers have pursued the question with other techniques, but so far this seems merely to create new grounds for apparently conflicting findings. Hulten and Schwab (1991b), for instance, found little infrastructure impact using a regional growth accounting technique, but Nadiri and Mamuneas (1994) and Lynde and Richmond (1992) found significant marginal benefits for public capital by adopting a cost-function approach.

Suppose the technical issues on which disagreement currently centers are eventually resolved, and future research produces consensus on the appropriate estimation and interpretation of the infrastructure/development link as portrayed, for instance, in specifications departing from Equation 2. What assistance might this offer practitioners? The answer is, not much. This is because the form of econometric models currently tested is unlikely ever to provide the fine-tuned information necessary for decision making.

We are a long way from obtaining parameter estimates that are precise enough to enable a government official to plug in values for new infrastructure stock and other variables in an estimated econometric model and obtain reliable estimates of future economic benefits. And although progress can surely be made along these lines, it is doubtful that econometrics represents the most promising strategy to obtain this more disaggregated information.

A government official facing a budget constraint might reasonably want to know the differential anticipated development effects of a dollar of investment in roads, wastewater treatment, or airports in his or her jurisdiction. How does the scale or amount of existing infrastructure stock affect the benefits from incremental additions to that stock? From the standpoint of economic development, what advantage is there to infrastructure spending (whatever the type) specifically targeted to serve business needs as opposed to mixed household and firm use? We also know very little about how the extent and nature of existing business development affect the benefits derived from infrastructure. How do infrastructure benefits vary across industries (cf. Nadiri & Mamuneas, 1994)? Even at the sectoral level, we know relatively little about how the long-term increase of services relative to manufacturing in aggregate economic structure affects the productivity of various types of infrastructure. How do the benefits vary across central-city, suburban, and rural areas? Given all the research on infrastructure, it is striking how little there is of this sort of disaggregated information.

A final word of caution pertains generally to methods based on extrapolating the experience of the past into the future. Over long time spans, technological change transforms the development consequences of any given type and vintage of infrastructure capital. The economic advantages associated with a central-city railroad terminal or access to an overland canal are a good deal less today than they were a century ago. Conversely, past estimates of development links may be largely irrelevant to predicting the future effects of, say, public infrastructure to facilitate an information superhighway.

■ Some Suggestions for Practitioners

Although past research offers little to assist practitioners in projecting the development consequences of potential infrastructure investments, observance of a few guidelines should help ensure that investment decisions are at least promising and prudent. What follows is a user's guide to some dos and don'ts of infrastructure investment.

Most of the following suggestions emerge from thinking about infrastructure broadly, beyond the conventional conception as accumulations of physical structures and equipment. Money spent on physical capital matters, but so do institutional elements relating to the administrative procedures for planning, constructing, operating, maintaining, and financing infrastructure systems. Infrastructure is one of the most important policy instruments available to state and local government officials for defining an area's residential and economic character. Wise use of this policy instrument can substantially benefit a community.

First, good planning and design can greatly enhance the flow of services from a given infrastructure investment. Public-capital investment should always be guided by and consistent with a jurisdiction's long-term development plans. Effective zoning and land use enhance an investment's payoff and help avoid needless duplication of capital stock. For example, the unit cost of providing infrastructure services to business will be greatly reduced if commercial and industrial development is concentrated in specified areas rather than permitted to spread helter-skelter. Separate infrastructure networks also can be designed to generate synergistic benefits. Water and wastewater pipes can be arranged coaxially, and both can serve as conduits for power and telephone lines and fiber-optic cables. These communication cables can also contribute to water and sewer service by transmitting information about fluid quality, flow rate, pressure variations, pipe condition, leaks, or contaminants that is derived from sensors in the pipe lining (Eberhard & Bernstein, 1984). Water and waste disposal systems can be designed to promote recycling by firms and thus greatly extend the services attainable from the infrastructure stock.

Infrastructure investment should also avoid "penny wise, pound foolish" pitfalls. Winston (1991) argued persuasively that building roads with better materials and thicker than required by prevailing engineering standards would produce great savings in repair costs. Eventually, even the best designed infrastructure wears out, and maintenance expenditures should not be neglected. The U.S. Congressional Budget Office (1988) estimated that the payback for filling potholes and other mundane repairs exceeds that of virtually any other kind of infrastructure investment. When an old section of pipe, road, or cable needs replacement, it should be done with a section that, though compatible with the existing system, incorporates the latest technology.

Second, proper pricing of infrastructure will encourage efficient use, increasing the flow of services attainable from a given stock of public capital (Gramlich, 1991, 1994; Winston, 1990, 1991). At a price of zero, excessive use is encouraged. Congestion results, exaggerating the need for new construction. The standard economic policy prescription is wider application of user fees, including, where relevant, peak-load pricing. Road congestion can be reduced by charging tolls on busy highways that are highest in peak periods. Technological advances, such as sensors on license plates, may permit fee collection without tollbooths. Congestion pricing would be easiest to implement at airports: Charge higher landing fees during peak hours rather basing fees on aircraft weight. Highway damage can be reduced by charging truck fees based on weight per axle rather than number of axles.

Third, it is important not to neglect infrastructure investment directed toward residential amenities. One dimension of a good "business climate" to which infrastructure makes an essential contribution is the establishment of an attractive environment for firm employees. Indeed, many of the most prized businesses, those that provide high-skill and high-wage jobs, are precisely those especially sensitive to this location determinant. In this sense, infrastructure investment directed to maintaining residential neighborhoods with first-rate roads, pipes, parks, recreational facilities, and libraries becomes an integral part of an economic development strategy.

Fourth, planners are advised not to engage in large-scale infrastructure investment that is premised on highly speculative and optimistic projections of future economic growth. For local governments, the literature offers no evidence to recommend a "Field of Dreams" infrastructure strategy of "If you build it, they will come." Infrastructure is not a dominant element in firms' location choice, although, for instance, many firms are sensitive to transportation costs. In particular, large firms making location decisions can expect provision of a full range of subsidized infrastructure services specially designed to meet their needs at any number of alternative sites. An agreement to provide infrastructure services subject to a firm's specifications is now a standard part of the negotiations through which government officials seek to attract private investment. No jurisdiction will gain a competitive edge from inappropriate and inflexible public capital constructed on the basis of government officials' past attempts to guess the infrastructure needs of future business.

Infrastructure investment should not outpace economic development by a wide margin. Though it is sensible for government officials to establish plans for desired development, the commitment of construction funds for more than demonstration or pilot infrastructure capital should await clear indications that it will be accompanied by business investment. Otherwise, given the range of other, relevant location determinants, the risk of low returns to public investment is very high.

Finally, all proposed infrastructure investments should be subjected to benefit-cost analysis. If one seeks the "state-of-the-art" research methodology for assessing the economic benefits of given infrastructure projects, then the best technique is surprisingly "low-tech" and familiar. Aggregate regression results, even if they uniformly suggest significant infrastructure/development correlations, can never guide actual investment spending. Only benefit-cost analysis can establish reliable estimates of an investment's merit or determine which among a variety of possible investments should be undertaken.

As explicated in textbook guides (e.g., Gramlich, 1990), benefit-cost analysis enables the practitioner to incorporate directly important considerations such as uncertainty, discounting, and the weighting of gains and losses to different groups. It also permits more precise evaluation of projects' net fiscal impact on residents, as determined by a jurisdiction's particular system of taxes and fees. And although the estimates may not be as precise as one would like, the benefit-cost framework is certainly open to the estimation and inclusion of a wide variety of indirect, general equilibrium or external effects.

Of course, simply using benefit-cost analysis does not preclude legitimate differences of opinion regarding the proper conceptualization and measurement of the relevant costs and benefits. But, again in contrast to regression analysis, at least these issues can be isolated, and the practitioner can and should explicitly assess the sensitivity of the results to alternative definitions and measurements.

Moreover, the benefit-cost approach enables one to avoid many of the measurement and conceptual problems noted in this chapter. In a case study, the analyst can

1. Develop measures of public-capital-stock service rather than simply past investment spending
2. Match tax and fee costs to firms associated with estimated infrastructure benefits
3. Identify a far richer and more pertinent range of economic benefits than the standard and unsatisfactory measure of development as simply output or productivity growth
4. Explicitly account for particular conditioning factors related to the existing stock of infrastructure and the local economic base
5. Help overcome nettlesome issues of causation; after all, if one really wants to pin down the direction of causation, how about just going out and surveying measures of economic development before and after the construction of infrastructure capital? Also, a systematic comparison of benefit-cost analyses applied to the same project ex ante and ex post is an important source of learning for methodological refinement.

Although benefit-cost methods have been around for a long time, the published literature contains surprisingly few examples of these techniques applied to public infrastructure. No doubt, many careful analyses have been conducted by practitioners at all levels of government, but such studies typically do not find their way into journal publication. There is a clear need for more accessible information that details actual examples of benefit-cost analysis applied to infrastructure. The federal government, a foundation, or a research institute might usefully sponsor an inventory of past studies to offer practitioners suggestions for addressing specific estimation problems as well as examples of benefit estimates across a range of different types of jurisdictions and infrastructure for comparison purposes.

REFERENCES

Aaron, H. (1990). Discussion. In A. Munnell (Ed.), *Is there a shortfall in public capital investment?* (pp. 51-68). Boston: Federal Reserve Bank of Boston.

Aschauer, D. (1989a). Is public expenditure productive? *Journal of Monetary Economics, 23*, 177-200.

Aschauer, D. (1989b). Public investment and productivity growth in the Group of Seven. *Federal Reserve Bank of Chicago Economic Perspectives, 13*, 17-25.

Aschauer, D. (1991). Infrastructure: America's third deficit. *Challenge, 34*, 39-45.

Aschauer, D. (1993). Genuine economic returns to infrastructure investment. *Policy Studies Journal, 2*, 380-390.

Costa, J., Ellson, R., & Martin, R. (1987). Public capital, regional output, and development: Some empirical evidence. *Journal of Regional Science, 27*, 419-437.

Duffy-Deno, K., & Eberts, R. (1991). Public infrastructure and regional economic development: A simultaneous equations approach. *Journal of Urban Economics, 30*, 329-343.

Eberhard, J., & Bernstein, A. (1984). A conceptual framework for thinking about urban infrastructure. *Built Environment, 10*, 253-261.

Eberts, R. (1986). *Estimating the contribution of urban public infrastructure to regional growth* (Federal Reserve Bank of Cleveland Working Paper No. 8610). Cleveland, OH: Federal Reserve Bank of Cleveland.

Eberts, R. (1990). Public infrastructure and regional economic development. *Federal Reserve Bank of Cleveland Economic Review, 26*, 15-27.

Eberts, R. (1991). Some empirical evidence on the linkage between public infrastructure and local economic development. In H. Herzog & A. Schlottman (Eds.), *Industrial location and public policy* (pp. 83-96). Knoxville: University of Tennessee Press.

Eisner, R. (1991, September/October). Infrastructure and regional economic performance: Comment. *New England Economic Review*, pp. 47-58.

Evans, P., & Karras, G. (1994). Are government activities productive? Evidence from a panel of U.S. states. *Review of Economics and Statistics, 76*, 1-11.

Garcia-Mila, T., & McGuire, T. (1992). The contribution of publicly provided inputs to states' economies. *Regional Science and Urban Economics, 22*, 229-241.

Gramlich, E. (1990). *A guide to benefit-cost analysis.* Englewood Cliffs, NJ: Prentice Hall.

Gramlich, E. (1991). How should public infrastructure be financed? In A. Munnell (Ed.), *Is there a shortfall in public capital investment?* (pp. 223-237). Boston: Federal Reserve Bank of Boston.

Gramlich, E. (1994). Infrastructure investment: A review essay. *Journal of Economic Literature, 32*, 1176-1196.

Holtz-Eakin, D. (1993a). Public investment in infrastructure. *Journal of Economic Perspectives, 7*, 231-233.

Holtz-Eakin, D. (1993b). State-specific estimates of state and local government capital. *Regional Science and Urban Economics, 23*, 185-209.

Holtz-Eakin, D. (1994). Public-sector capital and the productivity puzzle. *Review of Economics and Statistics, 76*, 12-21.

Hulten, C., & Schwab, R. (1991a, February). *Is there too little public capital? Infrastructure and economic growth.* Paper presented at the American Enterprise Institute Conference on Infrastructure Needs and Policy Options for the 1990s, Washington, DC.

Hulten, C., & Schwab, R. (1991b). Public capital formation and the growth of regional manufacturing industries. *National Tax Journal, 44*, 121-134.

Hulten, C., & R. Schwab. (1993). Infrastructure spending: Where do we go from here? *National Tax Journal, 46*, 261-273.

Lynde, C., & Richmond, J. (1992). The role of public capital in production. *Review of Economics and Statistics, 74,* 37-44.

Munnell, A. (1990a). How does public infrastructure affect regional economic performance? In A. Munnell (Ed.), *Is there a shortfall in public capital investment?* (pp. 69-103). Boston: Federal Reserve Bank of Boston.

Munnell, A. (1990b, January/February). Why has productivity growth declined? Productivity and public investment. *New England Economic Review,* pp. 3-22.

Munnell, A. (1992). Infrastructure investment and economic growth. *Journal of Economic Perspectives, 6,* 189-198.

Nadiri, M. I., & Mamuneas, T. (1994). The effects of public infrastructure and R&D capital on the cost structure and performance of U.S. manufacturing industries. *Review of Economics and Statistics, 76,* 22-37.

Reich, R. (1991). *The work of nations.* New York: Vintage.

Tatom, J. (1991). Public capital and private sector performance. *Federal Reserve Bank of St. Louis Review, 73,* 3-15.

Tatom, J. (1993). The spurious effect of public capital on private sector productivity. *Policy Studies Journal, 21,* 391-395.

U.S. Congressional Budget Office. (1988). *New directions for the nation's public works.* Washington, DC: Government Printing Office.

Winston, C. (1990). How efficient is current infrastructure spending and pricing? In A. Munnell (Ed.), *Is there a shortfall in public capital investment?* (pp. 183-205). Boston: Federal Reserve Bank of Boston.

Winston, C. (1991). Efficient transportation infrastructure policy. *Journal of Economic Perspectives, 5,* 113-127.

A COMMENT ON INFRASTRUCTURE
AND ECONOMIC DEVELOPMENT

Steven C. Deller

David Arsen has written an excellent review of the issues surrounding the question posed in the title of his chapter. Arsen identifies nearly all of the relevant points, ranging from the notion that the value of a stock of public infrastructure is derived from a flow of services as opposed to monetary investments to concern over causation to the notion that a region's growth and development hinge on a bundle of goods and services. He also correctly points out that the relevant literature has taken on two forms: firm-specific analysis (i.e., firm location modeling) and regional growth models.

Yet, despite the excellent review of the conceptual and empirical problems facing researchers studying the basic question outlined in the chapter title, he seems a bit too sure of the answer to the assigned question: "an unambiguous 'yes.' " I think a more truthful answer within the context of regional economic development is "It depends." My reasoning hinges on the issues surrounding causation: Does an investment in public infra-structure cause economic development and growth? Although this is not the direct question Arsen proposes to answer, it is in reality the relevant policy question facing many regional economic development decision makers. Part of the reason that I would back away from Arsen's firm response is that he does not develop a conceptual framework in which to pose the causation question. The literature has offered two such frame-works that help structure the relevant policy question.

As noted above, Arsen correctly observes that the economic and social conditions of the region in which the investment in infrastructure is to be made play a vital role in determining the level of future economic development and growth. To see this, Niles Hansen (1965) suggested that regions be classified into three types—"high," "medium," and "low." Hansen envisioned "high" regions as rapid-growth areas that can be characterized as congested and where current levels of public infrastruc-ture are a barrier or constraint to future growth. These regions tend to be rapidly growing urban centers that need new infrastructure investments to support new, higher levels of economic and social activity. Here, an investment in infrastructure is in response to congestion resulting from economic development and growth.

Hansen visualized "medium" regions as demonstrating the potential for economic development and growth. *Potential* was defined by Hansen as a sufficient level of social overhead capital, or social infrastructure—an abundance of quality labor, inexpensive energy, and social and cultural activities. Here, an investment in infrastructure *may* serve as a catalyst sufficient to foster rates of growth at levels such that the benefits derived by the public outweigh the costs.

Regions defined as "low" by Hansen are characterized by a low level of social overhead capital—a low standard of living, an untrained labor force, stagnant or declining industry, and a weak or ineffective institutional structure. Generally, these regions are viewed as having little potential for sustainable economic growth and development. Hansen suggested that in these "low" regions, investments in infrastructure may not have the desired outcome. In short, the costs will probably be greater than any benefits that might flow from the investment.

An alternative view of Hansen's concepts has been put in terms of *necessary and sufficient* conditions (Johnson, 1990; Sears, Rowley, & Reid, 1990). Three potential cases can be envisioned. One case involves situations in which infrastructure investment is *both necessary and sufficient* for stimulating economic growth and development. Here, an investment will guarantee economic development and growth. Unfortunately, probably few cases would fall into this category. An investment in infrastructure in and of itself will probably not lead to economic development (Fox, 1988; Sears et al., 1990).

Another case involves situations in which infrastructure investment is *necessary but not sufficient*. Here, more investment must occur prior to increased rates of economic development and growth, but it will not ensure that higher rates of development and growth will materialize. For example, the placement of communication lines is necessary for business location, but it will not ensure that businesses will locate in the area where the lines were laid. Sears et al. (1990) noted that "numerous cases probably exist where investments to fill gaps in a locality's infrastructure, made in conjunction with other community investments (e.g., investments in social overhead capital), lead to economic development" (p. 12).

The final case corresponds directly to Hansen's "low" regions; infrastructure investment is *neither necessary nor sufficient*. In this situation, there is some element or elements, such as an inadequate level of social overhead capital, that will serve as an impediment to higher rates of economic development and growth. Here the cost of the project incurred will be greater than any benefits that might flow from the investment. Unfortunately, many regions fall into this category.

In light of Hansen's conceptualization of the problem and the use of necessary and sufficiency conditions to view the problem, the conclusions of Fox (1988) seem perhaps more reasonable than Arsen's strong conclusion. The flow of services derived from the stock of infrastructure, both quality and quantity aspects, should be viewed as a constraint in the economic development and growth problem. If some part of the stock of infrastructure acts as a constraint to potential economic development and growth, then investment in infrastructure (i.e., a lifting of the constraint) may result in realized economic development and growth. On the other hand, if the stock of infrastructure does not act as the relevant constraint, but rather there is some other binding constraint (e.g., inadequate level of social overhead capital), investment in infrastructure will not result in economic development and growth.

The question of the role of infrastructure in economic development and growth is really not one question, but two. The first is the broader macro question: Is there a link? This question I believe Arsen answers in a strongly positive manner. The second is a more narrow micro-, or individual/community-level question: Is there causation? Given Arsen's suggestions for practitioners, I fear he has mixed these two fundamentally different questions. I concur with Arsen's answer to the first, broader, macro question. The answer to the second, more narrow, micro question is, I fear, "It depends."

REFERENCES

Fox, W. F. (1988). Public infrastructure and economic development. In D. L. Brown, J. N. Reed, H. Bluestone, D. A. McGranaham, & S. M. Mazie (Eds.), *Rural economic development in the 1980's: Preparing for the future.* Washington, DC: U.S. Department of Agriculture.

Hansen, N. (1965). Unbalanced growth and regional development. *Western Economics Journal, 4,* 3-14.

Johnson, T. G. (1990, October). *The developmental impacts of transportation investments.* Paper presented to the meeting of the Southern Regional Information Exchange Group 53 (SRIEG-53), Atlanta.

Sears, D. W., Rowley, T. D., & Reid, J. N. (1990). Infrastructure investment and economic development: An overview. In D. W. Sears, T. D. Rowley, & J. N. Reid (Eds.), *Infrastructure investment and economic development: Rural strategies for the 1990's.* Washington, DC: U.S. Department of Agriculture.

REPLY TO STEVEN C. DELLER

David Arsen

I am in full agreement with Deller's substantive observations regarding the contingent capacity of public investment to stimulate local or regional economic development. This is a central point of my chapter. The conclusion is important, for, as Deller correctly notes, development impacts are usually a key consideration for policy makers faced with public investment decisions. So our only apparent difference relates to a straw man that Deller has created.

Deller seems to have gotten stuck on a sentence in the chapter's introduction stating that the answer to the assigned question "Is there really an infrastructure/economic development link?" is an unambiguous "yes." Well, as the econometric literature demonstrates and as Deller acknowledges in his final paragraph, the answer *is* "yes." Deller then claims that this is also my answer to a different question: Namely, does public infrastructure investment cause economic development? There is the straw man. Of course, this is a more relevant question for policy. That is why the chapter pursued it, first conceptually, then in assessing the available empirical evidence, and finally in the policy recommendations. For a variety of reasons detailed in the chapter, the answer to this question is not an unambiguous "yes." Indeed, the answer is "It depends."

Deller writes that I do not propose to answer the relevant policy question noted above. He evidently missed the first paragraph of the paper's conceptual framework discussion, which states,

> There is little question that infrastructure investment is associated with economic development. There is a good deal less consensus about the direction of causation between these two variables. Economic development tends to call forth infrastructure improvements both by expanding demand for infrastructure services and by increasing fiscal resources to finance public investment. Though few would dispute this sort of linkage, most research on the infrastructure/development linkage, as well as the focus here, concerns the reverse causation: the potential for infrastructure investment policy to initiate or promote economic development.

Now, there are different ways to conceptualize the contingent development impacts of public infrastructure. Deller notes the useful taxonomies offered by Niles Hansen and others. I am happy to have these frameworks introduced into the discussion. They nicely complement, essentially recasting in different terminology, ideas that the chapter's conceptual framework develops. Given space limitations, I elected not to detail these taxonomies because they are not closely connected to the "state-of-the-art" empirical literature I was asked to survey. Instead, the chapter details infrastructure's contingent status in terms of variables that I believe are both more universal and more refined, relating to infrastructure's network characteristics, lumpiness, complementarities, and especially interactions with a particular area's existing economic base. In view of these and other considerations, I conclude that the capacity for infrastructure to initiate economic development is "highly contingent." For this reason, the chapter's suggestions to practitioners clearly caution against public investment premised on unfounded expectations that it will initiate private-sector development. Unless there are clear indications that public investment will remove a constraint to development and consequently be accompanied by private investment, the risk of low returns is very high.

Let me redirect attention to what I believe is the chapter's most important recommendation for policy makers. The net economic benefits or development potential of any proposed infrastructure project can only be assessed on a case-by-case basis at the state or local level. This assessment requires careful use of benefit-cost techniques that are sensitive to the features of the proposed public investment and an area's existing economic base. This information cannot be obtained from the sophisticated econometric studies that so many researchers are now pursuing. Nor can it be found in the stylized conceptual taxonomies that Deller evidently prefers.

5

What Is the Role of Public Universities in Regional Economic Development?

MICHAEL I. LUGER
HARVEY A. GOLDSTEIN

State-supported universities have always been used to further specific public policy goals. The first public university—the University of North Carolina at Chapel Hill—was chartered in 1789 by the state legislature as a means for "all useful learning [to be] . . . duly encouraged and promoted" (Powell, 1992, p. 3). Almost 50 years later, Congress passed the Morrill Act, which stipulated that the revenue from lands granted to each state should be used for

> the endowment, support, and maintenance of at least one college where the leading object shall be, without excluding other scientific and classical studies, and including military tactics, to teach such branches of learning as are related to agriculture and mechanic arts . . . in order to promote the liberal and practical education of the industrial classes in the several pursuits and professions in life. (Brickman, 1993)

Over the past 200 years, some 600 public 4-year institutions of higher learning, and 1,024 public 2-year institutions, have been established by state and local governments. Each year, these colleges and universities spend almost $100 billion on salaries, equipment, construction, and other items.[1]

To make a general statement, during the first two centuries of their existence, public universities were valued most as centers of learning and purveyors of knowledge, especially for the broad public (industrial classes). In the past several decades, the role of the public university has been undergoing redefinition due to both internal and external pressures. In short, public universities are relied on increasingly as engines of economic growth and development, particularly for their region.

■ External and Internal Pressures for Change

The external pressures operating on the university can be described by the following propositions:

1. We are living in an increasingly knowledge-based and global economy.
2. To sustain themselves economically in this environment, regions will have to be capable of generating and nurturing innovative individuals, businesses, and organizations on a continuing basis.
3. Universities are *the* institutions with the resources to provide the stream of knowledge, know-how, and human capital to their respective regions as the fuel for innovation, entrepreneurship, and regional synergy.

From the universities' perspective, internal pressures for change come from a number of sources, including falling enrollments due to demographic changes and increased competition from other institutions; reductions in revenue growth from government and private sources during economic downturns, and the incomplete restoration of funding during subsequent recoveries; institutions' limited ability or unwillingness to downsize when faced with budget austerity; greater competition for the noneducational uses of government funds; rising equipment and facility costs due to more sophisticated technology; rapidly rising salaries in the sciences and engineering due to competition with private industry and the drop in the number of PhDs awarded in those fields; and higher expectations from the public for free services.

■ New Roles, New Functions

As public officials and university administrators have redefined the role of the public university, they also have developed a new set of university-based programs. It is now common for universities to operate technology parks and research centers; to engage in collaborative research with industry scientists, often in shared space; and to encourage and support new product development and commercialization.

■ New Impacts, Different Measures

As universities develop into more complex organizations with multiple outputs, the types of impacts also change. This requires regional economists/planners/scientists to use a broader set of measures when exploring the regional economic impacts of universities. This chapter illustrates how that may be done. Specifically, we define different "outputs" from universities and then discuss how those can lead to different kinds of regional economic development. We use evidence from recent case studies to illustrate.

■ University Outputs and Impacts

In this section, we present a conceptual model of the university as a multiproduct organization that may help us understand why some universities are bigger engines of economic growth and development for their regions than others, and the trade-offs university administrators and economic development officials make regarding changes in the university's mission or particular mix of outputs. This characterization of the university departs from the traditional impact literature, which measures the regional consequences of higher education institutions mostly in terms of their spending multipliers.[2]

Universities as Multiproduct Organizations

Universities, in many respects, behave like any other economic organization. First, they are affected by conditions in their regional environment. However, unlike private businesses, they are not likely to move if local input prices rise or the local demand for services wanes. Universities find other ways to respond to economic changes, including alterations in the mix and quality of their outputs. In addition, they often attempt to shape and restructure their regional environment so as to improve the availability and quality of locally supplied inputs, including labor, infrastructure, amenities, and related knowledge-producing organizations.

Second, the mix and quality of universities' outputs have economic impacts that vary with distance. Some activities have a very local impact; others, if the "spatial gradient" of the impacts is relatively flat, can have a wider national, or international, impact. But the spatial impact gradient is not shaped the same for all universities, even for a given type of activity. Universities' size and certain internal characteristics, and the nature of the regional environment, affect the spatial distribution of economic impact for such functions as teaching or basic research.

In short, there is a two-way, dynamic relationship between what a university does and what happens in the regional economy. This relationship is important to understand, both for university decision makers who are concerned with the viability of their institutions and for local/regional officials who are interested in universities as potential stimulants of economic and business development.

We further conceptualize the modern research university as an organization that procures a set of inputs to produce a particular mix of outputs.

The typical inputs include supplies, equipment, services, external funding, students, and labor of varying skill levels. But the inputs also include the regional "milieu," which provides a set of amenities and other public goods more or less available to all individuals and organizations located within the region, as well as opportunities for interaction with other actors and institutions within the region.

Research universities have a much broader set of possible outputs today than they did 20 or more years ago. Today's set includes

1. The creation of new "basic" knowledge through research
2. The creation of human capital through teaching (i.e., knowledge transfer from faculty to students)
3. The transfer of existing know-how (i.e., technology) to businesses, governmental agencies, and other organizations
4. The application of knowledge to the creation and commercialization of new products or processes, or the improvement of existing ones (i.e., technological innovation)
5. Capital investment in the built-form and in the equity of private businesses
6. Leadership in addressing critical social problems
7. Coproduction of a knowledge-based infrastructure
8. The creation of a favorable "milieu"

We briefly discuss these outputs and their expected, or hypothetical, economic and business impacts below.

Knowledge Creation

The creation of basic knowledge through research is the raison d'être of the research university. Tratjenberg, Henderson, and Jaffe (1992, p. 3) noted that the scientific and technological literature defines basic research according to a focus on general scientific as opposed to particular technological questions. Those authors operationalized the concept by claiming that basic research is difficult for individual firms to appropriate (p. 4). For that reason, the geographic impact of basic research is likely to be rather wide. There is some debate in the literature about the extent of the spatial spillover of the benefits of basic research. Tratjenberg et al. concluded that "basic research results do not spill out as easily, as widely, or as quickly as the traditional view would suggest" (p. 1). Relatively speaking, however, the geographic impact is wide (see also Cohen & Levinthal, 1989; Jaffe, 1989; Jaffe, Tratjenberg, & Henderson, 1992).

Because basic research, by Tratjenberg et al.'s definition, is not directly appropriable, it is more likely to be funded by the government than by industry. Government funding can go directly to researchers for their time and equipment needs, and/or for the construction and operation of special government labs in which a mix of basic and appropriable research is conducted. We might conclude, then, that the basic research function of the university is not new but that some of the support structures for it are.

The fact that basic (and much applied) research is funded by the government leads to a different type of regional development effect due to the distribution of that funding as opposed to the ultimate impact of the research. The infusion of research dollars into selected local economies creates a wave of additional spending in the region that is an important source of regional growth.

Human Capital Formation

The knowledge transfer from faculty to students is regarded by many as the second of the two primary functions of research universities. And from a regional development perspective, there is no single more important ingredient for sustained economic development than a supply of creative, talented, and well-trained people.[3] An economic impact to the region is realized when the increase in human capital leads to an increased stock of labor skills in the region and then to increases in businesses' productivity.

Certainly, the transfer of knowledge from faculty to students, in classrooms, tutorials, advising, and supervised instruction, is not new. However, some new tools are broadening the spatial range of a university's teaching function. Specifically, interactive teleconferencing is used increasingly to bring specialized instruction to far-flung locations. That trend is expected to intensify in coming years. The economic impact of knowledge transfer through teaching depends on the "flow" of students: where they come from prior to matriculation and where they locate after graduation. Universities that draw talented students from a wide geographical area, so-called national or world-class universities, and that have regional labor markets offering abundant job and career opportunities would be expected to have the largest regional economic impacts; on the other hand, universities that mainly draw students locally but whose students tend to choose to leave the region for better career opportunities or perceived quality-of-life reasons may inadvertently contribute to a regional brain drain and have negative economic impacts.

The regional economic impacts accruing from the teaching and training function of particular universities are not usually estimated, in part because of the attribution problem and in part because of the large data requirements. There have been two traditions in the literature for estimating the aggregate economic value of higher education: (a) the estimated contribution of education to national income and (b) the rates of return on investment in schooling.

The return on investment in higher education has been measured as the increase in wages and salaries by the students undertaking the investment. That measures the *private* returns to the individuals, but it may also be considered as a reasonable proxy for the increment to marginal productivity accruing to businesses employing such individuals.[4]

A better measure of economic benefit would be the *social* rate of return, but that is even more difficult to measure. It takes into account that some of the costs of higher education are publicly subsidized (making the social rate of return lower than the private rate) but also that there are external benefits of higher education that society as a whole receives (which would make the social rate higher than the private). Whether the social rate of return is higher or lower than the private rate is an empirical matter that depends on whether the external social benefits of higher education outweigh the public subsidization of the per-student costs of teaching/training. Of course, the actual rate of return in a particular case fluctuates around the average rate of return, depending on academic field, gender, race, years completed, degree, and region. In any event, with data on the number of person-years of schooling and the average investment per student-year, appropriately disaggregated, the marginal economic benefit to society at large of the teaching/training function of a given university for a unit period of time could be estimated.

Because we are specifically interested in the economic benefits accruing to the *region*, however, we can count only those trained who have become employed and remain in the same region. The proportion of students who take jobs within the same region and the amount of time they stay vary considerably from region to region. This depends on the perception of the job opportunities for certain skill levels compared with other regions. The more specialized a job searcher's skills and educational training, the wider the geographic scope of his or her search and the less likely it is that he or she will stay within the region. Graduates of universities located within large metropolitan labor market areas will be more likely to remain in the region than those from universities in small and isolated regions, who will contribute to a brain drain.[5] This argument has an important policy implication: Because of the difficulty that disfavored regions will have in keeping university graduates within their labor market, locating universities in such regions may not generate the expected effects.

Transfer of Existing Know-How (Technology)

This output includes the outreach, extension, and public-service functions originally associated with land-grant universities in the United States but now solidly established in most public and private research universities. It is analogous to the knowledge transfer function discussed above, except that the "clients" for these services tend to be businesses, other private-sector and civic organizations, government agencies, and individual citizens, rather than enrolled students. In many cases, this technology transfer role can be combined with the teaching (and research) function. Administrators and faculty recognize that the local context can serve as a good laboratory for students to test the applicability of theory to real-world problems in engineering, agriculture, public health, the environment, social welfare, planning, education, economic development, and other fields.

As a general category, the "output" is the application of existing knowledge to the improvement of a product or process, or to problem solving. Universities provide these services institutionally by such organizations as industrial (and agricultural) extension services, small-business assistance centers, business schools, or economic and business research bureaus, and medical clinics, as well as by individual faculty through consulting (paid or pro bono). These activities are widespread. For example, in a recent survey, over 50% of responding U.S. research and doctoral-granting universities indicated that they sponsored technical assistance centers, and about 13% had small-business assistance centers (Luger & Goldstein, 1991). Matkin (1990) has counted "several hundred" technical assistance programs (TAPs) sponsored by universities.

Unlike teaching and basic research, the know-how transfer function typically has an explicit economic development objective. Accordingly, the spatial gradient of the resulting economic impacts is fairly steep. More generally, the extent to which the benefits are localized depends on the nature of the services provided and the geographic location pattern of organizations that need or seek that assistance. Relatively recent delivery and marketing mechanisms such as new modes of communication and information transfer have tended to widen the spatial range of university clients.

The magnitude of regional economic impacts from these activities depends, in large part, on the amount of university inputs devoted to technology transfer.[6] It also is a function of the density and receptivity of the pool of organizations within the region, the degree to which the expertise and know-how within the university are appropriate for the needs of the region's businesses and organizations, and the compatibility of the styles of communication and other forms of interaction between the providers of services and the recipient organizations.

Technological Innovation

This category refers mostly to the application of knowledge for the creation and commercialization of new products.[7] The range of programs includes patenting and licensing assistance offices, industrial liaison programs, centers for advanced technology (CATs), and joint university-industry research projects. From a regional economic development point of view, we would expect to see some combination of cost savings, increased sales, and increased employment within established businesses and an increased incidence, and higher rate of survival, of technology-based start-up companies.

This output may receive the most publicity, and draw the most controversy, of all the universities' nontraditional activities. One reason for the controversy is that this type of research is usually appropriable, in Tratjenberg et al.'s (1992) terminology, and thus runs counter to the tradition of a university in which knowledge creation is regarded as an open process and the products of research are located in the public domain.

Often this type of output from the university is not a "final" product but instead arrives at the marketplace by way of corporate labs. In any event, the application of knowledge to the creation and commercialization of new products normally requires ideas from university-based researchers to be patented and licensed to commercial producers. Intermediate steps include prototype development and testing.

The economic impact of this output typically has a steeper spatial gradient than that for basic research, for several reasons. First, some university researchers themselves spin off new firms, either while on the university payroll or as an alternative to university employment. The choice between these two options is determined, in part, by university policy. Some universities make such commercial activity very difficult for faculty. Others encourage academic entrepreneurship as an integral part of the university's official mission. In addition, firms in the university's region have a greater opportunity to learn about faculty research and enter into relationships than do businesses farther away. And during the prototype development and testing stages, the faculty inventor/researcher needs to be available. The extent of the regional effect will depend, in large part, on the composition of the local economic base. If the invention/research result is useful to an industry that has a large local presence, the effect will be large. However, there are many examples of the counterinstance: when research conducted in one region is important for industry located far away.

Capital Investment

State universities spend considerable sums of money each year building, renovating, and maintaining facilities, such as classroom buildings, laboratories, administrative offices, and activity centers, that are required to carry out their basic academic mission. Universities built on campuses, notably in the United States, also provide infrastructure, including internal roads, power stations, water delivery systems, athletic facilities, and student housing.

Increasingly, universities also invest in facilities that support the production of some of the "nontraditional" outputs, such as advanced-technology labs, small-business incubators, and research parks. And a number of universities in the United States have begun to invest in start-up business ventures and commercial and residential real estate in their regions, sometimes using endowment monies, as a potential (if risky) source of revenue and as a way to influence the type and quality of development in the university's environs. The multiplier effect of the outlays for these facilities, as well as for traditional capital projects—which can be substantial—typically is included in traditional analyses, but not the outputs of the facilities themselves.

The magnitude of the economic impact from construction alone depends on the extent to which it allows functions to occur that generate new economic activity at the university or in the region. The spatial gradient of the construction impact tends to fall off sharply because a large proportion of those investments tend to be located close to the university.

The construction and ownership of facilities and improvements by universities can have a negative effect on regions as well. Real property owned by public universities (i.e., the state), and facilities with academic uses, broadly defined, owned by many private universities in the United States, are exempt from local property taxes.[8] In-lieu-of-tax payments by the university to the local government sometimes are negotiated, but in many instances, local governments with large universities forego a considerable amount of tax revenue. The tax-exempt status of many of the new nontraditional facilities is the subject of some legal controversy.

Provision of Regional Leadership

Increasingly, top university officials and distinguished faculty are being asked by public officials to serve on commissions, boards, and committees to provide leadership in addressing critical social, economic, and environmental problems. What universities can contribute in this role, in addition to their technical resources, is a certain type of moral authority and political clout to help forge consensus among conflicting interests toward a plan of action. Typical problem areas might be affordable housing, crime and safety, literacy, governmental efficiency, environmental hazards, and economic development itself.

In providing regional leadership, universities can influence economic development by contributing to the region's quality of life—specifically, by making it more attractive as a place for the general population and business community to live, work, and invest. Of course, university interests are advanced by improvements in a region's quality of life. For example, faculty and student recruitment becomes easier, and the number of nonuniversity R&D facilities in the region is likely to grow. Publicly supported universities located in regions that are cited as "up and coming" in location surveys[9] are quick to take credit, whether or not they are actually responsible for the improved quality of life, as a way to curry favor in the state legislature.

Universities located in the larger and older cities in the United States historically have been the most active in community affairs and in consciously shaping community outcomes. However, in recent years, many universities in both North America and Europe, located in economically lagging smaller cities and nonmetropolitan areas, and in areas undergoing significant economic restructuring, have added this role to their missions.

Knowledge-Based Infrastructure

A region's knowledge-based infrastructure consists of the set of institutions and organizations, and their synergies, that support and increase the region's capacity for knowledge creation and dissemination, technological innovation, and entrepreneurship. The specific elements, in addition to universities, typically include the public schools, community colleges, training institutions, sources of venture capital and other forms of business finance, technical and managerial assistance, information service providers, research institutes, and the network of small and medium-sized R&D firms already in the region.

Just as a region needs a high-quality and efficient *physical* infrastructure—transportation and communication facilities, and various public utility systems—to have the capacity for future growth and development, so it needs a *knowledge* infrastructure to grow creative and innovative organizations and individuals that can lead to sustainable economic development.

A well-developed knowledge infrastructure has a high density of differentiated knowledge generators and a developed set of intermediary support services that provide possibilities for a high degree of interaction among them. Universities with policies that actively promote interaction of faculty and research units with other knowledge-producing organizations in the region will tend to contribute more to the productivity of the knowledge infrastructure. Also, we would expect that universities located in large metropolitan regions with higher densities of knowledge-producing organizations would be better able to generate regional synergy and entrepreneurial activity. Almost by definition, the economic impacts of the university's contribution to the knowledge infrastructure will be highly geographically localized.

Regional Milieu

Regional milieu is used here in a very inclusive way to encompass aspects of culture, community tastes and demand for public goods, political attitudes, and "entrepreneurial spirit." The regional milieu of the type considered here can stimulate economic development by attracting to an area creative and talented people who seek proximity to others like themselves. A research university, almost by definition, creates such a location dynamic. Through the production of the outputs discussed above, the research university inadvertently creates a set of externalities that may be shared by a large number of other actors and consequently can lead to regional economic benefits (and costs). We include here the production of a distinctive type of "milieu"—intellectual, cultural, social, and recreational—that certain individuals, businesses, and organizations, particularly those involved with research and development (R&D) functions, value highly. Universities affect the milieu of a region because the institutions themselves are often relatively large and employ a large concentration of highly educated and creative professionals who are willing to pay for high levels of public goods and amenities. That, in turn, attracts other highly educated professionals to the region who often have no direct ties with the university. The same dynamic, however, can lead to negative externalities as well. For example, a region that has experienced a high rate of growth based on attracting new R&D activity can suffer from commercial gentrification whereby higher input costs, such as labor, place traditional businesses at risk. That eventually can lead to significant out-migration and sharply reduced economic growth as a result of the inability of many types of lower-paid, but necessary, nonprofessional workers to afford to live in the region.

This last type of output, as well as the coproduction of a region's knowledge infrastructure, is not well defined, so the ensuing economic impacts are very difficult to measure. Because of that, and because they are not intentional but instead are by-products of what the university does, their "value" is often ignored or dismissed. Yet the long-term trend of an increasingly knowledge-based economy strongly suggests that those university outputs may be among the most important in contributing to economic development.

Summary

The various output categories and their expected major economic impacts are depicted in Table 5.1. Specific activities can overlap more than one of these outputs, and there are likely to be important complementarities between several of the categories. An obvious example is the production of new knowledge (research) and human capital creation in

TABLE 5.1 University Outputs and Their Likely Geographic Impacts

Activity	Traditional or New Activity?	Likely Geographic Impact	
		Broadest	*Narrowest*
Creation of new "basic" knowledge through research	Traditional	XX	YY
Formation of human capital through teaching	Traditional	XX	
Application of knowledge to the improvement of existing products and processes (transfer of know-how)	Traditional, new	XX	
Application of knowledge to the creation and commercialization of new products (technological innovation)	Traditional, new	XX	YY
Creation of physical capital and infrastructure	Traditional, new		XX
Leadership in addressing critical social problems	New		XX
Support services to other economic actors (knowledge-based infrastructure)	New		XX
Creation of a favorable milieu	Traditional		XX

NOTE: XX = spatial incidence due to activity's outcomes; YY = spatial impact of spending on research, regardless of research results.

the training of many graduate students. Also, there will be variation in the degree of opportunity costs for increasing production among types of outputs. For example, increasing technological innovation may require substantial additional capital and staff resources with attendant opportunity costs, whereas the investment in buildings and infrastructure requires the use of the university's capital resources but few of its limited human resources. The university's provision of leadership and its contribution to the region's knowledge infrastructure have very low costs and furthermore can be highly complementary with its teaching, research, and public-service activities.

■ Measuring the Regional Economic Impact of the Multiproduct University

In the previous section, we conceptualized the university as a multiproduct organization. By using a combination of inputs, the university pro-

TABLE 5.2   Some Key Questions to Ask About Outputs to "Measure"
University Impacts

Output	Key Questions to Ask
Knowledge creation via research	What is the effect of nonlocal research money (leveraged) on the university spending multiplier?
Human capital formation	To what degree does the university contribute to a reduction of brain drain (by region and type of worker) or induce brain draw?
Transfer of know-how	What kinds of technical assistance have business and government clients received from the university, and how has that affected their operation?
Technological innovation	How extensive is the university's licensing, patenting, and other commercialization activity? What links exist between university-based inventions and the local economic base?
Capital investment	In what range of capital projects is the university involved? Do those projects pay off? Are all projects appropriate activities for the university?
Regional leadership	To what degree are faculty, staff, and administrators part of the regional elite? How has their involvement improved public policy? What role do university alumni play in the political and business elite?
Knowledge infrastructure	How has local businesses' productivity been enhanced because of information, knowledge, and people flows among businesses, universities, and other institutions in the region?
Regional milieu	To what degree has the presence of a university, and the milieu it creates, induced businesses to move to the area (not specifically to have a business relationship with the university, but to enjoy the proximity)?

duces a set of outputs—some traditional, some fairly new—and, by doing so, influences its surroundings through external relationships as well. Through these backward, forward, and lateral linkages, the university is connected to the surrounding world, in particular to its regional economy.

Each of the outputs of the university can be measured, some more precisely than others, to give a comprehensive picture of the university's impacts on its region. To do that, the researcher must ask some key questions—some of which are listed in Table 5.2. These questions can be answered using university records, as well as interviews and surveys, of faculty, university administrators, alumni, and business owners at various distances from the university (to measure an impact gradient).

TABLE 5.3 Where Will Your Research Have Commercial Applications, by Discipline?

	Average Ranking (Lower Value = More Likely)			
	N.C.	S.E. U.S.	Other U.S.	Foreign
Computer, electrical, materials	3.4	4.2	2.9	3.5
Aero, mechanical, nuclear	3.0	2.1	1.9	4.6
Industrial, manufacturing, textiles	1.2	1.8	4.3	4.8
Chemical, environmental, civil	2.6	3.3	3.6	4.4

SOURCE: Goldstein & Luger, 1993.

In the remainder of this section, we present sample evidence about the regional impacts of universities in the United States. Given the broad scope of relationships between a university and its region, and the evident difficulty of measuring them, it is not surprising that the amount and quality of evidence differ considerably. In many areas, the available evidence is anecdotal and plagued by omissions and gaps.

The Regional Impacts of University Research

We have identified two types of impacts that occur as a result of universities' research activity: the economic growth that may result indirectly as a product of the research and the growth fueled by the injection of research dollars into the local economy.

Table 5.3 illustrates the geographic spread of the first type of benefit, based on responses by engineering faculty at North Carolina State University. Respondents were asked, "How likely will your research have commercial applications in the following four regions: North Carolina, the southeastern U.S., elsewhere in the U.S., or abroad?" We used a 1-to-5 scale, with 1 indicating the highest and 5 indicating the lowest likelihood.

In two of the four engineering disciplines, the faculty's inventiveness and entrepreneurial talent were considered to be more likely to benefit North Carolina than elsewhere, but for two other types of engineering, the reverse was true. This is not surprising, given the type of research that is produced: The clients for industrial, manufacturing systems, civil, and environmental engineering outputs tend to be locally distributed, whereas the users of computer, electrical, materials, aerospace, and nuclear engineering inventions are located all over the world. The result for textile engineering reflects the strength of that sector in North Carolina.

The University of North Carolina at Chapel Hill (UNC-CH) and other leading research universities bring considerable amounts of outside funds into their region that are then multiplied through the economy. Pings

(1994) estimated that the $10.8 billion in R&D funding provided by federal agencies to U.S. universities and colleges in fiscal year 1992 generated 366,867 jobs nationally in that year. In UNC-CH's case, external research funds almost equal the amount of state appropriations. This "leveraging" of those funds, illustrated in Figure 5.1, is important as a means to increase the multiplier effect of revenues originating within the state. Figure 5.1 shows that each dollar of *state* appropriations generates approximately $4 in net new private output. This multiplier is larger than for business spending because of the outside research funds that universities like UNC-CH capture and because universities tend to spend a larger share of revenues within the local economy.[10]

Regional Impacts From the Formation of Human Capital

The contribution of education (as a whole) to increases in national income has been estimated to be between 10% and 20%. Approximately one quarter of that can be attributed to higher education (Leslie & Brinkman, 1988). The return on investment in higher education also has been measured as the increase in wages and salaries by the students undertaking the investment. That measures the private returns to individuals but, as stated earlier, may also be considered as a reasonable proxy for the increment to marginal productivity accruing to businesses employing such individuals.

The definitions, methods, assumptions, and data vary considerably among the rate-of-return studies. Nonetheless, the general conclusion is that the private rate of return on investment in higher education in the United States is on the order of 10% to 20% (Leslie & Brinkman, 1988). The returns to graduate higher education are estimated to be somewhat lower than for undergraduate education (5%-15%).

From the perspective of universities lobbying legislatures for continued funding, university contribution to a slowdown or reversal of a brain drain is particularly important. Figure 5.2 summarizes evidence about the change in student flows among UNC-CH graduates between 1973 and 1983. Although the state still is a net exporter of graduates, the flow slowed considerably. We detected a similar trend for North Carolina State University students, with the ratio of destinations to origins falling from 2.06 for the 1979-1980 cohort to 1.19 for the 1987-1988 group. Extrapolating for both universities suggests that there no longer is a brain drain from those two campuses.

We also surveyed UNC-CH students and learned that 11% of the in-state undergraduates and 3% of the in-state graduate students currently enrolled would have attended schools in other states if they had not been admitted to UNC-CH. That amounts to some 1,500 current, highly quali-

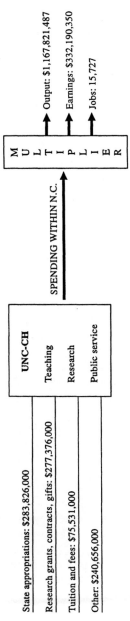

Figure 5.1. Revenues and Economic Impacts, University of North Carolina–Chapel Hill, Fiscal Year 1994
SOURCE: Adapted from Goldstein & Luger, 1992, p. 2.

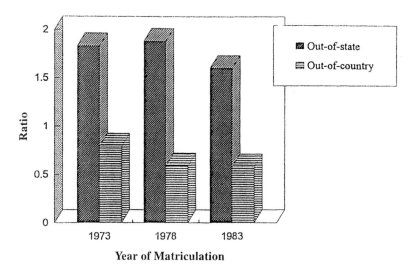

Figure 5.2.   Student Flows Among Graduates of the University of
North Carolina-Chapel Hill: Net Destinations to Net Origins
SOURCE: Goldstein & Luger, 1992, p. 8.

fied students who would not be in North Carolina at all if UNC-CH did
not exist. Similarly, almost 70% of out-of-state undergraduates and 80%
of out-of-state graduate students, or 4,700 students in total, would not be
studying in North Carolina if they had not been accepted by UNC-CH.
One might argue that if these students now at UNC-CH had studied in
another state, a similar number of other students would have come to the
university. However, those students presumably would have been less
qualified.

Other evidence suggests a significant brain draw to North Carolina
because of UNC-CH. For example, the number of graduate students who
petition the university for permanent state residency after their matriculation has increased in recent years. In calendar year 1990 alone, approximately 660 graduate students who were not native North Carolinians
changed their status from nonresident to resident, in part by convincing
the university that they intended to stay in the state permanently.[11] Certainly, the university was not always the reason these students came to
North Carolina to begin with, nor was it the only reason those students
decided to relocate here permanently. Undoubtedly, however, the university and the favorable milieu it helps create were significant factors in the
relocation decision.

The Regional Impacts From
the Transfer of Know-How

Estimates of the economic impacts of institutionally based technology transfer programs typically are based on a count of program outputs (e.g., number of clients assisted, or client-hours of assistance) and the recipient organization's self-assessment of the difference the assistance has made to the organization (for instance, changes in sales, employment, or productivity). Data generated from such self-assessments are subject to questions of both reliability and validity. As a result, there tends to be a lot of anecdotal information, which reduces one's ability to draw generalizable lessons.

Existing studies at least suggest that universities' and clients' perspectives on technology transfer do not coincide. Universities generally prefer to integrate technology transfer operations with their teaching/training and research activities as a way to demonstrate how responsive they are to external demands for public service and outreach and to economize on inputs. Many recipient organizations, on the other hand, prefer technology transfer to be detached from research and teaching/training activities (Matkin, 1990). Those clients' needs may not require PhD-level expertise, and the responsiveness, timeliness, and pertinence of the services provided might be compromised by the university's attempt to meet a number of objectives. If that is true, universities could be judged to be less appropriate than others (on efficiency grounds) for providing technology transfer services.

In our studies of UNC-CH and North Carolina State University, we asked businesses in the region and state how much they availed themselves of the universities' store of know-how and how important the transfer of know-how had been to their businesses. The frequency of interaction—around such activities as paid and pro bono consulting by faculty and provision of technical advice by industrial, agriculture, small-business, and other assistance centers—was substantial. For example, some 45% and 55% of businesses in the Research Triangle area used pro bono faculty consultants and free technical advice, respectively, from North Carolina State University's Engineering School. The use of most services was subject to a distance gradient; the exceptions were services targeted to specific industries.

Despite the widespread use of the universities' store of know-how, a very small percentage of respondents indicated that the interaction had affected their bottom line. (For example, 11% and 18% of respondents, respectively, agreed that North Carolina State's and UNC-CH's proximity helped make them more profitable. A much larger percentage disagreed with that statement.)

Regional Impacts of Universities'
Technological Innovation Activity

Among U.S. research-oriented universities, technology development has become almost as ubiquitous as teaching and research in the past 15 years. Peters and Fusfeld (1983) found that about one half of research universities surveyed had industrial liaison programs, though the vast majority had begun since 1978. In a 1990 survey of research and doctoral-granting universities, over 83% of the respondents indicated that they had licensing/patenting assistance offices, and 82% had joint university- industry research projects under way (Luger & Goldstein, 1991).

There is a dearth of generalizable studies that identify the particular models of university-based technology development that are most effective under different conditions. In part, this is because the very processes of technological innovation are still "black boxes," as Feller (1994) has noted. The existence of multiple models that vary widely—not only by technology field, but also by institutional environment, regional and national setting, particular human agents, and over time—would lead us to conclude, on the basis of available evidence, that the critical success factors tend to be more idiosyncratic than systematic.

Despite this rather harsh assessment, at least some suggestive findings have implications for what economic development role(s) universities should play. First, and not at all surprisingly, the regional impacts tend to be larger for universities located in larger metropolitan areas. This is simply due to a larger pool of organizations that potentially can realize economic benefit from technology developed within the university or jointly with nonuniversity personnel. On the other hand, there is more difficulty, particularly in macrolevel studies, in disentangling the effects of the university's activities from other putative regional influences.

Second, large corporations (or their R&D branch plants) are more likely than existing small and medium-sized businesses to benefit from the technology development activities in research-oriented universities. This is based on the observation that "academic research findings are seldom directly transferable into commercial innovations, requiring instead considerable technical and economic refining" (Feller, 1994, p. 23). Hence, those firms that are best able to keep informed of scientific and technological advances and that possess the resources and internal R&D capabilities to take advantage of these advances will be most likely to benefit from the direct affiliation, and research collaboration, with research universities.[12]

Third, the economic benefits accruing from the universities' technology development activities to their respective regions are small in comparison with the benefits those activities generate for society as a whole.

They also tend to be smaller than the share of regional benefits from technology transfer and teaching/training, as discussed above. The diffusion from universities and the adoption by industry of technological innovations are only minimally inhibited by distance, whereas collaboration between university and industry scientists and engineers can be increasingly accommodated in university labs and centers without the need for the separate location of corporate R&D facilities in the region. Also, just as large corporations enter into collaborative relationships with a portfolio of universities where researchers are doing cutting-edge work, research universities tend to be global rather than parochial in their search for the most fruitful collaborative arrangements. A survey of the research and technology development centers at the highly regarded College of Engineering at North Carolina State University in Raleigh, for example, indicates that the overwhelming majority of corporate and industrial affiliates of those centers were located out of state, despite the concentration of high-technology businesses in the same Research Triangle region.

The Regional Impacts
From Capital Investment

These impacts are typically the most spatially concentrated because the labor and materials used in the construction of plant and infrastructure are usually drawn from the local market, and a majority of university equipment purchases are from local vendors, if not local producers. Therefore, there is relatively little "leakage" of university spending out of the region, and the multiplier is large. Of course, the amount of leakage will depend on the breadth and depth of the local construction industry and the local network of suppliers and on the sophistication of the project being built. For example, a particle accelerator is more likely to be built by a specialized nonlocal firm than is a dormitory.

The methodology for estimating these impacts, in terms of direct and indirect job creation, and induced output and income, is well established. One traces an increase in final demand (e.g., the infusion of money to build a new facility) through rounds of spending, output generation, or employment in the regional economy, using patterns of interindustry trade compiled by the federal government. As noted previously, this approach is standard for university impact studies.

We can use our studies of UNC-CH and North Carolina State University to illustrate the nature of these impacts. In fiscal year 1990, UNC-CH spent $60,059,742 on construction projects. That translated into $139,767,845 in new output, $45,481,733 in earnings, and 2,346 new jobs (935 in construction, the remainder in sectors that sell to contractors or provide services).

Table 5.4 shows a similar effect for North Carolina State University, using a particular project now under construction—the Engineering Graduate Research Center (EGRC). That facility is costing the state approximately $55 million, $42.26 million of which is estimated to be spent directly within North Carolina. Of that amount, $30.8 million is for construction and related uses. That $30.8 million is projected to produce $96 million in output, $30.5 million in earnings, and 1,500 jobs. Most of the jobs are short term—as part of the construction. But several hundred (260) are more permanent, associated with maintenance and supplying and servicing the new facility.

Regional Leadership

University faculty, administrators, staff, and students commonly act as informed citizens and experts in their communities, frequently being appointed to boards, commissions, task forces, and working groups and serving in leadership positions in civic organizations. Some members of the university community also stand for election to public office. This involvement characterizes faculty, administrators, staff, and students in more than professional and public-service-oriented fields. Increasingly, public universities are documenting this activity and considering ways to encourage and reward it.

Alumni of the nation's public universities also become leading citizens in their respective states, forming both business and political elites. For example, 26.1% of business owners, plant managers, and CEOs surveyed in North Carolina identified themselves as alumni of UNC-CH, and 27% of professionals in engineering-related businesses in the state indicated that they were alumni of North Carolina State University's College of Engineering.

Universities' Contributions to a Region's Knowledge-Based Infrastructure and Milieu

These are the most indirect and the least well defined contributions to regional economic development that we consider, though they may very well end up being the most important. Knowledge infrastructure stimulates economic development by effectively lowering the costs faced by individual businesses for some of the critical inputs (e.g., specialized labor skills, information, ideas and intellectual stimulation) needed for successful innovations. It would be nearly impossible to attribute the regional development impacts to a single institution that forms part of the region's infrastructure or even to separate the impact due to the infrastructure as a whole (because of such high degrees of overlap with other contributing factors). However, we observe a strong positive correlation

TABLE 5.4 The Economic Impact of the College of Engineering in North Carolina: Economic Activity Directly and Indirectly Attributable to the College of Engineering

	Construction			Ongoing (Annual) Operation		
	Output (Sales)	*Earnings*	*Employment*	*Output (Sales)*	*Earnings*	*Employment*
College of Engineering	$96.0 mil.	$30.5 mil.	1,500 jobs	$122.1 mil.	$35.3 mil.	2,000+ jobs
Eng. Grad. Research Center				$16.5 mil.	$4.8 mil.	260 jobs
Total				$138.6 mil.	$40.1 mil.	2,260+ jobs

SOURCE: Goldstein & Luger, 1993.

between the degree of development of a region's knowledge-based infrastructure, including the prominence of one or more research universities, and its capacity for generating innovative businesses and sustained economic development.[13]

The best known examples of this positive correlation are in California and Massachusetts. The cases of Stanford University and MIT spawning huge numbers of spin-off businesses in the Silicon Valley of California and along the Route 128 corridor in Massachusetts, respectively, able to take advantage of the explosion of national defense spending for applied research in the sciences and engineering in the late 1950s through the 1970s, have defined standards of success that almost all other universities and their regions cannot realistically emulate (Dorfman, 1982; Krenz, 1988; Luger & Goldstein, 1991; Saxenian, 1985; Sirbu et al., 1976). Yet other research universities in the United States have stimulated more modest, but still significant, numbers of technology-based spin-offs, including the University of Utah, the three research universities of the Research Triangle region of North Carolina (Duke University, University of North Carolina at Chapel Hill, and North Carolina State University), and the University of Texas, Austin (Brown, 1987; Ko, 1993; Luger & Goldstein, 1991; also see Brett, Gibson, & Smilor, 1991).

Measuring the effect of the regional milieu on economic development is an even more heroic task, in part because the concept itself is so slippery and in part because of the usual, but even more complicated, attribution problems. The best evidence is from studies of the location decisions of R&D and other high-technology firms, which show that the presence of research universities in the region was a very important factor even when the business did not expect to have any direct relationships with the university (e.g., Ko, 1993; Malecki, 1987).

Our studies of UNC-CH and North Carolina State University again can serve as illustrations. Almost 15% of surveyed businesses now located in the Raleigh-Durham-Chapel Hill metropolitan area agreed that the nearby location of UNC-CH significantly affected their decision to relocate to the region. Similarly, 8% of businesses in the Piedmont region (stretching from Wilson-Rocky Mount east of Raleigh, through Durham, Chapel Hill, Winston-Salem, and Greensboro, to Charlotte-Rock Hill, South Carolina) indicated that the close proximity of North Carolina State University's College of Engineering was a factor in their decision to relocate or start their business in North Carolina.

■ Normative Considerations and Conclusions

In this chapter, we have argued that public research universities in the United States have to cope with a growing set of demands. External pressures—specifically, the unfolding of a knowledge-based global economy—are positioning universities, as traditional centers of knowledge creation, also to be engines of growth for their regions. To fulfill that role, universities are required to undertake new activities. Some of those activities create strains within the academy. Some also are better than others in generating economic growth.

It remains to be determined whether the changes we have documented are desirable, at least from an economic development perspective. In part, the answer depends on whether the stresses and conflicts that are created within the university are manageable and whether universities have comparative advantages in stimulating economic development. If they do not have comparative advantages or if the resulting conflicts cannot be managed, then other institutions may be more appropriate for these roles. If they do have comparative advantages, for what specific set of activities are universities best suited, and what kinds of internal organizational rearrangements might be needed to enhance their effectiveness?

For policy-making purposes, we want to know whether and to what extent the various university activities lead to regional economic development and whether those activities are best conducted by universities, as opposed to other types of institutions (private businesses, not-for-profits, etc.). We address the second of these questions here. That is tantamount to asking whether a legislature would maximize the return on a given appropriation, in terms of economic development results, by channeling the funds to a university as opposed to another type of institution.

To answer that question, we should ascertain the unique strengths of universities, in general, that give them comparative advantage. The first of these is the concentration of technical knowledge and expertise. That is not to say that professionals in industry are not equally intelligent and inventive. The difference is that the faculty at research universities are rewarded not only for creating new knowledge in specialized areas but also for keeping abreast of scientific and technical advances in broader areas of knowledge.

The second comparative strength of universities is their credibility, based on the perception that research is nonpartisan and scientifically rigorous. This perception is not always borne out, but to the extent that scholarly research must pass through peer review before it is published, it is generally more credible than industry or special-interest-group research (see Long, 1992).

A third comparative advantage for universities is their pool of talented, but inexpensive labor: the students.

These advantages are not granted to universities automatically. Universities need to develop and cultivate them. Technical knowledge and expertise in broader areas of knowledge require interaction and cooperation among faculty members who are experts in their respective fields. Credibility of a university disappears easily when its research does not meet a certain level of quality. Because a few bad examples can cause major damage, universities are well advised to develop mechanisms that guarantee a minimum quality of their output. Students as qualified but inexpensive labor will only be a comparative advantage for a university when it puts its students to work.

Provided that universities develop and cultivate their comparative advantages, what do these suggest about the use of public funds for various activities? First, knowledge transfer—from faculty to students in the classroom and from faculty (and research staff) to other economic actors via the provision of technical and managerial assistance, expert testimony, and consulting—draws directly on the first unique strength of universities. Private consultants certainly do the second type of knowledge transfer as well, and in many cases with good results. However, the second unique strength of the university discussed above would suggest that university-based knowledge transfer generally would be more credible (in terms of quality and neutrality).

The concentration of technical expertise also supports the basic research function. This is research that has many potential applications. Industry is at least as well suited to conduct applied research that results in improvements to its products or specific new products that the business can commercialize. In fact, the best research marriage between universities and industry is the combination of basic research strength with industry knowledge of applications. This point was made, as well, by Rosenberg and Nelson (1994), who warned universities away from research with specific industry applications. An exception could be made for R&D in industries or countries dominated by small and medium-sized firms that do not have their own R&D capacity. Then universities can fill important gaps and "prime the pump" of private R&D. Rosenberg and Nelson provided numerous examples of how state universities developed productive research programs affiliated with small-sized local industries.

The availability of willing students who do not yet command a market wage provides unique opportunities for universities to engage in labor-intensive public-service activities—in particular, those that would serve as a training ground for professional (or preprofessional) students.

The preceding discussion suggests that there is a legitimate basis for university involvement in all of the activities that we have identified, both traditional and new. That does not mean, however, that all are appropriate for every university. Like private business, a university may find it preferable to specialize and pursue only a certain set of activities. As we have shown, however, not all activities are equally likely to lead to regional economic development.

In summary, university spending on supplies and facilities and on employee wages and salaries, and related student and visitor spending, have significant regional development consequences via multipliers. Knowledge transfer also has considerable regional development effects, especially in regions that have a large enough job base to retain graduates. Students trained in the local university are more productive. So are businesses and governments in the region that have received technical and managerial assistance from the university. The third set of activities with highly localized effects are those relating to public service and leadership. That is perhaps most pertinent to universities in larger and more urban settings.

The creation of new technologies and improvements in existing technologies also has an impact gradient, but it is flatter than for the activities mentioned above. Spin-offs are likely to be located in the same region, but the creation of specialized knowledge can be appropriated by businesses anywhere in the world.

The mechanisms for effective knowledge and technology transfers require cooperation from local and regional businesses, government institutions, and policy makers, as well as time to develop. For example, businesses have to adjust to the demands of the university, its faculty, and its students to make the multiplier effect work. Similarly, when there is no tradition of technical and managerial assistance to government at the university, government officials will have to spend some time with university representatives explaining their needs and constraints. The initial stages will be characterized by uncertainty and misunderstanding until the partners have developed a basis of trust and a common language.

Taking on new responsibilities is not without risk for universities. As we have seen, the set of requests and expectations universities face has broadened considerably in recent years. Because of limited resources, no university can fulfill all these requests simultaneously. Therefore, universities have to decide their product mix and analyze questions such as "Which requests are essential for the long-run development of the institution?" "Are there market niches the university can occupy?" and "What are the short- versus long-run implications of a certain strategy?"

However, when the university chooses a certain product mix and targets a specific market—such as serving the local or regional economy—the adjustments and resource allocations it must make will have implications for its structure and development potential. To serve the targeted market most efficiently, the university will develop structures that are suitable for this specific need but may be ill suited for other functions. Structures and instruments that prove to be adequate under today's set of conditions may be hampering the university's future ability to adjust to changes in its dynamic environment.

As a result of fundamental changes in society and the economy, universities today face a broader and more complex set of demands than they did a few decades ago. In addition to the traditional teaching and research functions, universities today are expected to provide leadership and infrastructure, to stimulate the economy through technology transfer and expenditures, and to create a "milieu" that is favorable for economic and societal development. Our analysis suggests that there is a legitimate basis for universities to get involved in all of these activities, although it may not make sense for each institution to take on the whole set of responsibilities. Whether a certain university should become actively involved in a specific function depends not only on the university and its structure but also on the region and its economy.

Taking on new functions may bring challenging new management tasks to university administration. At best, administrators will need to target the appropriate new areas for the university and to balance new and old requests. It is likely, however, that taking the necessary steps will meet opposition and cause stress within and outside the organization. Managing these periods of crisis by communicating the aims and long-term perspective of restructuring and by carefully balancing its costs and benefits will be a major task for university leaders. To drive the development process rather than be driven by it, university administrators need a long-term perspective in addition to considerable managerial competence. To take the necessary steps to meet the challenging new set of demands, universities need sufficient flexibility in and control over their internal structure and operation (Neave, 1994; Williams, 1994).

NOTES

1. See the *Chronicle of Higher Education*'s Almanac Issue, September 1, 1995. The institutions counted above do not include 6,961 vocational institutions, many of which also are public.

2. A large number of universities in the United States have conducted this analysis using regional input-output modeling techniques. Caffrey and Isaacs (1971) provided the standard methodology for studies of this type. Gran, Mulkey, and Malecki (1995) cited 12 studies of universities or university systems, completed between 1991 and 1994, that employ this approach.

3. The key role of universities in supplying such a labor force for a technologically rapidly changing economic environment was articulated as early as 1969 by the National Academy of Sciences: "A key requirement in the attainment of social and economic objectives for a given region lies in the development of human capabilities and talents, and the attraction or retention of the most gifted and innovative segment of the population."

4. These studies, by Hansen (1963), Hanock (1967), Griliches and Mason (1972), Raymond and Sesnowitz (1975), and others, typically disaggregate individuals by level of schooling and demographic characteristics. In rate-of-return studies, producers' surplus is

usually neglected in the measurement of productivity gains realized by the producing firms. Also, the use of wage and salary measures rather than total compensation (i.e., including fringe benefits and working conditions) underestimates the rate of return on the order of 10% to 40%. Other non-market-induced benefits of investments in higher education, such as improved health, are usually ignored. For a complete typology of market and nonmarket impacts of investment in schooling, see Haveman and Wolfe (1984).

5. Amenity level may affect the decision to remain or move on graduation. We are prone to ignore the two-way, intimate relationship between a university and its regional environment. Although this chapter is focused on the impacts of the university on its regional environment, the regional environment is a critically important input, or determinant, of the set of outputs and impacts that a university can potentially generate.

6. Besides dollars in resources to pay for staff time, the degree of commitment by the university's administration, the incentives provided to faculty and staff, and the effectiveness of the organization of service provision within the university (e.g., decentralized along essentially departmental lines, highly centralized through a public service, or having an extension division with its own staff devoted to service provision) have often been cited as important internal factors.

7. It would also include the application of knowledge to the transformation of existing products. Measurement is difficult because commercializable products sometimes result unexpectedly from research, not as an explicit objective of the activity.

8. Most private universities in the United States are extended tax-exempt status as charitable organizations, as defined in §501.C.3 of the Internal Revenue Code.

9. Livability comparisons are published periodically by *Money* magazine, the *Places Rated Almanac*, and other books and journals.

10. See Goldstein and Luger (1992) for a complete presentation of the methodology and results. Other researchers estimate even larger multipliers. For example, Purdue University researchers report in *Perspectives* (Spring 1995) an eightfold multiplication of state appropriations into new output in Indiana.

11. Data were provided by Mary Seachrist, Associate University Counsel, from her annual report to the Chancellor. Phone conversation, August 9, 1991. Our survey results are consistent with these numbers. Fifty-three of the 167 graduate/professional/special student respondents were granted in-state status after their matriculation.

12. This assessment was made on the basis of the experiences of what Feller (1994) called the "university-industry model." It may be less true for the "product development model" and for models that we place in the category of technology transfer in this chapter.

13. One can argue that the implied direction of causality may be the opposite direction: The generation of a number of innovative businesses (brought about by other factors) makes it possible to build a more highly developed infrastructure because one of the contributors to the infrastructure is a pool of successful entrepreneurs, talented scientists and engineers, R&D shops, and so on.

REFERENCES

Bania, N. R., Eberts, R., & Fogarty, M. (1987, July). *The role of technical capital in regional growth*. Paper presented at the meeting of the Western Economic Association, Santa Barbara, CA.

Becker, G. (1975). *Human capital: A theoretical and empirical analysis*. New York: National Bureau of Economic Research.

Brett, A. M., Gibson, D., & Smilor, R. W. (1991). *University spin-off companies.* Bollman Place, MD: Rowman & Littlefield.

Brickman, W. W. (1993). *Land-grant colleges.* New York: Funk & Wagnall's [online encyclopedia, Microsoft Encarta 97].

Brown, W. S. (1987). Locally-grown high technology business development: The Utah experience. In W. S. Brown & R. Rothwell (Eds.), *Entrepreneurship and technology: World experiences and policies.* Harlow, UK: Longmans.

Caffrey, J., & Isaacs, H. H. (1971). *Estimating the impact of a college or university on the local economy.* Washington, DC: American Council on Education.

Chronicle of Higher Education Almanac Issue. (1995, September 1). Vol. XLII, p. 7.

Cohen, W. M., & Levinthal, D. (1989). Innovation and learning: The two faces of R&D. *Economic Journal, 99,* 569-596.

Cote, L. S., & Cote, M. K. (1993). Economic development activity among land-grant institutions. *Journal of Higher Education, 64,* 55-73.

Dorfman, N. S. (1982). *Massachusetts' high technology boom in perspective: An investigation of its dimensions, causes, and of the role of new firms.* Unpublished manuscript, Massachusetts Institute of Technology, Center for Policy Alternatives.

Feller, I. (1990). Universities as engines of R&D-based economic growth: They think they can. *Research Policy, 19,* 335-348.

Feller, I. (1994, March). *The university as an instrument of state and regional economic development: The rhetoric and reality of the U.S. experience.* Paper presented at the Conference on University Goals, Institutional Mechanisms, and the Industrial Transferability of Research, Stanford University, Stanford, CA.

Felsenstein, D. (1994). *The university in local economic development: Benefit or burden?* Unpublished manuscript, Hebrew University of Jerusalem.

Florax, R., & Folmer, H. (1992). Knowledge impacts of universities on industry: An aggregate simultaneous investment model. *Journal of Regional Science, 32,* 437-466.

Geiger, R., & Feller, I. (1994). The dispersion of academic research in the 1980s. *Journal of Higher Education, 66*(3), 336-360.

Goldstein, H. A., & Luger, M. I. (1992). *Impact Carolina: The University of North Carolina at Chapel Hill and the state's economy.* Report prepared for the UNC-CH Bicentennial Observance, The University of North Carolina, Chapel Hill, NC.

Goldstein, H. A., & Luger, M. I. (1993). *The economic impact of North Carolina State University's College of Engineering on the state.* Report prepared for the College of Engineering, North Carolina State University, Raleigh, NC.

Goldstein, H. A., Maier, G., & Luger, M. (1994, June). *The university as an instrument for economic and business development: U.S. and European comparisons.* Paper presented at the Symposium on University and Society: International Perspectives on Public Policies and Institutional Reform, Vienna, Austria.

Gran, S., Mulkey, D., & Malecki, E. (1995, April). *The economic impact of the University of Florida on the state of Florida.* Report presented at the meeting of the Southern Regional Science Association, San Antonio, TX.

Griliches, Z. (1984). *R&D, patents and productivity.* Chicago: University of Chicago Press.

Griliches, Z. (1986). Productivity, R&D and basic research at the firm level in the 1970s. *American Economic Review, 76*(1), 141-154.

Griliches, Z., & Mason, W. (1972). Education, income, and ability. *Journal of Political Economy, 80,* 74-103.

Hanock, G. (1967). An economic analysis of earnings and schooling. *Journal of Human Resources, 2,* 310-346.

Hansen, W. L. (1963). Total and private rates of return to investment in schooling. *Journal of Political Economy, 71*, 128-140.

Haveman, R. H., & Wolfe, B. L. (1984). Schooling and economic well being: The role of nonmarket effects. *Journal of Human Resources, 19*, 377-407.

Jaffe, A. (1986). Technological opportunity and spillovers from R&D: Evidence from firms' patents, profits and market value. *American Economic Review, 76*, 984-1001.

Jaffe, A. (1989). Real effects of academic research. *American Economic Review, 79*, 957-970.

Jaffe, A., Tratjenberg, M., & Henderson, R. (1992). Geographic localization of knowledge spillovers, as evidenced by patent citations. *Quarterly Journal of Economics.*

Klausner, S. Z. (1988, April). *Dilemmas of usefulness: Universities and contract research.* Paper presented to the American Bar Foundation.

Ko, S. (1993). *The incidence of high technology start-ups and spin-offs in a technology-oriented branch plants complex.* Unpublished Ph.D. dissertation, University of North Carolina at Chapel Hill.

Krenz, C. (1988). *Silicon Valley spin-offs from Stanford University faculty.* Unpublished manuscript. Stanford, CA: Stanford University Business School.

Leslie, L., & Brinkman, P. T. (1988). *The economic value of higher education.* New York: American Council on Education and Macmillan.

Long, E. L., Jr. (1992). *Higher education as a moral enterprise.* Washington, DC: Georgetown University Press.

Luger, M. I., & Goldstein, H. A. (1991). *Technology in the garden: Research parks and regional economic development.* Chapel Hill: University of North Carolina Press.

Malecki, E. J. (1987). The R&D location decision of the firm and "creative" regions. *Technovation, 6*, 205-222.

Marsden, G. M. (1994). *The soul of the American university: From Protestant establishment to established nonbelief.* New York: Oxford University Press.

Matkin, G. W. (1990). *Technology transfer and the university.* New York: American Council on Education and Macmillan.

National Academy of Sciences. (1969). *The impact of science and technology on regional economic development.* Washington, DC: Author.

Neave, G. (1994). The stirring of the prince and the silence of the lambs: The changing assumptions beneath higher education policy, reform, and society. In D. D. Dill & B. Sporn (Eds.), *Through a glass darkly: Emerging patterns of social demand and university reform.* Paris: Pergamon, for the International Association of Universities.

Noble, D. (1977). *America by design.* New York: Knopf.

Peters, L. S., & Fusfeld, H. I. (1983). *University-industry research relationships.* Washington, DC: National Science Foundation.

Pings, C. J. (1994). *Memorandum to university and college presidents and chancellors.* Washington, DC: Association of American Universities.

Powell, W. S. (1992). *The University of North Carolina: A pictorial history* (3rd ed.). Chapel Hill: University of North Carolina Press.

Purdue University. (1995, Spring). State money returns high yield: $2 billion. *Perspectives*, p. 13. West Lafayette, IN: Office of University Relations, Purdue University.

Raymond, R., & Sesnowitz, M. (1975). The returns to investments in higher education: Some new evidence. *Journal of Human Resources, 10*, 139-154.

Romer, P. (1990). Endogenous technological change. *Journal of Political Economy, 98*, S71-102.

Rosenberg, N., & Nelson, R. (1994). American universities and technical advance in industry. *Research Policy, 23*, 323-348.

Saxenian, A. (1985). Silicon Valley and Route 128: Regional prototype or historic exceptions? In M. Castells (Ed.), *High technology, space, and society*. Beverly Hills, CA: Sage.

Sirbu, M. A., Treitel, R., Yorsz, W., & Roberts, E. B. (1976). *The formation of a technology-oriented complex: Lessons from North America and European experiences*. Unpublished manuscript, Massachusetts Institute of Technology, Center for Policy Alternatives.

SRI International. (1992). *New York State Centers for Advanced Technology Program*. Final report to the New York State Science and Technology Foundation. Menlo Park, CA: SRI.

Tassey, G. (1991). The functions of a technology infrastructure in a competitive environment. *Research Policy, 20*, 345-361.

Thomas, M. (1985). Regional economic development and the role of innovation and technological change. In A. T. Thwaites & R. P. Oakey (Eds.), *The regional economic impact of technological change*. London: Francis Pinter.

Tratjenberg, M., Henderson, R., & Jaffe, A. (1992). *Ivory tower versus corporate lab: An empirical study of basic research and appropriability* (Working Paper No. 4146). Cambridge, MA: National Bureau of Economic Research.

University of North Carolina at Chapel Hill. (1994). *The first state university: A pictorial history of the University of North Carolina*. Chapel Hill, NC: University of North Carolina Press.

Williams, G. (1994). Reforms and potential reforms in higher education finance in Europe. In D. D. Dill & B. Sporn (Eds.), *Through a glass darkly: Emerging patterns of social demand and university reform*. Paris: Pergamon, for the International Association of Universities.

COMMENTARY ON THE ROLE
OF PUBLIC UNIVERSITIES

Donald F. Smith, Jr.
Robert E. Gleeson

Universities have wide-ranging effects on the regions they inhabit. About that there is no debate. Yet the last decade has seen renewed controversy about the nature, scale, direction, and utility of interactions between universities and their surrounding communities.

Of particular interest has been the impact of higher education on regional economic development. Motivated largely by the destabilizing effects of pervasive economic restructuring, governments, corporations and philanthropies in the United States that for decades provided routine, generous financial support to higher education are rethinking their priorities.

For the first time since the 1930s, universities find it necessary to justify themselves to skeptical external audiences. Historians have long pointed out that education is America's secular religion. If that is so, then universities are America's civic cathedrals. Yet that status no longer shields universities from Americans' favorite pastime: getting their dollar's worth. It is good politics today for congressional committees and state legislatures to cut funding for higher education. And cuts go well beyond allegations of waste. The 1996 Republican budget proposal, for example, called for the total elimination of National Science Foundation funding for most social science research. Although funds were restored in the end, such proposals reveal how deeply skeptical American society has become about the contributions universities make to broader society.

Within this context, Michael Luger and Harvey Goldstein provide an interesting and very thorough review of the various ways that researchers have tried to conceptualize and measure the relationships between universities and their surrounding regions. The questions outlined in Table 5.2, for example, are both central and informative. Very importantly, Luger and Goldstein remind us that modern, industrial societies have become knowledge economies and that universities are key organizations that support the creation and dissemination of knowledge. Though they may partly play the role of civic cathedrals, universities still have great practical utility in the global economy. This is especially true for regions because it is within regions that knowledge-based organizations can operate in close proximity to each other and thereby create and transfer new knowledge that can be turned into economic growth.

Universities play essential roles in creating and supporting regionally based clusters of knowledge-based organizations. But universities are not a panacea for regional economic development. The authors hit directly on the oft-overlooked point that universities and their host regions provide reinforcement for each other. Building a major university in a region that lacks the other building blocks of a knowledge-based infrastructure not only is *not* a guarantee of economic development but can actually have negative consequences. This is most notably true in the case of students from a region who go through a university, build human capital, and then migrate to other regions to seek employment, thus creating a "brain drain."

Although Luger and Goldstein give us an excellent first step in grappling with these issues, we see several ways of enhancing their approach. First of all, their review reinforces our concern about how to measure the outputs of universities better. We simply do not yet have reliable measures for many of the most important contributions universities make to supporting their region's cluster of knowledge-based organizations. Most measures instead gauge processes, such as educational attainment, research grants awarded, technology transfers programs/centers, patents, commercial licensing, equity positions in start-ups, and acres of technology parks. Although all of these measures give useful information about traditional—and new—activities on the part of universities, they do not directly measure the ultimate effects universities have on regional economic growth. Nor do they allow us to untangle the directionality of effects: That is, do universities develop regions or vice versa? Though the authors offer elements of measurement needed to operationalize an examination of impacts, these are never tied together into a coherent approach or methodology that would facilitate an analysis of the impacts.

In addition, despite the insightful discussions of the many new roles universities are playing or attempting to play, it remains true that the primary mechanisms by which universities affect their region's economic development are the two traditional means: providing human capital and acting as major employers/purchasers within the region. New roles, though important, do not approach the scale of these long-accepted university functions.

Another concern is the authors' choice of public universities for their study. Though public universities are certainly more numerous, and their aggregate spending dwarfs private universities, an argument can be made that private universities may play more important roles in most regional networks of knowledge-based organizations. One need only point at the literature on Silicon Valley and Route 128 to make the point that private universities must be included in any general research regarding universities and regional economic development. In our own region, two private universities, Carnegie Mellon and Duquesne, have been central players in initiating and staffing the ongoing Regional Economic Revitalization Initiative.

A more minor flaw is that the authors provide no perspective on the very dubious nature of impact estimation and multiplier effects. Impact assessments are usually based in input-output tables for the nation (occasionally modified in some way for the region), with coefficients that are many years out of date. No one in academe actually believes the point estimates, and they can best be described as being generally indicative. Better methods and data will be needed to provide any semblance of accurate estimates.

One final issue also deserves attention. Universities are important organizations within regional networks of knowledge-based institutions. But they are supportive organizations. They cannot alone create sufficient activities to spur regional economic development. In their role as supporting organizations, universities are shaped in fundamental ways by how other knowledge-based organizations go about their work of creating economic growth from new knowledge. Universities organize their contributions in relation to the environment around them. For most of the last century, the general business/social environment in the United States has organized itself around the notions of increasing specialization and the rigorous separation of thinking and doing. Whether one calls this social organization Taylorism, Fordism, or some mixture of the two, universities have shaped themselves accordingly. Yet increasing evidence suggests that the networks of knowledge-based organizations that inhabit the world's most productive regions are moving away from this traditional form of business/social organization and are moving toward the type of ideas expressed by advocates of total quality and high-performance work organization. If this is true, one can expect universities gradually to restructure themselves in response to the larger social reorganization going on in the regions they inhabit.

In closing, Luger and Goldstein provide us with an important and useful exposition of the changes taking place in the roles played by public universities in regional economic development. Although they detail quite well the types of questions one needs to assess and the measures currently available to researchers, we encourage them and others to take the debate to the next level in formulating comprehensive methodologies for understanding the complex relationships between universities and their regional economic environments.

RESPONSE TO THE COMMENTARY

Michael I. Luger
Harvey A. Goldstein

Donald F. Smith, Jr., and Robert E. Gleeson provide a thoughtful review of our chapter and make several critical observations. Here we respond briefly to their comments.

Their first observation is that we offer only elements of measurement needed to operationalize an assessment of universities' impacts and do not tie those elements together into a coherent approach or methodology that would facilitate analysis. That is a point worth pondering. Presumably, Smith and Gleeson are imagining something like a "cost-benefit analytic framework" that organizes impact assessment in other fields (e.g., in environmental policy). In essence, our approach is along those lines, though we do not identify it as such. The problem, however, is that so many of the benefits of universities are indirect and intangible. And although costs of education are straightforward to calculate, the proper assignment of costs to population subgroups (the incidence question) is knotty because of the social nature of knowledge creation.

Smith and Gleeson recognize that our intention, however, was simply to broaden the way we think analytically about universities. The literature focuses on just one of several potentially important economic development effects. Smith and Gleeson point out that "despite [our] insightful discussions of the many new roles universities . . . play, it remains true that the primary mechanisms by which universities affect their region's economic development are the two traditional means: providing human capital and acting as major employers/purchasers within the region." That is correct if we are using traditional measures of economic development. But if we are using different yardsticks—for example, the creation of new technology for the world—some of the other roles are also of primary importance. Moreover, human capital creation for a region is achieved by several university activities, not just teaching. Similarly, the size of the employment/spending multiplier depends on other characteristics of the university, such as the amount of externally funded research. We do not agree with Smith and Gleeson's categorical statement that "new roles . . . do not approach the scale of [the] long-accepted university functions." The reality is more complex than that.

With regard to the multiplier effects of universities, Smith and Gleeson observe that we provide no perspective on the very dubious nature of impact estimation and multipliers. That is, indeed, an important point. As they indicate, most studies of university impacts use multipliers generated by others, often for different economic regions and time periods. Undoubtedly, that compromises the accuracy of the university impact analyses. For that reason, we generate our own multipliers in the case studies used for our chapter, using recent state input-output tables. That enables us to conclude that generally research universities have larger multipliers than comparably sized private business because of the nature of their funding (much from external sources) and their purchasing patterns (generally more localized).

Another observation Smith and Gleeson make is that we overlook the important economic development role played by private universities and focus too much on state-supported institutions. We agree fully that private universities also can play an important role in regional economic development. In other research, we are studying private universities as well and are comparing the motivations for their activities, their internal structure and incentives, and their regional impacts with those of the public universities. Here, we made a strategic decision to limit our scope, for two reasons. First, the chapter is part of a larger project in which we compare U.S. and European universities, and most of the latter are state supported. Second, this chapter is about economic development policy, and it is the public institutions that are used most by states as explicit instruments of policy for economic development.

Smith and Gleeson's final comment is that we should say more about how "universities are shaped . . . by how other knowledge-based organizations go about their work of creating economic growth from new knowledge." Their point is well taken, and in future work we will augment our discussion of the internal pressures operating on universities with insights from the business literature on total quality and high-performance work organization.

6

Taxes and Economic Development in the American States: Persistent Issues and Notes for a Model

PAUL BRACE

For 60 years, the American states have ridden a roller coaster of federal expansion and contraction. In the first half of this era, states languished in a relatively passive role, politically and economically, as the federal government expanded. In the 1950s and 1960s, state governments were the "fiscal stepchildren" (Van Horn, 1989) of the federal system. Beginning in the 1970s, this situation changed, placing state governments under pressure to address new political and economic issues. State capitols became centers of innovation and leadership, rather than politically moribund backwaters (see, e.g., Osborne, 1990). By the 1990s, however, new changes have descended on the states that may spell a totally new direction in state politics and economics.

Though many events can be highlighted as sources of political and economic change in the states (e.g., the Depression in the 1930s, redistricting and civil rights legislation in the 1960s), the ones of primary interest here, and of significance in the dynamics of contemporary state political economies, concern fiscal changes. Since the 1960s, the states have had to pick up an ever-growing piece of the revenue pie, as illustrated in Figure 6.1.

In a trend initiated during the Carter years, federal revenues to the states have diminished. The Omnibus Reconciliation Act of 1981 reduced federal aid payments to the states and localities by 7%, the first cut in constant dollar terms in 25 years (Nathan & Doolittle, 1987). Against this backdrop of decreasing federal aid has been an increase in federal mandates requiring the states to provide such things as health care to the poor and services for the disabled.

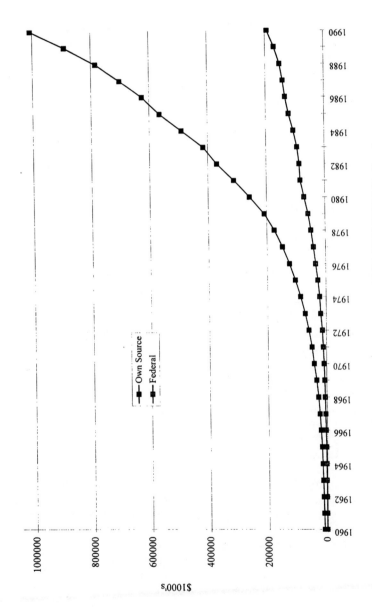

Figure 6.1. Average Annual State Revenues by Source, 1960-1990 (1982 Dollars)
SOURCE: Computed from data in U.S. Department of Commerce, Bureau of the Census, Annual Report, Compendium of State Government Finances (various years).

141

The states have adapted their fiscal policies to cope with these growing needs. In the 1960s through the mid-1970s, states added or increased taxes to finance new or expanded services in health, education, transportation, and environmental protection. These taxes supplemented the generous contributions of the federal government (40% of state expenditures came from the federal government in 1975; Advisory Commission on Intergovernmental Relations [ACIR], 1984). Tax increases in the states dropped off precipitously after the tax revolt in California, as can be seen in Figure 6.2. No new taxes were added from 1978 to 1980, and taxes declined in many states (ACIR, 1984, p. 71).

A new brand of leadership in state capitols was being heralded in the midst of the federal fiscal retreat and the recession of the early 1980s. Increasingly, states engaged in economic entrepreneurism, investing heavily in the development of new technologies, providing capital to new businesses, pursuing expanded export markets, and training workers (Eisinger, 1988; Osborne, 1990). Many states also raised taxes in this period to cope with the national recession of the 1981-through-1983 period and further reductions in federal aid. This increase in tax changes is illustrated in Figure 6.2.

When the national economy grew in the mid- to late 1980s, states had an abundance of revenues as a result of these tax increases that allowed many states to engage in spending programs aimed at stimulating their economies. States as diverse as Mississippi and New York created venture-capital corporations. A majority of states created high-tech development and technical assistance programs to stimulate technological growth. By the mid-1980s, all but a handful of states had invested in programs to stimulate foreign trade and market development.

Armed with new revenues, several state governors embarked on bold economic development programs. In Michigan, James Blanchard created a Strategic Fund and a Modernization Service, intended to use state financial resources to help the state modernize and diversify its manufacturing base. In Massachusetts, Michael Dukakis helped initiate an Industrial Service Program to provide state funds intended to help distribute growth throughout the state and provide businesses with capital and services. In New Jersey, Governor James Florio initiated a public-works investment program to improve transportation and create 150,000 jobs. As the economic boom of the 1980s ended, so too did the political fortunes of these and other economically innovative interventionist governors.

With the national recession of the 1990s, another era of state political economy has emerged. With the shortfall in revenues resulting from the recession, and the unrelenting pressure of new federal mandates, many states found themselves teetering on the brink of fiscal disaster in the 1990s. Fifty percent of the states could not fund their fiscal 1992 budget,

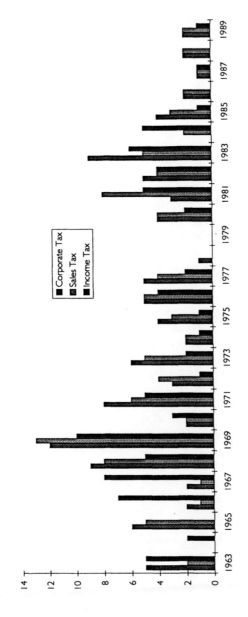

Figure 6.2. Total State Tax Increases by Type and Year

Figure 6.2. Total State Tax Increases by Type and Year

SOURCE: Created with sales and personal tax data supplied by Professor Richard Winters of Dartmouth, who compiled these data using various issues of the *Tax Administrators News*. These data were supplemented and corporate tax changes were added by reference to "Recent Trends in Taxation" in biennial volumes of *Book of the States*.

despite tax increases totaling over $16 billion in fiscal 1991. To address these financial exigencies, states have had to cut programs or raise taxes.

The latest developments in state taxation and state economies are at the heart of any discussion of the role of taxation in shaping state economic performance. The salient features of the current political economic milieu: a national recession, reduced federal aid, increased federal mandates, and political groups in many states that remember with fondness the tax revolts of past decades. The states are caught in a fiscal vise, and there are telling signs that this is influencing taxation and economic development strategies in the states.

With unemployment a concern, job growth remains a popular item on the agendas of most state governments. In the midst of a recession and in the wake of tax increases, pursuing jobs with state expenditures has become a less practical approach to development. The pursuit of jobs in a tax- and revenue-constrained environment has led to a renewed emphasis on tax incentive packages and general tax cuts to stimulate economies. Recent efforts by the states to attract firms have been high-visibility bidding contests with astronomical stakes that have left deep gouges in the revenues of the victorious contestants. Only time will tell if the high bids will be repaid over the long haul with sufficient growth to offset these generous outpourings of revenues and tax inducements.

Consider the cases from only the last few years. The bidding began in the 1980s as Japanese automakers obtained then-record subsidies for new plants in Ohio, Tennessee, and Kentucky. More recently, the bidding wars have expanded to include domestic and European firms, and the subsidies have grown astronomically. For example, when United Airlines announced its intention to locate a maintenance facility, a high-stake auction among the states resulted. The "winner," the city of Indianapolis, hooked the prize with a bid of $291 million worth of subsidies. The cost per job in this contest worked out to almost $50,000. Northwest Airlines recently secured $270 million in loans from the state of Minnesota to prop up the company's shaky finances. In the automobile industry, South Carolina scored big by luring BMW, but not without a large price tag. The state offered Bayerische Motoren Werke AG at least $130 million in incentives to build its factory there. The state also paid to move 100 families so the company could have the exact site it wanted, offering to lease the reclaimed land to the prestige automaker for a dollar a year. No doubt the success of its primary German competitor in securing favorable treatment in America fueled expectations at Mercedes Benz, for the firm drove an extremely hard bargain before settling in Alabama. Benz considered more than 20 states. North Carolina, site of a Mercedes-owned Freightliner truck plant, was an initial front-runner, and Governor Jim Hunt reportedly pursued Benz harder than he had ever pursued any other potential investor

(Browning & Cooper, 1993). Mercedes was a very skilled negotiator in this process. After receiving offers for certain things from several states, they would add these to their "ideal" contract proposal. All the main competitors were eventually persuaded to offer a $5-million welcome center next to the factory where visitors could pick up cars, have them serviced, and visit an auto museum. It also received commitments for free 18-month employee training programs. Competing states also promised to buy large quantities of the four-wheel-drive vehicles that the new factory was to produce. The kicker for luring this plant came at the last minute. Mercedes asked the remaining states if they would be willing to pick up the salaries (roughly $45 million) of the workers during their first year on the job, when they were in state-financed training, not yet actually producing anything. North Carolina said no. Alabama said yes to this and many other demands, and the resulting bid, at least that which has been made public so far, pushes the envelope of state generosity.

The initial disclosure of $253 million in incentives to attract the 1,500 jobs offered by Mercedes was later revealed to be a $325-million package. Some of the incentives were symbolic and almost comical, but they reveal the length to which states have been driven to capture these plums. For example, Alabama officials agreed to place the Mercedes emblem atop a local scoreboard in time for the big, nationally televised Tennessee-Alabama football game at no charge to the company. Mercedes, furthermore, was designated the official automobile of the state of Alabama, and the state agreed to purchase no fewer than 100 of the company's new sport utility vehicles, with a 10-year commitment to purchase 2,500 vehicles. The financial core of the deal was very serious business, however. Among the incentives were commitments by the state to spend $1.5 million per year for 5 years on image advertising and public relations promoting Mercedes' decision to locate in Alabama, to contribute $5 million to Mercedes if it decides to build the previously mentioned welcome center, to make the U.S. Navy port at the Alabama State Docks available to Mercedes if it decides to ship cars through the port of Mobile, to build a $30-million permanent training center at the plant site as well as to contribute $5 million a year toward the center's operation, to make a tax credit available in advance in the form of an interest-free loan, and to provide an additional $60 million to educate and train Mercedes employees, including sending people to Germany to study at the company's headquarters. In all, Alabama will spend approximately $216,000 per direct job. What is truly remarkable is that Alabama's bid for Mercedes is not a record. Through a combination of job training grants, local site improvements, and income tax credits, Kentucky provided a steel firm with a $140-million package to secure 400 direct jobs, for a $350,000-per-job subsidy.

If the pressure to create sweetheart deals were not enough, many states currently find themselves under pressure to cut existing tax rates. This pressure comes partly from a desire to appear more business friendly and partly from a pervasive antitax mood in the country. It is not difficult to find overblown claims about the deleterious effects of taxation on state economies. Malcolm Forbes, Jr., heir to *Forbes* magazine, described the state personal income tax as destructive and pointed to it as the source of New Jersey's recent economic woes (Redburn, 1994). This thinking, furthermore, is translating into action in many states. Christine Todd Whitman proposed slashing state income taxes by 30% over 3 years in her campaign for governor of New Jersey. Arguing that this measure would keep businesses from being lured away to states with lower tax rates (Redburn, 1994), this newly elected governor succeeded in passing a package of bills that cut the state income tax by 5% across the board (Blumenthal, 1994). Governors Fife Symington of Arizona, Zell Miller of Georgia, and William Weld of Massachusetts have announced planned tax cuts. Even liberal Democrats are pushing for tax cuts under the assumption that these cuts can boost growth. Most recently, Mario Cuomo of New York proposed easing business taxes to make his state more business friendly. Preliminary signs for a cut in the state of Washington are also positive as Democratic Governor Mike Lowry has pledged tax breaks for high-tech industries.

To be economically successful, the present flurry of tax cuts in the states will have to stimulate substantial growth to compensate for lost revenues. Federal mandates have not subsided, and federal expenditures have not risen appreciably. Although there have been promising signs of economic recovery at the national level, state tax cuts in the 1990s may isolate state coffers from the inflow of dollars such as those that followed the 1980s recovery. In New Jersey, the initial 5% tax cut will reduce revenues by as much as $285 million (Blumenthal, 1994). Given a constitutional amendment requiring a balanced budget, and a projected budget deficit of between $1.5 and $2 billion, it is unclear how Governor Whitman can follow through on the additional 25% cut she proposed in her election. As a stimulus to the state's economy, the 5% cut does not appear likely to bring about dramatic economic change in the state. The reduction was estimated to translate into a net change in income of $24 per year for the largest category of taxpayers in the state (Blumenthal, 1994).

A new era of "go-it-alone" or "fend-for-yourself" federalism is upon us. With pressing federal mandates, diminished federal dollars, increased concern about jobs, and heightened animosity among voters about increased taxes, state politicians are threading their way through these often conflicting cues and developing policy proposals that strike at the heart of taxation and development. Many states appear poised to engage in predatory competition to lure employers and placate voters' antitax sentiments. In the emerging global economy, firms appear well suited to capitalize on the probusiness, antitax sentiments in the states. From automobile manufacturers to theme parks, firms are succeeding in obtaining extremely favorable deals from governments eager to lure jobs and income to their states. Though it is too early to say with certainty what the effects of these new bidding wars will be, it seems that the thirst for new income may be leading some states to subsidize profits for firms instead of requiring payment for the costs of their development.

■ Do Taxes Influence State Economic Development?

With competition from a globalizing economy, escalating tax wars between the states, and the public outcry for jobs, questions about the state tax/development nexus are more salient than ever. The answer should be simple: High taxes harm growth! Yet despite this seemingly sensible expectation and the elegance of some economic theories pointing to the detrimental consequences of higher taxes, empirical evidence about the effects of taxes on state economic performance has often been contradictory or vague. The developmental consequences of taxes in the American states raise a number of significant questions about policy making in a federal system and the interplay of national and subnational political economies. Fiscal policy in the states reflects both exogenous changes and endogenous demands. Economic performance in the states, similarly, may be viewed as resulting from a blend of external and internal forces. In an analytically simple world, tax considerations would reflect endogenous demands and resources, and economic effects would be manifest in growth determined strictly by endogenous economic forces. In the real world, changing policy directions in Washington have greatly influenced state tax policy. State economies, in like manner, have reflected booms and busts in the national economy. Because of the enormous influence of external forces on both tax choice and economic performance, the role of taxes in the process of development at the state level has been ambiguous.

Economists have generated an extensive body of research on the effects of taxation on states, and these studies vary widely in their methodologies and their findings. A dominant conclusion of early studies was that state tax levels had no appreciable impact on state economies (Bloom, 1955; Campbell, 1965; Due, 1961; Sacks, 1965; Struyk, 1967; Thompson & Mattila, 1959). The findings of many of these early studies may be suspect because they employed very simple and questionable research designs. Simple correlational analyses were often used. However, many later studies with more sophisticated designs and methodologies have failed to show state taxes as having any notable effect on state economic performance (Genetski & Chin, 1978), the birth of firms (Carlton, 1983), or business location decisions (Kieschnik, 1981; Wasylenko, 1981). Some research provides mixed results on the impact of taxation on the states. Helms (1985), for example, argued that the economic consequences of taxation are contingent on the use to which the tax revenues are put, with income transfers hurting growth. Others have made similar observations (Plaut & Pluta, 1983; Romans & Subrahmanyam, 1979; Wasylenko & McGuire, 1985). Still other research has argued that taxes have unambiguous negative effects on state economies (Brace, 1990; Brace & Mucciaroni, 1990; Grierson, 1980; Kleine, 1977; Newman, 1983; Vedder, 1982).

Although more recent studies employing more sophisticated estimation procedures have tended to show the detrimental consequences of taxes on growth, the overall picture remains ambiguous. Eight of 14 studies reviewed by Eisinger (1988, p. 211) revealed little or no effect of taxes on growth, with 6 of those 14 finding some or great influence. Of the 58 interstate or intermetropolitan studies conducted between 1979 and 1991 surveyed by Bartik (1991, pp. 216-247), a majority found significant negative effects. It is worthy of note, however, that 22 of these studies report positive or nonsignificant economic effects.

Despite an abundance of research, we remain uncertain of the effects of taxation on the long-term economic development of the states. Considering the evidence to date, the best we can say is that state taxes probably hinder growth on some economic indicators, and in some periods, under certain conditions, in the short run. Though impressive advances have been made, we lack sufficient confidence to make general assertions about these effects.

Three central issues have hindered a more satisfactory and general understanding of the effects of taxation on state economies: theoretical ambiguities, measurement problems, and period effects. Each of these methodological/technical issues is tied to basic substantive concerns about the nature of political economy in a dynamic and competitive federal system. These issues are discussed in detail below.

Growth, Development, and
Taxation: Theoretical Issues

At the theoretical level, there are several basic and compelling reasons for expecting higher taxation to retard economic growth. From Adam Smith to Milton Friedman, economists have reasoned that excessive taxation removes incentives to invest by lowering rates of return. By removing capital from the ostensibly more efficient market economy, and placing it in arguably less efficient government expenditures, taxation also serves to undermine the allocative effectiveness of the marketplace by shifting resources from more to less productive use. From the standpoint of historical evolution (Olson, 1982), or in comparison to other jurisdictions (Peterson, 1981), taxation in support of redistributive policies has been argued to retard economic growth. Though some economists argue for the stimulative economic effects of government expenditures, at the state level these expenditures may spill over into other jurisdictions, and the demand-stimulating benefits of expenditures for roads, facilities, or equipment may go partly or mostly outside the state. Payments to state employees or workers contracted to the state, similarly, often find their way to commodities produced outside the state or nation. Largely, then, whatever short-term stimulative effects state expenditures might have on state economies would appear to be more than outweighed by the detrimental consequences of higher tax levels.

State expenditures may have long-term economic benefits, whereas short-term effects may be meager or nonexistent. As one account noted, "Irrespective of the prevailing low wage levels, absence of unionization, availability of energy, and politically hospitable environment, no firm will locate in a region unless the requisite social and physical overhead capital has already been installed" (Watkins & Perry, 1977, pp. 47-48). These and other observers pointed to the importance of massive public outlays for physical or educational infrastructure for securing employers. Thus, although state expenditures may not have a Keynesian-demand-stimulating effect in the short run, these outlays may shape the attractiveness of states to firms and workers in the long run. From this perspective, short-run growth from lower taxes may come at the cost of long-term economic development if expenditures for education and infrastructure cannot be maintained.

A major shortcoming of many studies of state political economy is a failure to come to grips with the complex, multifaceted nature of taxation *and* expenditure and of short-term growth and long-term development. Remarkably, economic growth and economic development are not necessarily compatible objectives. Growth typically creates new demands for public services that often outpace new revenues created by that growth. Job growth and tax-base growth are treated as mutually reinforcing goals by many policy makers, but, as one study observed, in-migrants attracted by new employment will usually more than absorb additional revenues created by the accompanying economic growth. This problem is magnified when new jobs are of the low-wage variety.

In the late 1970s, some came to acknowledge that simply increasing the quantity of jobs within a state might not affect the overall well-being of its citizenry as well as focusing on the quality of jobs (Eisinger, 1988). From this perspective, economic development involves a qualitative increase in collective well-being. Employers attracted primarily by low taxes and low wage rates are unlikely to produce revenues, or a workforce that produces revenues, sufficient to offset the increasing drain on public services created by that growth. Thus, from this perspective, job gains from low taxation could diminish and not increase the economic development of a state by decreasing the collective well-being of its citizenry. In short, the "beggar-thy-neighbor" low-taxation approach to growth does not necessarily lead to economic development when qualitative as opposed to quantitative facets of economic change are factored in. An embarrassment of riches in new jobs may also lead to a paucity of revenues for public programs with long-term economic benefits.

Emerging empirical evidence suggests that increasing state taxes to finance improved public services can have a stimulative economic effect. Expenditures for physical infrastructure and education have been revealed to enhance a state's economic performance, at least under the right conditions (Brace, 1990, 1993; Jones, 1990). State and local tax increases to finance increased public capital have been shown to increase the growth rate of private employment (Munnell, 1990). Increased taxes that improve state and local public services to business have also been shown to increase growth (Bartik, 1991). One study (Helms, 1985) contended that when revenues from increased state and local taxes were used for anything but welfare expenditures, growth in personal income resulted. A complete picture of the taxes-development nexus must allow for the positive long-term effects of expenditures as well as any short-term loss of growth.

Measuring Tax Effects

Any discussion of taxation at the state level must consider several notable features of the operating environments of subnational political economies. The states represent political institutions that operate in a politically and economically constrained environment. The environment prohibits deficits and often limits taxation and expenditures. Of the 50 states, only Vermont has no type of balanced-budget restriction (Fisher, 1988). Since 1976, 18 states have adopted taxation or expenditure limits generally restricting growth in state-source revenues or expenditures to growth in income, population, or inflation.[1] By law, the revenue options of the states are greatly circumscribed.

Beyond constitutional and statutory restrictions on taxation, the structure of American federalism may act to inhibit efforts to increase taxes. The Tiebout (1956) theory of local expenditures, designed for local governments, also highlights features of federalism that could retard taxation and expenditures by states. The argument posits that mobility of residents or firms to locales where governments best satisfy their preferences will force governments to compete and thereby become more efficient. This mobility, according to Thomas Dye (1990), may make taxation even more economically salient at the state level than it is at the national level. At least in theory, mobility should accentuate disincentive effects of higher taxes in one jurisdiction. When confronted with higher federal taxes, people can work less, save less, and avoid new entrepreneurial activity. When confronted with a higher state tax, Dye reasoned, people can pursue these options as well as move to a lower tax jurisdiction, and thus mobility should highlight the growth-retarding effects of high taxes in any single jurisdiction.

Decisions to move, of course, are not taken as lightly as a normal consumption decision, and a host of locational factors other than taxation may affect decisions about movement. Nonetheless, the typical rhetoric surrounding tax bills in state capitols would suggest that the competitive position of states loom large in the thinking of many politicians. As will be seen below, firms seeking preferable tax settings can exert enormous pressure on state governments, inducing them to engage in fierce economic competition. Mobility might accentuate the growth-retarding effects of an outlying state, but this mobility might also preempt such an outlier by inducing states to adopt fairly similar tax policies.

Taxation at the state (and local) level is noteworthy in that regressive tax instruments are relied on most heavily. As Paul Peterson (1981) has observed, the federal government may tax on the ability-to-pay/equality criterion, but residential competition forces states and localities to levy taxes on the benefits-received principle, taxing residents according to the level of services they receive. Concerns about the health of local economies lead state and local governments to rely on regressive tax instruments, accounting for why sales or consumption taxes are the single largest source of revenue to state governments, producing almost 30% of total state governmental general revenue.

The sales tax does produce the anticipated economic consequences along borders where consumers can adjust their spending with relative ease. Several studies have demonstrated that a disadvantageous sales tax rate leads to a statistically significant reduction in sales in the higher-tax jurisdiction (see, e.g., Fox, 1986). Potential or actual border effects are likely to serve as restraints on the levels of sales taxes within the states. In 1987, for example, these rates varied from 3% in Colorado, Georgia, North Carolina, and Wyoming to 7.5% in Connecticut, and Delaware, Montana, Oregon, and New Hampshire had no sales tax (ACIR, 1987). Comparisons, of course, are made difficult by statutory exemptions, but in general there is remarkably little variation in sales tax rates within regions.

Though merchants may complain about sales taxes, they are probably an inconsequential issue to most producers. Indeed, higher sales taxes may reduce the need for reliance on a progressive income tax or a corporate tax. The domain of choice realistically available to states in their tax policy considerations is greatly limited by economic concerns resulting from the potential and actual mobility of residents and businesses. These concerns must at least partly account for the heavy reliance of the states on the sales tax and for the relatively low variance in this tax between states.

The states tend to rely on taxes that may be of least concern to producers and to adjust rates to be economically competitive with one another. In addition, revenue structures tend to converge regionally (Fisher, 1988, p. 19). States within regions, with similar tax structures, may offer similar factor costs or access to markets. The result may be tax differentials among the states that do not, on average, provide sufficient incentives to overcome other differences such as labor costs, energy costs, or access to markets. Like Willie Sutton, some employers are where they are because "that's where the money is." New York's taxes and labor may seem high compared with Texas and Oklahoma, but neither of those states can offer the unique benefits and market access of New York City. Northern New Jersey or southern Connecticut might seem like reasonable alternatives to some employers, but intraregional tax differences may be insufficient to offset the costs of being away from New York City. In general, unique intraregional resources may tie firms to certain areas of the country, and minimal intraregional differences in tax burdens may make moving economically unfeasible.

Another measurement issue concerns the scope of the impact of taxes on growth and development. Different taxes may have differing effects on alternative segments of state economies. David Osborne (1990) observed,

> The truth is that tax levels are far less important than most of us assume. Extremely high tax rates obviously scare some investors away, and extremely low taxes just as obviously inhibit government's ability to pay for quality schools, highways, and the like. Between these extremes, the important questions do not concern the level of taxation. They concern the kind of taxes a government levies: Do they encourage investment or consumption? Speculation or productive investment? They concern the stability of tax rates over time. (pp. 255-256)

From this perspective, the type of taxes, and changes in taxes, may be the more important indicators of their effect on economic performance.

Just as individual taxes may have differential effects on alternative economic indicators, growth in one economic indicator may come at the expense of another. The United States has had considerable success in generating jobs, but unions have weakened, and the minimum wage has failed to keep pace with inflation. The result has been a growing class of "working poor." The decline in demand for low-skilled workers in industrialized democracies has translated into unemployment in Europe and lower wages in the United States. Thus, job growth may come because of low incomes, and low taxes that stimulate job growth may be associated with suppression of income growth. Conversely, per capita income growth may retard job growth (Brace, 1993). The differential effects and interconnectedness of taxation require that a complete picture of the effects of taxation consider the direct and indirect effects of taxes on alternative measures of growth.

Period Effects and the Dynamic
Nature of State Political Economies

There are cycles in American history that affect state and national economies (Brace, 1993). The states must be viewed within the national context that affects them, but this context is itself a shifting one. For evaluating the consequences of state tax policy on economic performance, what this may ultimately mean is that the effects are variable through time, depending on the impact of a host of exogenous factors.

Since World War II, there have been a host of notable changes in the political economy of the United States, too many to recount here. Several general trends and recent developments, however, are worthy of note. There has been a steadily increasing trend in state and local taxation. The weight of this increasing taxation in this era has been falling more and more on individuals and less and less on businesses (Eisinger, 1988, p. 142). Property taxes on business as a proportion of state and local revenues fell from 20.3% in 1957 to 12.3% in 1977 (ACIR, 1981, p. 62). During this period, the proportion of revenues raised from individuals rose from 63.2% to 69.4%, with increases in sales and particularly income taxes accounting for most of the increase.

Until the mid-1980s, the federal government indirectly assisted economic development in the states through its grants and loan programs. In addition, the federal tax exemption on industrial development bonds provided stimulus for economic development, and throughout the 1970s and early 1980s tax-deductible economic development bonds were a primary avenue for local economic development. This all changed with the Tax Reform Act of 1986, which greatly curtailed this avenue of economic development and also marked the effective demise of many other federally funded development-related programs. In the wake of these changes, state and local governments have had to dig into their own pockets to provide inducements to economic development.

Against this backdrop of ever-increasing pressures on state economies and fiscal resources have been booms and busts in the national economy that have influenced greatly the levels of growth experienced in the American states. Recessions in the 1970s, 1980s, and 1990s created fiscal crises that the states had to address. Similarly, national booms in the 1960s and 1980s stimulated growth in state economies and revenues, reducing pressures for additional taxes. State economies are open and enormously sensitive to exogenous economic influences (Brace & Cohen, 1989). Analyses that fail to control systematically for national-level influences are seriously underspecified, and this underspecification could mask the effects of taxation. Over time, state taxation and state economic performance are only imperfectly connected and are often bounced around by enormous national influences over which they have little or no control. Satisfactory analysis of the effects of taxation on economic performance in the states must control for the changing features of fiscal federalism embodied in period effects and for the effects of the national economy.

■ Taxes and Development: Notes for a Model

The above discussion highlights substantive and technical issues concerning tax effects at the subnational level that must be addressed for a more satisfactory understanding of the role of taxation in economic development. In this section, these substantive and technological concerns are incorporated with the specification of research design/methodology, variable measurement, and a model for better addressing the overall effects of tax choices on state economies.

Research Design/Methodology

As has often been noted, the states present an excellent comparative laboratory in which to evaluate the effects of many public policies. Questions about policy are captured in unit (or state) differences. States with different policies may be compared, and inferences may be derived about the effects of policy on the observed differences between states. The comparative effect of taxation, by definition, raises questions about cross-sectional differences between states.[2] To make such comparisons, one normally constructs a data set consisting of observations on multiple units (e.g., states) at a single point in time. Economic performance, again by definition, raises questions about longitudinal change in economies. To examine change, we typically construct a numerical (temporal) sequence of observations on a single unit (e.g., state).

Taxation and economic performance together blend comparative (cross-sectional) and over-time (longitudinal) questions: How do economies of states with different policies change over time? The most suitable methodology for such research questions is pooled cross-sectional time series. In this research design, variables for a number of different cross sections are observed over time, allowing for the estimation and comparison of *unit* (i) effects over *time* (t) change (see, e.g., Stimson, 1985).

In simplified form, a basic pooled model of state economies can be expressed as

$$\text{Economic Performance}_{it} = a_0 + b_k X_{kit} + e_{it}$$

where Economic Performance$_{it}$ is a measure of economic output or employment change, as discussed below; a_0 is an intercept; b_k is a parameter estimate of the effects of k independent variables on Economic Performance$_{it}$; X_{kit} is the value of k independent variables on state i in time t; and e_{it} is the error term.

In such a model, taxation effects are modeled as an independent variable (X) and hypotheses about taxation and economic performance are evaluated substantively and statistically by scrutinizing the estimates of associated slope coefficients b. The typical expectation is that the slope coefficient should be negative and statistically significant in a model with appropriate controls. Pooling raises a host of technical issues that are best addressed by reference to one of many articles (e.g., Stimson, 1985) or texts (e.g., Judge, Hill, Griffiths, Lutkepohl, & Lee, 1988). However, several substantive and theoretical issues are of special importance. The relatively small differences in tax policies between the states, and the expectation that the impact of taxation is probably quite small on year-to-year changes in state economic performance, requires that special emphasis be given to estimator efficiency. An efficient estimator is one with minimum variance. Hypothesis tests about the estimated effects of taxation or any other variable are likely to support the null (i.e., no effect) if effects are comparatively small or estimators do not have minimum variance. In studies of economic performance of states, efficiency is often threatened because error variances across states are typically not constant but vary as a function of state size. For example, annual growth in a large state is likely to be quite stable, whereas it may fluctuate relatively dramatically for a small state. This problem, described as heteroskedasticity, leads to estimates with larger than minimum variance and can lead to the erroneous conclusion that an independent variable is not statistically significant when in fact it is.

Given the time-series component of the data, another problem that warrants concern is autocorrelation. When error terms are serially correlated, the estimated error of parameters is underestimated, leading to erroneous conclusions about the significance of independent variables (Ostrom, 1990). Estimation procedures such as GLS-ARMA that correct for each problem jointly in pooled data provide yield estimates that are more efficient and unbiased than models that correct for only one or neither problem (Sayrs, 1989). With data on economic change in the states over time, and employment of suitable estimation procedures to correct for likely threats to inference, the first steps are taken to give hypothesis tests about the effects of policy on economic performance a fighting chance.

The Left Side of the Equation: Dependent-Variable Measurement

As noted above, there is reason to expect taxation to have differential effects on differing sectors of the economy. Changes in one sector, furthermore, could lead to indirect changes in other sectors. Given these concerns, multiple indicators of economic performance should be examined and the interrelationships between these indicators explored.

One set of measures could consider economic outputs, such as those contained in the gross state product measures. Total output, output from private sources, and outputs from manufacturing (durable and nondurable) and service could be considered. Another set of measures would consider employment change. Here, overall and manufacturing employment change could be examined. Other measures, such as the creation or loss of firms, or export growth, might also be incorporated in an analysis.

A wide variety of transformations could be imagined for these variables, some dealing with rather complex technical issues such as stationarity that will not be addressed here. However, some simple substantive concerns can highlight a simple set of transformations of the dependent variables. First, because state size will greatly influence aggregate output figures, each measure should be per-capitized to control for this systematic bias. The variables could then be computed as net differences from the national average to help control for enormous national-level influences that are no doubt exerting an influence on state-level fluctuations. These net differences from the national average may then be expressed as first- differences (i.e., $Y_t - Y_{t-1}$), allowing one to examine the rate in change of a state from the national average on a given economic indicator. This variable would be positive for states that were growing *and* outpacing the national average, zero for states with economies mirroring the nation's economy, and negative for states that were falling behind the national level of growth.

Once methodological issues have been diagnosed and satisfactory estimates of the effects of taxation obtained, interrelationships between each of these output measures should be explored to consider if economic losses/gains in one sector are associated with losses/gains in another. As noted above, policies that nurture income growth might actually harm job growth, or, stated more simply, policies that keep labor prices down might stimulate job growth but suppress income growth. Other such interrelationships should be explored.

The Right Side of the Equation:
Independent-Variable Measurement

As noted above, different taxes may have different effects. One solution to this problem is to incorporate multiple taxes in the analysis. Tax levels, furthermore, are enormously difficult to compare because of wide variability in the states. In personal income taxes, for example, the deductions allowed and the income categories to which these taxes are applied vary widely among the states and over time, thus making systematic comparisons of rates exceedingly difficult. Corporate taxes, similarly, exhibit wide variability in what is taxed and under what conditions. Even sales tax comparisons are made difficult by extensive lists of exclusions that vary appreciably between the states.

Problems with the comparability of tax rates can be avoided by focusing on tax changes. If it is assumed that, other things being equal, a state economy is moving along, driven by some underlying process of economic forces, a tax change should affect this process by altering the operation of some (or many) economic processes. If higher tax levels have the hypothesized negative effects, there should be significant pre- and post-tax change differences. Also, states that change (raise) their taxes should experience lower growth than states that do not change their taxes. These tax changes may be modeled as simple dichotomies, equaling 0 before the change and 1 thereafter.

Of course, a wealth of factors other than taxation could be reasoned to affect economic change. To keep analysis manageable, a lagged dependent variable can be employed in analysis, narrowing the focus of analysis to variation in economic output at time $_t$, controlling for the level of output at time $_{t-1}$. This specification assumes that the effects of the multiple influences that could be affecting economies are mostly captured in the lagged value of the dependent variable, avoiding the necessity of incorporating a host of alternative explanatory influences.

Growth is only one side of the story, as the discussion above made clear. Tax increases may also enhance fiscal resources necessary for expenditures that may attract or retain business over the long run. The direct consequences of tax change, and the indirect effects of revenue increases on development, should also be considered. The effects of expenditure on growth in the states may be measured using distributed lag estimation (Brace, 1990). To capture the two faces of tax effects, a comprehensive model of taxation on economic performance should include tax change variables and a distributed lag component for expenditures as independent variables. In combination, these components should tell us something about the short-term effects of tax changes on growth and expenditure and the effects of expenditure on growth over the long run.

Period Effects

Even with a suitable estimation technique, reasonable and well-measured variables, and important interrelationships addressed, it still may be the case that these models are not stable over time because of the dynamic nature of American fiscal federalism and the changing global economy. The effects of taxation in different eras may vary, depending on the fiscal policies of the federal government or the nature of the international economy. One reasonable hypothesis is that state-level policies may have mattered less in periods when the nation's economy was more insulated from global fluctuations and when the fiscal reach of the federal government was long. In such an era, state economies may have flourished regardless of their policies. In the increasingly competitive global economy, when the federal government has reached its fiscal limits, state policies could be much more important (Brace, 1989, 1990, 1993).

To consider variability in the effects of state taxation over time, models such as those outlined above must be estimated in different time periods. This is not simply a technical matter because it can reveal important insights into the interplay of national and subnational political economies and provide a clearer understanding of the incongruity of findings concerning the economic effects of state tax and other policies. For example, many state-level attributes such as taxation, government capacity, and economic development policy were found to be unrelated to growth in per capita personal income in the 1960s and 1970s. In the 1980s, however, in the context of dramatic changes ushered in by the Reagan administration in the fiscal environment of states, state-level attributes emerged as significant and plausible influences on personal income growth (Brace, 1990, 1993). These results suggest that national-level retreat and state-level advances have tangible consequences for what states do and the effects they have.

■ Conclusion

The model proposed above employs a widely available estimation procedure with known advantages to readily available data, using variables that are easily and plausibly constructed. The benefits of analyzing such a model are multiple. First, the significance and magnitude of tax changes on economic performance could be evaluated. Second, short-term effects on growth and long-term effects on, and of, expenditure, may be reasonably ascertained. Third, the effects of different taxes may be estimated. Fourth, the effects of taxation on growth *and expenditure*, and resulting effects of expenditure on long-term performance, may be considered. Fifth, the responsiveness of differing sectors of the economy may be gauged. Sixth, the trade-offs between tax changes and growth/decline in one sector may be weighed against the indirect consequences of taxation in other sectors. Finally, this model may be used to examine the differential effects of state tax policies through time, yielding clearer insights into fiscal federalism and the economic salience of state tax policy.

In a setting where attention is focused on taxation, where international and intranational competition is forcing states to examine and reexamine their competitive positions, and where highly visible tax incentive bidding wars have become commonplace, the information gleaned from such an analysis is more than a simple academic exercise. Though political realities will probably play a prominent role in the unfolding debates about economic development in the states, the quality of this debate will be elevated and the resulting policies improved if sound, solid information about the short- and long-term effects of taxation is part of the discussion. At present, it would seem, much of the debate and policy emerging in the states is premised on the notion that the economic effects of taxation are palpable. What may be more tangible, at least in the long run, is the diminished infrastructure and education capacity of states. Short-term growth may come at the expense of long-term development. In such circumstances, it would be typical to say, "Only time will tell." However, data and methodology are available at present to answer many of the important questions left unaddressed in the current debate. Substantively and theoretically, there are many sound reasons for pursuing a clearer understanding of the effects of state taxation on short-and long-term economic performance in the states.

NOTES

1. The 24 states are Alaska, Arizona, California, Colorado, Connecticut, Delaware, Hawaii, Idaho, Louisiana, Massachusetts, Michigan, Missouri, Montana, Nevada, New Jersey, North Carolina, Oklahoma, Oregon, Rhode Island, South Carolina, Tennessee, Texas, Utah, and Washington (ACIR, 1995).

2. Taxation can also raise longitudinal questions about a single state before and after a tax change.

REFERENCES

Advisory Commission on Intergovernmental Relations. (1981, March). *Regional growth: Interstate tax competition* (Advisory Commission on Intergovernmental Relations Rep. No. A-76). Washington, DC: Author.

Advisory Commission on Intergovernmental Relations. (1984). *Significant features of fiscal federalism.* Washington, DC: Author.

Advisory Commission on Intergovernmental Relations. (1995). *Fiscal discipline in the federal system: National reform and the experience of the states.* Washington, DC: Author.

Bartik, T. J. (1991). *Who benefits from state and local economic development policies?* Kalamazoo, MI: W. E. Upjohn Institute.

Bloom, C. C. (1955). *State and local tax differentials and the location of manufacturing.* Iowa City, IA: Bureau of Business and Economic Research.

Blumenthal, R. G. (1994, March 7). As New Jersey slices taxes, it's unclear how state will maintain cuts in future. *Wall Street Journal*, p. B5A.

Brace, P. (1989). Isolating the economies of states. *American Politics Quarterly, 17*, 256-276.

Brace, P. (1990). The changing context of state political economy. *Journal of Politics, 53*, 297-313.

Brace, P. (1993). *State government and economic performance.* Baltimore: Johns Hopkins University Press.

Brace, P., & Cohen, Y. (1989). How much do interest groups influence economic growth? *American Political Science Review, 83*, 1297-1308.

Brace, P., & Mucciaroni, G. (1990). The American states and shifting locus of positive economic intervention. *Policy Studies Review, 10*, 173-181.

Browning, E. S., & Cooper, H. (1993, November 24). States' bidding war over Mercedes plant made for costly chase. *Wall Street Journal*, pp. A1-A6.

Campbell, A. (1965). State and local taxes, expenditures, and economic development. In C. F. Conlon (Ed.), *State and local taxes on business.* Princeton, NJ: Tax Institute of America.

Carlton, D. W. (1983). Location and employment choices of new firms: An economic model with discrete and continuous variables. *Review of Economics and Statistics, 65*, 440-449.

Due, J. F. (1961). Studies of state-local tax influence on location of industry. *National Tax Journal, 14*, 163-173.

Dye, T. R. (1990). *American federalism: Competition among governments.* Lexington, MA: Lexington.

Eisinger, P. (1988). *The rise of the entrepreneurial state.* Madison: University of Wisconsin Press.

Fisher, R. C. (1988). *State and local public finance.* Glenview, IL: Scott, Foresman.

Fox, W. F. (1986). Tax structure and the location of economic activity along state borders. *National Tax Journal, 39*, 387-401.

Genetski, R., & Chin, Y. (1978). *The impact of state and local taxes on economic growth.* Chicago: Harris Bank.

Grierson, R. E. (1980). Theoretical analysis and empirical measurements of the effect of the Philadelphia income tax. *Journal of Urban Economics, 8*, 123-137.

Helms, L. J. (1985). The effect of state and local taxes on economic growth: A time series cross section approach. *Review of Economics and Statistics, 67*, 574-582.

Jones, B. D. (1990). Public policies and economic growth in the American states. *Journal of Politics, 52*, 219-233.

Judge, G. G., Hill, R. C., Griffiths, W. E., Lutkepohl, H., & Lee, T.-C. (1988). *Introduction to the theory and practice of econometrics* (2nd ed.). New York: John Wiley.

Kieschnik, M. (1981). *Taxes and growth: Business incentives and economic development.* Washington, DC: Council of State Planning Agencies.

Kleine, R. J. (1977). State and local tax levels and economic growth: A regional comparison. In S. J. Bowers & J. L. Staton (Eds.), *Proceedings of the Annual Conference on Taxation of the National Tax Association*. Princeton, NJ: Tax Institute of America.

Munnell, A. H. (1990, September/October). How does public infrastructure affect regional economic performance? *New England Economic Review*, pp. 11-33.

Nathan, R. P., & Doolittle, F. C. (1987). *Reagan and the states*. Princeton, NJ: Princeton University Press.

Newman, R. J. (1983). Industry migration and growth in the South. *Review of Economics and Statistics, 65*, 76-86.

Olson, M. (1982). *The rise and decline of nations*. New Haven, CT: Yale University Press.

Osborne, D. (1990). *Laboratories of democracy*. Boston: Harvard Business School Press.

Ostrom, C. (1990). *Time series analysis: Regression techniques* (2nd ed.). Newbury Park, CA: Sage.

Peterson, P. (1981). *City limits*. Chicago: University of Chicago Press.

Plaut, T., & Pluta, J. (1983). Business climate, taxes and expenditures, and state industrial growth in the United States. *Southern Economic Journal, 50*, 99-119.

Redburn, T. (1994, January 20). Whitman and Cuomo seek to cut different taxes for the same reasons. *New York Times*, p. 1.

Romans, T., & Subrahmanyam, G. (1979). State and local taxes, transfers and regional economic growth. *Southern Economic Journal, 46*, 435-444.

Sacks, S. (1965). State and local finances and economic development. In C. F. Conlon (Ed.), *State and local taxes on business*. Princeton, NJ: Tax Institute of America.

Sayrs, L. W. (1989). *Pooled time series analysis*. Newbury Park, CA: Sage.

Stimson, J. (1985). Regression in time and space: A statistical essay. *American Journal of Political Science, 29*, 914-947.

Struyk, R. (1967, Winter). An analysis of tax structure, public service levels, and regional economic growth. *Journal of Regional Science, 7*, 175-184.

Thompson, W. R., & Mattila, J. M. (1959). *An econometric model of postwar state industrial development*. Detroit: Wayne State University Press.

Tiebout, C. (1956). A pure theory of local expenditures. *Journal of Political Economy, 64*, 416-424.

U.S. Department of Commerce (various years). *Government finances*. Washington, DC: Government Printing Office.

Van Horn, C. (Ed.). (1989). *The state of the states* (2nd ed.). Washington, DC: Congressional Quarterly.

Vedder, R. (1982, Fall). Rich states, poor states: How high taxes inhibit growth. *Journal of Contemporary Studies*, pp. 19-32.

Wasylenko, M. (1981). The location of firms: The role of taxes and fiscal incentives. In R. Bahl (Ed.), *Urban government finance*. Beverly Hills, CA: Sage.

Wasylenko, M., & McGuire, T. (1985). Jobs and taxes: The effects of business climate on states' employment growth rates. *National Tax Journal, 38*, 497-511.

Watkins, A. J., & Perry, D. C. (1977). *The rise of the Sunbelt cities*. Beverly Hills, CA: Sage.

COMMENTARY ON TAXES AND ECONOMIC DEVELOPMENT

Judith Kossy

As someone who has worked in the field of economic development for more than 15 years, I welcome the renewed focus on the economy at national, regional, and local levels. However, the simplistic agenda of cutting taxes, regulations, and public expenditures as the primary avenue to economic growth and development is a woefully inadequate way to address the complex issues of a 21st-century economy. Such approaches fail to recognize the function of expenditures as investments in a community's economic and social development and the necessity of investments to the maintenance of a region's long-term competitiveness. They also fail to consider jurisdictional distribution of costs and benefits.

Paul Brace's chapter "Taxes and Economic Development in the American States: Persistent Issues and Notes for a Model" is therefore a welcome departure from the emotion-laden symbols, taxes and deregulation, that are used in much political and ideological rhetoric. His proposal for a comprehensive analytical and methodological framework places questions regarding the long-term value of tax cuts and expenditures where they should be, in the context of their impact on the quality of development, income growth, and distribution of costs and benefits. His model reflects the real complexity of the relationships between taxes, expenditures, and development and the importance of exogenous variables and time-lag factors. Thus, he offers a potentially valuable tool in evaluating the economic impacts of state-level policy.

The editors therefore have asked me to review Brace's model from the perspective of local and regional development. In what seems to be an inextricable trend away from federal control, state policies will play increasingly critical roles in establishing the framework for private-sector investment and operations. In the end, however, all economic development activity is carried out locally. It is affected by and affects local factors, which mediate the impact of state policy.

In that context, I find that the proposed model identifies four important factors that serve as a checklist for considering the direct impact of taxes and expenditures on business and investment. First, different taxes and tax incentives have differential impacts on different classes of business, individuals, and local taxing jurisdictions. Second, taxes are rarely the exclusive or even dominant factor in business decisions to invest or disinvest in a particular jurisdiction or business. Third, over time the availability and quality of services and infrastructure may be at least as important to business as tax rates, if not more important. Finally, the model acknowl-edges a critical distinction between "economic growth," which focuses primarily on levels of economic activity, and "economic development," which considers the character of activity, quality of jobs created, and distribution of income.

The strategic value of the model, however, is limited because it does not go far enough in recognizing the role of local factors in the location of investment and business operations. In reality, complex regional and local finance and tax structures made up of overlapping services and tax jurisdictions mediate the locational impacts of state policy. Businesses may take advantage of state incentives and use state facilities in a variety of communities, each with its own tax structures, resources, and cost base. Significant inequalities among the jurisdictions foster substantial competition for public and private investment. This competition reduces the choices of many poor communities designing a tax and spending plan that encourages high-quality development in the long term.

Western New York provides a good example of this dynamic. Local and state taxes in the five-county region are among the highest in the country, and public bureaucracies are considered to be among the most inefficient and costly. Communities with a declining industrial base, aging infrastructure, and high rates of unemployment and poverty have been forced to increase taxes and fees on the remaining businesses just to cover costs of maintenance. As Brace acknowledges, these costs may be out-weighed by other factors such as proximity to suppliers, customers, a high-quality workforce, and efficient, cost-effective transportation systems. However, if the base of suppliers, workforce, and quality of the infrastructure declines relative to other jurisdictions, the business finds less justification for continuing to pay high costs.

Over time, these fiscally poor, declining communities lose their ability to maintain their existing base, much less attract new investment, because they have neither low costs nor high-quality assets. They must work within the delicate balance of minimizing costs while maintaining quality, but they have little financial flexibility to do so. Competition with other lower-cost localities further narrows their choices and constrains their ability to act strategically.

The city of Niagara Falls, for example, felt that it had little choice in granting a substantial property concession to Nabisco to supplement state incentives. Nabisco promised to retain 500 jobs and modernize the facility in return for state financing and a lower tax appraisal. Without the tax deduction, they would consolidate operations in a newer plant in the

Midwest. The city knew that because of the extreme difficulty of replacing Nabisco and the jobs, it would lose far more revenue if the plant left than it would from the tax concessions. At the same time, the concessions would strain even further the limited city budget for other services, such as water and sewerage. There is almost no opportunity for communities such as Niagara Falls to invest in the type of infrastructure that will generate new businesses in the future. The cities feel that they have little economic ability to adopt to any strategy other than a concessionary one. These concessions, unlike those in growing areas, actually decrease public revenue in order to maintain economic viability rather than expand it.

Similarly, state expenditures in infrastructure and facilities, such as a state university, a highway, or fiber-optic lines, can serve business in a number of localities within a regional market. Investors who may be attracted to specific facilities can choose where to locate. All things being equal, new investment gravitates to newer, growing jurisdictions with lower costs, more efficient services, modern facilities, green field sites, wealthier residents, and healthier businesses rather than to older, declining communities within a region, state, or nation.

A recent site decision typifies this dynamic. New York State recently offered $13 million of incentives to a grocery chain to build an $80-million distribution center in western New York in an effort to save 650 jobs. The project involved moving out of three sites, two in the city of Buffalo, and consolidating operations in a single state-of-the-art facility. Pennsylvania and Ohio had competing sites with demonstrably lower development and operating costs. State incentives could have been used anywhere in western New York. The company chose a site in the town of Lancaster, a small growing community about 20 miles from the city of Buffalo, that has substantial greenspace, close proximity to major high-ways, and a 10-minute commute to the airport. The city of Buffalo will lose tax revenue, whereas Lancaster will expand its revenue base and cover many of its additional costs with fees. Therefore, the state incentives were provided to keep the company and jobs in western New York, but the costs and benefits were distributed unevenly within the region.

As states assume greater control of federal resources and programs, their policies will have greater impact on the framework in which businesses operate and invest. They will affect the patterns of development within regions and the pattern of disparities among them. If they do not acknowledge inequalities among communities, they may very well exacerbate them. A strategically balanced development policy must account for local differences if it is concerned with the distribution of outcomes across space, as well as the impact on growth and income. Paul Brace's model could help states develop a more balanced framework by including local factors.

COMMENTARY ON TAXES
AND ECONOMIC DEVELOPMENT

Brett W. Hawkins

Professor Brace brings to his task the expertise of a scholar who has himself published research on the economic effects of state taxation. No one has a more valid license to criticize the literature reporting that state taxes have damaging economic effects than a scholar, such as Brace, who has himself found evidence of damaging effects. Probably no one appreciates more that the literature is difficult to sum up, except by noting contradictory findings or saying that there is no unambiguously clear weight of the evidence.

Of course, because the overall picture remains ambiguous, a reviewer of the literature could just as easily select particular studies ("one study observes") and use the subjunctive mood ("might not," "could," "may,") to criticize the assertion that high taxes *do not* retard growth as to criticize the assertion that they do. A reviewer could as appropriately argue that the effect of high taxes may be harmful as to argue that it may not be. Other valuable reviews of the evidence conclude that higher tax burdens have significant negative effects. Some reviews include studies of communities as well as states (Blair & Premus, 1987; Wasylenko, 1985).

Quite properly, Brace observes that research on the subject of the effect of taxes is not just an academic exercise; it has critically important implications for policy choice. In that light, there is surely as much reason in 1997 to caution policy makers *not* to assume that tax burdens are trivial to business location decisions as to advise them to disregard the tax burdens they impose. Ambiguous as it is, the evidence is sufficiently clear to disallow policy makers the luxury of assuming, as they once did (Brownlee, 1974), that the taxes they levy bring about better services with no negative effects on the behavior of individuals or firms.

As for services, if state policy makers ever assumed that education and physical infrastructure are irrelevant to location decisions by businesses and skilled employees, they are not entitled by the evidence to assume it any longer.

The cost of assuming that tax burdens, or particular tax types (those incorporating a progressive rate structure, for example), are not a significant impediment to economic health if in fact they are—as Brownlee (1974) and Vedder (1982) observed about progressive income taxes—is economic stagnation and decline. New York City lost 44,500 manufacturing jobs for each 1% increase in city business income taxes, according to a commission appointed by Mayor Beame (Bartlett, 1980). Stagnation and decline are, of course, not just matters of profit, output, and expansion —or even of employment. They also involve personal income. After decades of scholarly argument that taxes are not a significant impediment to business location or expansion, and a shorter, more recent period arguing that they are, the most appropriate context for presenting a research model designed to improve our knowledge of the subject (and better guide policy) is, it seems to me, one arguing that policy makers cannot proceed as if taxes were not an impediment. The possibility exists, of course, that research employing the Brace model might end the debate. But it is not likely, however admirable the model is. And it is very admirable.

The model attractively synthesizes many desiderata in the study of tax effects on economic health. I have two small suggestions and one reservation. One suggestion is that the model explicitly include per capita personal income as a dependent variable, which it does not seem to do even though Brace does mention it in his essay. In other research (Brace, 1993), he has used it as a dependent variable. A second suggestion is that explicit consideration be given to the indirect economic effects of tax burdens, such as their effects on the residential choices of mobile families, and thereby on the location of skilled labor.

My reservation concerns the units of analysis. It seems to me that even if analysis using the Brace model produces the strongest possible findings about the effect of state taxes on state economic health, there will still remain the question of where local taxes and their effects fit into research-based generalizations about the economic effect of taxes on state economic health. Local jurisdictions use varied tax instruments in the same state (some counties tax sales and some do not), and of course the rates of local taxation vary greatly among jurisdictions within the same state and across states. Several studies conclude that the economic health of the states is affected by property and other taxes levied by local jurisdictions. Local taxes appear to be especially important to business location decisions within metropolitan areas and for high-technology firms (see Blair & Premus, 1987; Wasylenko, 1985). Because of evidence that different factors are involved at different levels, Blair and Premus recommended a two-stage approach to the study of business location decisions, the first being the state or region and the second being the specific community and site. Naturally, covering another level of phenomena would worsen an already big problem of noncumulative research. It might therefore be wise to take state taxes by themselves, as Brace perhaps believes, and then turn to local ones for their added effects, if any. I wish that he had commented on this issue in his thoughtful, informative, and commendable essay.

REFERENCES

Bartlett, B. (1980, Winter). Higher taxes may deliver lower revenues: New York City revisited. *Taxing and Spending, 3,* 51-57.

Blair, J. P., & Premus, R. (1987). Major factors in industrial location: A review. *Economic Development Quarterly, 1,* 72-85.

Brace, P. (1993). *State government and economic performance.* Baltimore: Johns Hopkins University Press.

Brownlee, E. W. (1974). *Progressivism and economic growth: The Wisconsin income tax 1911-1929.* Port Washington, NY: Kennikat.

Vedder, R. K. (1982, Fall). Rich states, poor states: How high taxes inhibit growth. *Journal of Contemporary Studies, 5,* 19-32.

Wasylenko, M. (1985). *Business climate, industry and employment growth: A review of the evidence* (Occasional Paper No. 98). Syracuse, NY: Syracuse University, Metropolitan Studies Program.

REPLY TO JUDITH KOSSY AND BRETT W. HAWKINS

Paul Brace

I appreciate the able and thoughtful readings provided by Judith Kossy and Brett Hawkins. Having labored in the area of state economic development for a number of years, it is heartening to see a convergence among scholars and practitioners over the central issues. Each of us supports a broader view of the impact of economic development efforts in the subnational setting, and each of us agrees that these issues are of paramount importance to policy makers in states and communities. Although it appears we have some minor differences, I think the main points of difference are quite subtle rather than major.

To the extent that there is common criticism from these two reviewers, it is in my lack of focus on local economic development. Kossy and Hawkins would like to see greater emphasis on localities. In large measure, the logic of my model can be extended to localities. My choice of states as units of analysis stems partly from my training and interests as a political scientist and partly from my familiarity with the widely available and readily comparable data for the states. My essay seeks to highlight some of the generic analytical issues that surround the study of taxation and development in subnational economies, and on this level I think both would agree that the discussion succeeds.

Although I focused on states to keep the discussion manageable and to highlight what can be done with state-level data, it was not my intent to suggest that tax choices by municipalities are unimportant. Future studies should consider the interplay of state tax policies with the fiscal choices made by municipalities and their effects on economic development. For example, some state tax incentive programs may actually harm local employers by placing them at a competitive disadvantage to newly lured firms. Alternatively, some incentive packages might have the effect of merely moving existing employers around within a state, contributing nothing to a state's overall economic output but perhaps diminishing its revenues in the process. Communities, furthermore, may have to alter their tax policies to correct for problems created by state tax changes. Early experiences with New Jersey's tax cuts suggest that some of the burden is merely shifted to local governments. These and other important issues should draw the attention of students of urban and local political economy.

Professor Hawkins quite correctly asserts that politicians at the subnational level cannot proceed as if taxes did not matter. Where we differ, I think, is that he mixes this behavioral pattern on the part of politicians with the empirical findings on the effects of state taxes on economic performance. Where I supply ample anecdotal evidence to support his first conclusion, I do not think the scholarly literature can support a conclusion that taxes have an unambiguously detrimental effect on economic development, despite the behavior of these politicians. As my review indicates, the jury remains out on the impact of taxes on economic growth.

To underscore my support for his first point, I note in the essay that "the pursuit of jobs in a tax- and revenue-constrained environment has led to a renewed emphasis on tax incentive packages and general tax cuts to stimulate economies" and that "many states currently find themselves under pressure to cut existing tax rates. This comes partly from a desire to appear more business friendly and partly from a pervasive antitax mood in the country. . . . This thinking, furthermore, is translating into action in many states." I then supply six brief examples from across the country and the ideological spectrum, ranging from Christine Todd Whitman in New Jersey to Fife Symington in Arizona, Zell Miller in Georgia, Mario Cuomo in New York, and Mike Lowry in Washington.

In sum, Hawkins and I agree that policy makers are *not* proceeding as if taxes were not an impediment. I do not think, however, that we should confuse the course of action being pursued by these and other politicians with actual support for the proposition that taxes harm growth or that tax cuts will stimulate growth. *This course of action could be motivated by political pragmatism rather than economic virtue.*

Hawkins points to some additional literature that suggests that high taxes harm growth. One source he cites is an occasional paper by Wasylenko (1985). I rely on Wasylenko's major published work with McGuire (1985) in the *National Tax Journal* indicating that the impact of taxation is highly contingent in nature. I also cite Plaut and Pluta (1983), which is to many the definitive work in this area. Also, considering that 8 of the 14 studies cited by Eisinger (1988) revealed little or no effects of taxes on growth (and only 6 of 14 found some or great influence), or that 22 of the 58 interstate studies conducted between 1979 and 1991 surveyed by Bartik (1991) found negligible or *positive* effects from increased taxation, I am entirely comfortable with my assertion that the effects of taxes on growth are and remain ambiguous. I think my interpretation of the empirical support for the hypothesis that taxes harm growth as ambiguous is entirely consistent with the thrust of literature on taxation and economic development.

Please note that I am not guarding a hypothesis that taxes do not affect growth: My book (Brace, 1993) and several articles (Brace, 1990; Brace & Mucciaroni, 1990) provide limited empirical evidence that taxes can have a detrimental influence on economic performance. I could drop the subjunctive "might not" "could" or "may" and write as if this were a well-established empirical pattern. This would not be consistent with the highly varied findings in the literature, and although it would be desirable to leave readers of a review essay of this type with a clear-cut and simple interpretation of the impact of taxes on growth, the waters are much too muddy to support such a conclusion.

Hawkins recommends inclusion of per capita personal income as a measure of economic output. In the essay, I call for future studies to broaden the economic indicators that they examine, and certainly looking at per capita income would contribute to this. In fact, since writing this essay, I have become aware of another data set. The Regional Economic Information System (REIS) provides measures of income for very refined categories across a wide variety of types of employers in states over time.

Hawkins also suggests that the model address the effects of tax burdens on the residential choices of families and the location of the labor force. Although it might be possible to apply the model I have developed to population migration patterns, I have not given much thought to this. Instead, I argue for examining job creation and employment levels. Though tracking the impact of state tax policy on interstate migration patterns is an admirable and productive pursuit, I have limited my discussion to changes in economic outputs and jobs. The interpretation is perhaps narrower than Hawkins would like, but I think that for most students of economic development at the state level, this is a sound approach.

REFERENCES

Bartik, T. J. (1991). *Who benefits from state and local economic development policies?* Kalamazoo, MI: W. E. Upjohn Institute.

Brace, P. (1990). The changing context of state political economy. *Journal of Politics, 53,* 297-313.

Brace, P. (1993). *State government and economic performance.* Baltimore: Johns Hopkins University Press.

Brace, P., & Mucciaroni, G. (1990). The American states and shifting locus of positive economic intervention. *Policy Studies Review, 10,* 173-181.

Eisinger, P. (1988). *The rise of the entrepreneurial state.* Madison: University of Wisconsin Press.

Plaut, T., & Pluta, J. (1983). Business climate, taxes and expenditures, and state industrial growth in the United States. *Southern Economic Journal, 50,* 99-119.

Wasylenko, M. (1985). *Business climate, industry and employment growth: A review of the evidence* (Occasional Paper No. 98). Syracuse, NY: Syracuse University, Metropolitan Studies Program.

Wasylenko, M., & McGuire, T. (1985). Jobs and taxes: The effects of business climate on states' employment growth rates. *National Tax Journal, 38,* 497-511.

7

Is Industry Targeting
a Viable Economic
Development Strategy?

KENNETH VOYTEK
LARRY LEDEBUR

You seek a final portrait—or a small gallery of related and refracted portraits—which can be held beside the living face of your subject in the light of day without the work losing its integrity. . . . It's not a bad methodology for those prone to coffee and facts, rather than wine and imagination. You hang around for the details.

—Thomas Boswell, *The Heart of the Order* (1989)

This chapter explores

- The utility of industry targeting as an economic development strategy to guide urban economic development efforts
- The basic steps involved in conducting an industry-targeting analysis so that those choosing to purchase or provide such a product will better understand the methods, use, and limitations of industry targeting
- The development of industry targeting as an explicit economic development strategy
- The rationale underlying industry targeting and the current state of the art (or "best practice") in industry targeting
- The empirical information evaluating the impact of industry targeting on economic development outcomes

Economic development has become an important activity for local communities. Development practitioners are searching for an "edge" in marketing their communities to potential industry prospects and in

171

guiding economic development. One tool that has emerged to guide economic development efforts is industry targeting.

Industry targeting is a way to focus the use of economic development resources. As practiced currently, industry targeting is a process of identifying those industries that are performing well according to some dimension of industry performance and that are compatible with the attributes or assets of the local community that are important in influencing industry location decisions. Industry targeting is also interesting in that it blends theory and practice (Bowden, 1971; Smith, 1981). It is essentially a process for modeling the location decision of a firm and comparing those factors important in influencing location decisions with an assessment of how a community "stacks up" in terms of these factors.

Industry targeting borrows from the industrial location theory literature as a way to improve the effectiveness of economic development programs and practice. Finally, industry targeting is attractive to economic developers because they can assess how industries have performed over time, how they are forecast to perform, what factors they value as they make location decisions, and how these industries help the community achieve its economic development objectives.

■ The Context

From the mid-1970s and into the early 1980s, the U.S. economy experienced changes in the distribution of jobs across industries and regions, a regional redistribution of population, stagnation of income growth, a decline in productivity, a reduction in investment, and increased competition from industries abroad. Against this backdrop, several leading policy analysts suggested that the federal government should adopt a national industrial policy and the key features of such a national policy (Reich & Magaziner, 1982).

One of the key features of a national industrial policy involved targeting industries of particular importance to the national economy and allocating investments to promote their growth. A range of opinion, however, existed as to the criteria that should be used to identify these industries and the general sorts of industries that should be targeted within a national industrial policy. Although the discussions over whether the federal government should adopt such a policy have remained largely confined to a debate over the best way to address the long-term problems facing the national economy (Krugman, 1983), local and state governments stepped into this void and have focused on revitalizing their economies as well as planning and implementing effective economic development strategies (Eisinger, 1988; Hansen, 1989).

Economic development has become an important function for local governments and is characterized by increasing competition among localities. In response to this competition and the limited resources available for economic development, localities have sought out methods that would maximize their chances of success in spurring economic development and growth. An important innovation that many localities have adopted is the emphasis on a more strategic approach to planning and implementing policies and programs aimed at stimulating economic development.

Strategic planning for economic development emphasizes several important steps, including a careful analysis of the current economic situation, analysis of the strengths and weaknesses of the local economic environment, identification of key strategic issues, and, finally, development of strategies based on the prior steps (Blakely, 1989; Kotler, Haider, & Rein, 1993; Luke, Ventriss, Reed, & Reed, 1988). This information is also a critical input into the industry-targeting process.

The practice of economic development has also undergone rethinking. Economic development practice has moved away from simple focus on recruitment (i.e., "smokestack chasing") to encompass activities including industry retention, business formation, expansion, and offers of a variety of assistance to firms (i.e., labor training, modernization assistance, financial assistance). One problem that economic development professionals face, given limited economic development resources, is what types of activities they should focus on.

■ Industry Targeting Defined

One strategy to guide economic development practice that has received wide attention is industry targeting (Growth Strategies Organization, 1992). In its simplest form, industry targeting can be thought of as a process of "picking winners" from a range of alternative industries on the basis of some criteria relating to industry performance (e.g., job growth, increasing productivity, increasing investment levels). In some circles, industry targeting is equated with identifying industries that are performing well. However, industry targeting, in its more sophisticated development or form, is more than simply identifying "winners" (Thompson, 1987; Thompson & Thompson, 1987). It is a process of carefully assessing the locational needs of a particular industry and reconciling them with the location attributes available within a locality (Klassen, 1967; Moriarty, 1980). Thus, industry targeting is a process of modeling the location decision process of a firm. Moreover, Porter (1990) argued that targeting is a process of singling out particular industries for support and development and that it requires a set of government policies that work in tandem with the determinants of competitive advantage.

Industry targeting is appealing to economic development professionals for several reasons (Lipman & Miller, 1987). It is a potentially attractive economic development strategy, given the desire to maximize the "return on investment" of scarce economic development resources. By assessing the competitive advantages and disadvantages of a particular area, it also serves as a tool in devising a comprehensive set of economic development strategies to capitalize on the area's competitive advantages and address its competitive disadvantages (Boyle, 1987; Gregerman, 1984). Thus, the process of industry targeting also provides a basis for more general economic development strategic planning.

Industry targeting also provides a means of focusing economic development marketing efforts. Thus, it has been referred to as a "pinpoint bombing" or "rifle" rather than "saturation bombing" or "shotgun" approach to economic development marketing. Information generated from a targeting study may, for instance, identify industries that could diversify a local economy, fill a particular niche within a local economy by providing goods or services previously purchased from outside the area, provide critical inputs into existing industries, and identify industries most likely to find the community potentially attractive as a location or simply industries that are compatible with the economic development goals of a community. Industry targeting, then, provides critical input into economic development strategic planning, as well as potentially serving as an explicit economic development strategy itself. As Boyle (1987) observed, "Building a sound economic growth strategy . . . [requires a] . . . careful and honest self-analysis" of the local community (p. 12).

Several different approaches to targeting have been discussed in the literature. Hansen (1989) and Vaughan and Bearse (1981), for instance, suggested that economic development efforts could be targeted along several dimensions: sectoral, social, and spatial. Sectoral targeting involves targeting specific industries or sectors of an economy, social targeting involves targeting activities that would benefit particular population segments, and spatial targeting involves targeting specific geographic areas. Though any of these dimensions could be used, sectoral (or industry) targeting is the one most often discussed in the economic development literature.

■ The Evolution of an Idea

Industry targeting as an explicit economic development strategy has evolved over time. A guiding principle in its evolution has been the desire to develop a more systematic and rational approach to economic development planning. An important step in devising an economic development strategy is the selection of goals and targets. As economic development has emerged on the policy screen for many communities, interest in identifying potential "targets" for economic development has increased as well. However, as with many ideas in urban economic development, the ideas embodied in industry targeting have been borrowed from many disciplines and many different contexts. Nearly 50 years ago, Edgar Hoover (quoted in Smith, 1981), in his classic work on industry location, suggested that

> locational policy should be carefully selective in its application to different industries. No locational policy can be intelligently implemented unless its effects on different types of industry are taken into account in advance. Different industries vary widely in their locational responsiveness to possible controls and in the leverage they exert by passing on the effects of locational change to still other industries. If the aim of a policy is to increase or stabilize employment in a specific area, concentration of effort on a carefully chosen group of industries may be essential for success (p. 385)

The observation by Hoover suggests a rationale for industry targeting but was not to be acted on for several years.

■ The Diffusion of Industry Targeting

By the 1960s, increased interest in economic development as a means of addressing the problems of underdeveloped and depressed regions and countries began to emerge. Klassen (1967), for instance, provided guidance on employing several different methods to identify (or target) prospects for economic development in less developed countries. An important outcome of the process outlined by Klassen (1967), in addition to identifying industry targets, was the logic of the analysis: It provided useful information on the particular attributes of an area that are important in influencing industry location decisions and suggested actions to capitalize on an area's competitive advantages. As these "methods" were worked out in the context of depressed regions and countries, growing interest in the problems and lagging economic growth in urban areas and regions in the United States emerged as well. In reviewing the rationale and approaches to increasing employment opportunities in Philadelphia in the 1950s, Petshek (1968) observed that the city economist was deciding what industries and firms should be targeted as a way to guide the city's development efforts. An important issue raised by this work was that economic development could be systematically approached and that industries could be identified as a way to focus economic development efforts.

In the early 1970s, several articles were written outlining alternative approaches to a more systematic approach to economic development planning. These articles attempted to change the nature of how economic development practitioners conducted their business. The approaches outlined within these articles reflected increased technical sophistication of planners and analysts, the development of better information and tools and methods to analyze local economies, and a desire to address critiques of the efficacy of economic development efforts (Hansen & Munsinger, 1972; Laird & Rinehart, 1979; Sweet, 1970a, 1970b, 1971).

Hansen and Munsinger (1972), borrowing from the framework of systems planning, suggested that economic development could be more systematically approached and outlined an approach that capitalized on the ideas embodied within system planning. This approach first reconciles a community's economic development goals and its resources and assets (competitive advantages/disadvantages) and then proceeds to identify industries to target on the basis of their compatibility in helping the community achieve its economic development goals and whether they will respond to the locational assets the community offers them.

Sweet (1970a, 1970b, 1971) outlined two complementary approaches to industry targeting in a series of articles to guide economic development decision making. One approach suggested that communities approach industrial development by using a "screening matrix" that would evaluate industries "in terms of their desirability and potential for locating in the region under study" (Sweet, 1970a, p. 124). This approach entails arraying industries along one axis of a matrix and a series of attributes of the local area (including both factors influencing location decisions and attributes the community is seeking) on the other axis.

By the mid-1970s, the foundations of industry targeting had been laid. Building on this work and given the growing interest among communities in economic development, several economic development consulting firms began to market their services to communities to aid them in their economic development planning.

■ The Rationale for Industry Targeting

Communities and economic development professionals are faced with an almost bewildering array of industries and business on which to focus in their economic development efforts. Millions of business establishments and hundreds of industries exist from which to choose. Thousands of additional businesses come into existence and go out of existence in a year. Faced with these choices, a community must be able to find a way to sort through this mass of information and identify industries and firms that would find their community an attractive location for their investment dollars. Industry targeting is a way for communities to plot their efforts more rationally, given limited resources.

At its core, industry targeting is a method for guiding decisions on how best to use economic development resources to generate economic growth and development. It also provides a community with a forum to discuss its economic development strategy. The process—including a careful and candid assessment of a community's strengths and weaknesses and potential industrial targets—provides a rational and analytical basis for developing an economic development strategy. Industry targeting assists communities to determine how best to focus economic development resources on the most promising development opportunities. Given limited resources, communities must be judicious in how they spend these resources.

Economic development has been criticized for being "a relatively untargeted intermunicipal sales or marketing competition" (Levy, 1990, p. 158). Just as private firms must target their marketing efforts, communities must also focus and rationalize their economic development efforts. Industry targeting also gives a community an opportunity to remove some of the randomness associated with economic development (Koebel & Bailey, 1993). Instead of making economic development a reactive process, responding to external forces, the community can make it more proactive (Kotler et al., 1993).

Industry targeting is not without its critics, however (McGahey, 1990). These critics argue that industry targeting is a flawed approach to determining an economic development strategy for several reasons. Problems include factors related to how industry targets are defined, the impossibility of accurately replicating the location decision of a firm, the political problems of favoring some industries and not others, and the missing of opportunities in untargeted industries.

Some argue that industry targeting cannot and should not be done (Vaughan, Pollard, & Dyer, 1985). These critics argue that it is unlikely that communities can "outguess" the market and that rather than targeting scarce resources on particular industries, they are better advised to use their resources to undertake more general efforts that improve local conditions (e.g., improving education or access to financial capital).

Second, critics argue that many of the factors affecting industry performance are beyond the scope of local (or even state) control. Thus, local efforts to influence industry performance or location decisions are likely to be futile. The general tools of economic development policy are not cost-effective, are not important in influencing location decisions, and result in a focus on using these subsidies rather than the more general resources available to local community.

Third, target industries are typically identified on the basis of a single criterion or a very narrow set of criteria to assess industry performance. For instance, focusing only on industries increasing employment may miss industries that provide other important benefits to the local economy. Thus, the process of attempting to identify industry targets could lead to ignoring opportunities in untargeted industries (Hansen, 1989; Vaughan et al., 1985).

Fourth, the process of industry targeting is a potentially politically charged issue (Bartik, 1989; Hansen, 1989). By identifying specific industries for attention, industries and firms not targeted may be unable to access the special incentives or other tools offered to particular industry targets, and charges of unfair favoritism may result.

Finally, as Smith (1981) argued, "It may be seldom possible to identify with confidence anything approaching the optimum pattern of industrial location for a development plan, because of the shortcomings of existing explanatory models and the lack of numerical data on many of the relevant variables" (p. 392). Thus industry targeting is difficult because the data used to "drive" the process are unavailable or inadequate, the models of industrial location are imperfect, and many of the factors influencing business firms are "invisible" (Doeringer, Terkla, & Topakian, 1987).

■ The Basic Contours of Industry Targeting

The basic idea behind industry targeting is to identify industries that match a community's comparative advantages. Two approaches to industry targeting have been used.

The first is the economic screening method. In this approach, industries are passed through a series of "screens" or "filters" and are divided into those that pass these screens and those that do not (Lee, 1987). Those not passing a screen are dropped from further analysis. Those passing through the screens may then be further examined according to additional criteria not explicitly included in the screens. Screens may represent particular traits of the industry being examined (e.g., growth rates, wages paid, or cyclical sensitivity) or particular location factors (e.g., market access, transportation facilities, or labor force availability) that a community offers to a potential industry. This approach is particularly attractive when a wide array of potential industries are being examined and the list must be winnowed down to a more manageable number. It is flexible (the screening criteria can be easily modified), relatively inexpensive to deploy, and easily understood by the user. For instance, McLean and Voytek (1993) presented a particular approach using this screening method. A series of screens identify industries on the basis of their recent growth patterns and their relative concentration in the local economy. The goal is to find those industries that would find the community an "optimal" location. Industries passing through these "screens" are then identified as "target industries" and examined further.

The second approach used in industry-targeting studies is called the "simultaneous" or "combined-scoring" method. The combined-scoring method, unlike the screening method, which passes industries through successive screens and identifies those passing all screens as industry targets, ranks potential industries simultaneously and considers all the variables important to the industry's location requirements or the particular traits that the community is seeking. This latter approach has the advantage that it does not drop from the analysis industries that satisfy some, but not all, of the criteria established as important in identifying a particular industry target. Whereas the screening approach identifies optimal industries, the combined-scoring approach identifies industries that may satisfy "many but not all conditions" (Lee, 1987, p. 126). This results in a ranking of industries based on their scores, in which the industry with the highest score represents the best fit for the community. Industries with lower scores are less likely to find the community attractive. This particular approach is, however, much more computationally demanding than the screening approach as the number of industries and factors considered increases.

■ Getting Under the Hood: The Mechanics of Industry Targeting

Industry targeting is a sequential process. The basic approach used in identifying industry targets is outlined in Figure 7.1. The steps in this process can shift and need not occur in the order presented in the figure. For instance, the community goal fit analysis can be done before the community and industry resource assessments. Moreover, the process is not as neat or as linear as Figure 7.1 suggests. Industry targeting is a messy, time-consuming, data-intensive, and potentially vexing process.

The first step in undertaking industry targeting is to assemble the necessary data. This is the most important step in industry targeting, but it is often the one given the least attention in the literature. To be implemented, industry targeting requires that a wide array of data be assembled. Indeed, assembling the data requires that the analyst be creative and resourceful in pulling together the information. The success and quality of industry targeting are dependent on the quality of the data used (Kraushaar, 1986).

In some cases, statistical data to assess a particular factor may be available; in other cases, quantitative data are unavailable, and qualitative data must be used that require the analyst to make an informed assessment on the basis of his or her knowledge and expertise. In some cases, secondary data are readily accessible, whereas in other instances, original data can or should be collected (e.g., through a business survey). In addition, trade associations sometimes collect information that can be

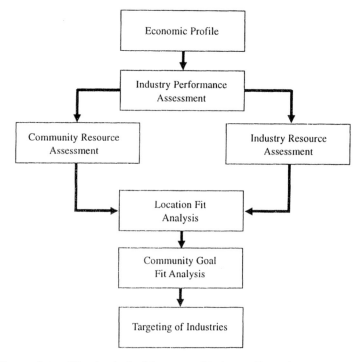

Figure 7.1.   The Analytical Process of Industry Targeting

used by the analyst to assist in the assessment. These may be trade associations of particular industries (e.g., office furniture manufacturers) that can be identified as part of the targeting analysis, or they may be associations of particular industries that provide critical services (e.g., the Edison Electric Institute for comparative information on electricity costs, the Society of Office and Industrial Realtors for data on the comparative costs of land, or the National Association of Regulatory Utility Commissioners or state departments that regulate particular industries for comparative data on residential gas bills). In addition, previous studies of the local area should be consulted. A state transportation department may have a wealth of information available on the availability of air, truck, and rail services in an area.

The Economic Profile

An industry-targeting exercise includes assembling data profiling the local economy and provides information on the basic structure of the local economy and how the local economy has changed over time (Jones, Manifold, & Spencer, 1980; McLean & Voytek, 1993; Syverson, 1985). The importance of undertaking this economic base analysis is downplayed by some practitioners (Boyle, 1987), but an economic profile provides useful information on the local economy and can answer several key questions, including

- What industries (and firms) are important to the local economy
- How the local economy and these key industries have changed over time
- How diversified the local economy is
- Which industries are growing and which industries are declining

This information can help a community in thinking about the goals that it wishes to achieve through economic development efforts. For instance, a community may wish to build off its existing industry base (Siegel, Reamer, & Hochberg, 1987; Syversen, 1985), to offset the impact of economic cycles on the local economy by attracting less cyclical industries that diversify the local economies industry mix (Cho & Schuermann, 1980; Conroy, 1974), or to identify employment opportunities for displaced workers or job training efforts (Gillis & Shaffer, 1985; McKee & Froeschle, 1985).

Industry Performance Assessment

The next step involves an assessment of industry performance. The purpose is to identify industries that have the potential for future growth and that present opportunities for further economic growth and expansion (Thornton, 1984). There are several rationales for identifying growing industries. First, industries growing nationally are those most likely to be searching for new locations as they expand to meet increasing demand for their goods or services. Second, an important objective of an targeting analysis is to improve the performance of the local economy, and thus it is important that industries doing well are targeted to ensure that a local economy is stable and growing. However, it may also be important to target industries that are particularly important to a local economy.

Two approaches to identifying growth industries have been used in conducting industry-targeting analysis. One approach is to examine past patterns of industrial growth and decline in the local community and to compare these growth rates with national and regional standards or benchmarks. Growth industries can be identified on the basis of employment, establishments, value-added, or other criteria that measure industry performance (Klassen, 1967). Identifying growth industries requires analyzing data in a detailed breakout of industrial categories (at least three-digit Standard Industrial Classification [SIC] level) within the local economy and comparing the industry growth rate in the local area to trends regionally and/or nationally. But there are problems with the way in which industries are classified and the level of detail available that may hamstring the ability to identify growing industries (Boyle, 1987; Kraushaar, 1986; Kraushaar & Feldman, 1987; Thompson, 1987).

This comparison of industry growth rates in the local economy can be carried out by using simple comparisons or by using economic analysis tools such as shift-share analysis or location quotients (McLean & Voytek, 1993; Regan, 1988; Stevens, Treyz, Arvidson, Ehrlich, & Delmar, 1984). Using tools such as shift-share or location quotients can help to identify the reasons for differences in industry performance and industries that are performing well in a particular region but not nationally. The goal is to identify industries performing well. Thus, one can identify industries growing nationally, regionally, and in the local community and focus on them in terms of the latter analysis. This particular approach is "backward" looking in that it relies on past performance as a basis for picking these industries, and it must be used with this limitation in mind.

Another approach to identifying growth industries is to use national projections or forecasts of future industry performance. These projections are available from several federal agencies, such as the Department of Commerce (*U.S. Industrial Outlook*), the Department of Labor (annually in the *Monthly Labor Review*), or private forecasting groups such as DRI, WEFA, or Predicasts. These projections can serve as the basis for identifying potential growth industries. Using national projections of future growth is not without its problems, however. Industry performance varies across regional and local economies; thus, industries projected to perform well nationally may not be doing well in a local economy or vice versa (Garn & Ledebur, 1982). Finally, identifying possible growth industries is only one step in the process. It is not enough simply to identify and assess industry performance; these growth industries must examined in terms of the likelihood that they will find the community attractive and that they will help the community achieve its economic development goals.

Community Resource Assessment

Information must also be assembled on the locational features of a local economy to help in evaluating the local economy's competitive advantages and disadvantages. There are two components to this part of the process: developing an inventory of community resources and assessing or evaluating the community resources.

The first step is to develop an inventory of community resources (Boyle, 1987; Gregerman, 1984; Hunker, 1974; Minshall, Miller, & White, 1976; Moriarty, 1980; Ryans & Shanklin, 1986; Vaughan, 1976). This inventory consists of information on the natural, human, institutional, physical, and economic resources that a community offers to potential industries. Many of the factors examined are those that have emerged from the vast literature on industry location theory and have been continually refined as the business location decision process has been examined. Ideas on what location factors or community resources should be collected can be obtained from research studies examining business location decisions; results of surveys that are periodically published, such as *Area Development* or *Industrial Development*; discussions with or surveys of local businesses; discussions with industry experts and industrial realtors; or other compilations of data, such as those contained in the *Places Rated Almanac*. Information on the following categories should be assembled:

- Labor market characteristics, including skills, cost, and availability
- Access to markets, including suppliers and markets/customers
- Access to support services
- Land and facilities, including cost and availability
- Transportation resources, including cost and availability
- Utilities, including cost and availability
- Capital availability
- Quality-of-life factors, including housing costs, cultural resources, climate, and recreation resources
- Business climate factors, including the cost and quality of public services, taxes, and other government-imposed costs and regulations

Each of these broad categories can be further subdivided. For instance, an area's utilities can be examined in terms of the cost, quality, and supply (or availability) of gas, electric, telecommunications, water, and wastewater services. However, to be useful, the process must go beyond a simple inventory to include a critical self-assessment of these community resources, in terms of strengths and weaknesses and possibly in comparison to other areas seen as competitors.

A community's assessment of its resources in terms of strengths and weaknesses (Ryans & Shanklin, 1986) establishes a community's competitive advantages and disadvantages. The process of identifying these advantages and disadvantages highlights resources that present economic development opportunities to capitalize on and factors that also present barriers to economic development.

This assessment involves the assignment of scores to particular factors. Scores can be assigned on the basis of the availability and other characteristics of the factor. These scores are based on some sort of scale, with higher scores reflecting factors that represent advantages/strengths and lower scores reflecting weaknesses. The choice of how to scale these assessments is often left up to the user. Some analysts use a 3-, 5-, or 6-point scale. Whatever scale is chosen, the analyst must be able to rate a characteristic as excellent, good, adequate, or poor. Further, the rating must be defensible.

An example of an advantage that other communities cannot easily replicate is that of transportation access for Baltimore. Baltimore provides access to an international airport for both passengers and cargo, a state-of-the-art port facility, and a wide range of trucking firms and rail lines servicing the area. The availability and range of transportation resources in Baltimore constitute a very important asset for the area that might well be very crucial to certain industries.

It is critical that a community do a realistic and honest analysis of its strengths and weaknesses. Only then can realistic candidates for targeting be identified. Although this assessment is clearly subjective, some model purveyors have attempted to offer guidance in undertaking it and in differentiating rankings among the factors (i.e, providing standards or benchmarks) to highlight an area's strengths and weaknesses (Minshall et al., 1976). In addition, this assessment can be done through either a comparative approach (how do other areas stack up against this community in terms of the factors examined) or in isolation.

Although the comparative approach is probably best to use in such an assessment, it requires that data on the comparison communities also be collected and adds to the time and complexity of the approach. Thus, most assessments are typically undertaken only for the community under study and are based on standard benchmarks.

Industry Resource Assessment

The next step is to determine the relative importance of a particular factor to the location decision of an industry because industries vary in those factors that are important in influencing their location decisions. Scales are also developed to highlight the relative importance of particular factors to an industry's location decision. The same scales can be used as were developed during the community resource assessment. Higher scores indicate factors that are very important in influencing an industry's location decisions, and lower scores represent factors that are unimportant to that industry. For example, a medical instruments manufacturer may place more emphasis on access to research and development opportunities represented by hospitals, higher education institutions, and cultural resources and may place less emphasis on access to an area's transportation resources. Thus, although transportation is a strength of the Baltimore economy, it is not an important factor in influencing the location of medical instrument manufacturers.

This particular step is difficult because finely detailed data on the importance of location factors to a particular industry are difficult to find. Indeed, Klassen (1967) argued that a study of the location requirements of all growth industries is an important contribution for making such an analysis workable at a local level. With regard to some factors important to an firm's location decision, quantitative benchmarks are available that can be used to gauge the importance of particular production inputs (energy, transportation, labor, supplies from other industries) from input-output tables or publications of the Department of Commerce that show how goods and services are used in the production of a particular good or service. In the early 1970s, the Economic Development Administration (EDA) published the results of a survey of the importance of a range of location factors for a series of industries at the four-digit SIC code level. However, the data are now nearly 25 years old, and the coverage of industries is limited to those industries that were growing at the time of the survey (Lipman & Miller, 1987).

Other surveys, such as those published in *Site Selection Handbook, A Guide to Industrial Site Selection, Area Development,* or *Industrial Development,* are aggregated so that it is difficult to determine how the industry location factors vary from industry to industry. In other cases, econometric studies of business location decisions have pointed out the importance (or unimportance) of particular factors to the decisions of firms within particular industries. Another potential source of information is trade associations representing industries that are preliminary industry targets. Not only can they provide information on the prospects for the industry, but they may also have insights into the factors that are important influences on investment decisions within their industry.

Boyle (1987) has proposed that rather than looking at industries according to SIC codes, such an analysis should classify industries along five broad categories (office, research and development, production, distribution, and consumer service operations) and that each of these five categories should be further subdivided according to ownership type (large public corporations, companies that are closely held but whose owners actively manage the business, and companies that are just emerging). This is useful because even within an industry, the relative importance of location factors will vary according to the type of facility being considered. Within each of these categories, industries can then be arrayed and examined.

Location Fit Analysis

This step involves combining the results of the community resource assessment and the industry resource assessment to determine how well the communities' attributes match up with the attributes important in influencing location or investment decisions of particular industries (Minshall, 1976; Thornton, 1984). In some variations, this step follows directly after the industry performance assessment and is not broken into two separate steps. However, it is useful to separate out each of these assessments so that the resources of the community and the needs of industry can be examined in turn. Further, the processes of developing community and industry resource assessments permit the development of weights for those factors most important to the industry.

For each factor, the community resource assessment score is multiplied by the industry resource assessment score to produce a weighted score for that particular factor. Thus, if for Baltimore the availability of transportation resources is a strength of the local area (score = 5 on a 5-point scale) but is not very important to medical instrument manufacturers (score = 1), the weighted score will be 5 (5 × 1). These scores are summed across all factors to produce an overall score that reflects the compatibility of the resources offered by a community with those needed by an industry.

Community Goal Fit Analysis

The industries identified then must be assessed in terms of their compatibility with the goals and the objectives that the community—local public officials, economic development professionals, and local citizens —has defined for itself in terms of its economic development efforts. For instance, a community may identify two primary goals for its economic development programs: increasing the number of jobs in the local economy and increasing the income in the area. Thus, industries will be examined in terms of their contributions to achieving these two goals.

Whereas in the prior step, industries were assessed in terms of what the community could do for them, in this step, they are assessed in terms of what they can do for the community in terms of its economic development objectives (Minshall, 1976). Here, industry targeting begins to assist the community in thinking about its goals and objectives for economic development. As Hansen and Munsinger (1972) suggested, "The objective of industrial development efforts should be to fully exploit resources in a manner consistent with the long-term goals," and "Development activities are more than marketing efforts" (p. 77). Although the community will probably not screen out industries because of their incompatibility with the community's development goals, it may use this information to prioritize industry targets on the basis of degree of compatibility with the community's economic development goals.

In part, the goals can be identified prior to an industry-targeting analysis. Indeed, Smith (1981) argued that the starting point for any strategy should be the setting of goals. However, more recent thinking about economic development planning has suggested that the establishment of goals should come after a thorough analysis of a local economy (Jones et al., 1980; McLean & Voytek, 1993).

Overview

As stated earlier, industry targeting can be carried out using one of two methods. The first approach, the screening method, uses the output from each screen (or step) to narrow successively the list of potential industries examined and ultimately to end with a small number of potential industries to target. Because of the bluntness of this approach and problems of imperfect information about possible industries, the screening method is not recommended.

In contrast, the combined-scoring approach does not eliminate any industries from consideration. Rather, it retains all industries and ranks and scores them in terms of their overall fit with the local community. These "scores" are then used to categorize industries in terms of their priority for economic development efforts, with higher-scoring industries receiving more attention and lower-scoring industries receiving less attention.

A third approach might combine aspects of the screening and combined-scoring approach. For example, only those industries passing some critical performance criterion might be examined. The industries that passed through this screen would then be examined in terms of their weighted score in the location fit analysis. Thus, in this step, industries would be rank-ordered in terms of their overall compatibility with the local community.

■ Alternative Approaches to Industry Targeting

Undersupplied Industry Analysis

The above discussion suggests that the basic approach to industry targeting is data dependent, time-consuming, and dependent on the quality of the judgments made and decisions rules adopted. Further, it suggests that a community is unlikely to be able to undertake an industry-targeting analysis on its own. Undersupplied industry analysis is an alternative approach that may be attractive to many communities and economic development professionals because the data are often readily available through secondary sources, the analysis is not overly complex and is easy to use, and the expense is relatively low in terms of both time and resources. The approach has been popularized by the Council for Economic Action (CEA) in Boston (Koebel & Bailey, 1993) and was first used in the early 1980s. It was developed to assist neighborhoods to identify potential development opportunities, primarily in the retail and service sector, but in more recent years, it has been expanded for use by communities, and it focuses on other sectors as well (e.g., manufacturing).

The objective of undersupplied industry analysis is to identify industries that are underrepresented (or undersupplied) in the local community compared to other similarly situated communities. The basic approach is similar to that embodied in developing location quotients or in some of the variations of location quotient analysis (i.e., minimum requirements approach or the K value technique), although the unit of analysis is different.

Undersupplied industry analysis first requires that a small number of similar local economies (approximately 6-12) be selected for comparison on the basis of a set of characteristics such as population size, industry structure, economic characteristics (e.g., income), social characteristics, and other factors that could influence the structure and composition of the local economic base. Detailed data on the distribution of the *number of business establishments* in a range of industries (at least at the two-digit SIC code level, or more detailed if practical) are then compiled, and the average number of business establishments across these areas for each industry is computed. One then looks for instances in which the number of business establishments in each industry in the local economy being examined differs markedly from the average computed for the comparison areas. The analysis focuses on those industries that have a smaller number of business establishments than would be "expected" on the basis of these comparison areas. These industries then represent potential industry targets. Finally, one examines industry structure and past and projected growth rates to refine further the identification of target industries.

Although this approach is intuitively appealing, it is fraught with problems. First, the selection of the comparison areas must be carefully thought through. For instance, even if areas are selected on the basis of the characteristics outlined above (i.e., population, industry structure, and other socioeconomic characteristics), care must be taken to ensure that these selection criteria represent the most important determinants of an area's local economy. Work reported in Wolman and Voytek (1989) and in McLean and Voytek (1993) showed that proximity to other, often larger, metropolitan areas needs to be considered in identifying comparison areas as well.

Second, the analysis relies on counts of establishments rather than employment to identify undersupplied industries. It is not clear why establishments are chosen for the comparison (unless, of course, this is because detailed employment data are unavailable because of disclosure problems). Thus, a community could be the home of one large discount hardware chain employing 75 people, whereas in the other comparable areas, the retail hardware business could be dominated by several very small corner hardware stores. The community with the one large chain (representing one establishment) would be considered "undersupplied" in comparison to these other areas.

Further, the tenuous assumptions that are mustered in criticisms of location quotients apply to undersupplied industry analysis as well (i.e., that demand and consumption patterns are uniform, that labor productivity does not vary, that each industry produces a similar product, and that no cross-hauling is present).

Input-Output Analysis

Another approach that has been used either as a stand-alone approach to industry targeting or as a critical input into industry-targeting exercises is input-output analysis (Bohm, Herzog, & Schlottmann, 1983; Lee, 1987). Unlike undersupplied industry analysis, input-output analysis is a complex and sophisticated approach to identifying industry targets. Its objective is to identify, on the basis of interindustry linkages, industries whose input or output needs are not being met in the local economy as potential targets for economic development opportunities. An analysis of the interindustry linkages can identify industries that are importing important inputs used in producing their product (import substitution opportunities); industries that use goods produced within the local economy but are located outside the area (export promotion opportunities); and industries exporting goods to other areas that may have a particular competitive advantage in the local economy and whose growth and development should be encouraged. An analysis of these linkages may also suggest opportunities for economic development efforts aimed at encouraging industry clusters based on these interrelationships (Pascal & Gurwitz, 1983; Porter, 1990).

The first step in conducting an input-output analysis is to develop a chart depicting the linkages among the industries in the area that provides information on the industry demand for goods and services used to produce its product(s) and the ways that products of this industry are used in producing products in other industries. Developing this "localized" input-output chart is not easy, and a range of methods have been suggested, including "regionalizing" national input-output tables and using survey techniques to identify the linkages. Each industry's links to other industries are then examined to determine whether they differ from some standard or benchmark. These differences are then examined to determine whether these "gaps" represent potential industry targets for economic development efforts.

Input-output analysis is informative because it highlights how industries are linked, revealing the potential opportunities presented by identifying supply-and-demand relationships and specifically those that are not in equilibrium. It also presents information to help guide efforts focused on encouraging agglomeration economies or industry clusters that are linked in terms of supply and demand for goods and services. Finally, it provides useful information to examine the market orientation of an industry and whether an industry "fits" with a local economy. The early work of Sweet (1970a), further extended by Bohm et al. (1983) and Lee (1987), shows how input-output analysis can be integrated into a more comprehensive industry targeting model. However, input-output analysis is unlikely to be useful as a stand-alone approach. First, as Chinitz (1961) observed, it focuses only on the demand side of a regional economy and does not help to understand the supply side of a local economy. Second, it is fairly complex and cannot be easily used by someone not trained in this sort of analysis. Third, localized input-output charts are difficult to develop, and many of the techniques available to customize national or regional input-output tables are imperfect at best.

■ Issues in Industry Targeting

Industry targeting, by any measure, is an art, not a science. In many ways, it is unproved. There are no careful evaluations of alternative approaches that assess their effectiveness in identifying "high-probability" targets for local industrial development initiatives. Most information available on "outcomes" is highly anecdotal. The strongest argument in support of industry targeting may well be its usefulness in helping planners to avoid doing "wrong things," often very costly, rather than in helping them to identify clear courses of action for local economic development efforts.

Four key issues surrounding industry targeting must be considered. First is the range of methodological issues in using this approach. We are learning more about these issues. Existing models and approaches, though limited, are growing in sophistication (Gorin, 1991). A critical limitation remains, however. Most industry targeting focuses on manufacturing. This is where we have the best information on how location decisions work. We know very little about other industries, such as services, especially business information and financial services (Gillis, 1987). As the economy and patterns of global interaction continue to evolve, the significance of this limitation grows.

The second issue is how industry-targeting analysis is used in actually guiding the design of comprehensive economic development strategies. We do not know very much about this critical issue. Clearly, industry targeting has become part of the conventional wisdom of local economic development. It has strong appeal to economic development professionals. However, we do not have useful information on how to integrate industry targeting into comprehensive economic development plans and strategies, and we have no clear guidelines for development professionals on how to do this effectively.

Third is the issue of what actions follow the identification of priority industry targeting. The contribution of these techniques to a deepening understanding of dimensions, trends, and underlying dynamics of economic change is singularly important. The usefulness of this information in preventing costly, wasteful, and unproductive initiatives from being undertaken is great. The risk that industry targeting may engender ill-conceived, misdirected, zero-sum competitive strategies is also great. In many, and perhaps most, cases, industry targeting is viewed as a tool of industry recruitment and interjurisdictional and interregional competition, rather than as one component, clearly important, of a multidimensional economic development strategy.

Finally, effective industry targeting requires expertise, talent, experience, and knowledge. It is not an endeavor for the uninitiated or the innocent. As an underdeveloped art, it is not a mechanical process. It requires a high degree of intuition. It cannot be taught or learned in a few weeks. The potential for abuse and misuse is great by either the untutored or the unscrupulous.

REFERENCES

Bartik, T. J. (1989). The market failure approach to analyzing regional economic development policy. *Economic Development Quarterly, 4*, 361-370.

Blakely, E. J. (1989). *Planning local economic development.* Newbury Park, CA: Sage.

Bohm, R. A., Herzog, H. W., Jr., & Schlottmann, A. M. (1983). Industrial location in the Tennessee-Tombigbee corridor. *Review of Regional Studies, 10*, 28-37.

Bowden, E. V. (1971). The theory and the practice of regional development economics. *Land Economics, 47*, 113-121.

Boyle, R. M. (1987). *Developing strategies for economic stability and growth.* Washington, DC: National Council for Urban Economic Development.

Chinitz, B. (1961). Contrasts in agglomeration: New York and Pittsburgh. *American Economic Review, 51*, 279-289.

Cho, D. W., & Schuermann, A. C. (1980). A decision model for regional industrial recruitment and development. *Regional Science and Urban Economics, 10*, 259-273.

Conroy, M. E. (1974). Alternative strategies for regional industrial diversification. *Journal of Regional Science, 14*, 31-46.

Doeringer, P. B., Terkla, D. G., & Topakian, G. C. (1987). *Invisible factors in local economic development.* New York: Oxford University Press.

Eisinger, P. J. (1988). *The rise of the entrepreneurial state.* Madison: University of Wisconsin Press.

Garn, H. C., & Ledebur, L. C. (1982). Congruencies and conflicts in regional and industrial policies. In M. E. Bell & P. S. Lande (Eds.), *Regional dimensions of industrial policy* (pp. 47-73). Lexington, MA: Lexington.

Gillis, W. R. (1987). Can service producing industries provide a catalyst for regional economic growth? *Economic Development Quarterly, 1*, 249-256.

Gillis, W. R., & Shaffer, R. (1985). Targeting employment opportunities toward selected workers. *Land Economics, 61*, 433-444.

Gorin, D. R. (1991). POINT: Potential Oklahoma Industrial Targets. *Economic Development Review, 9*, 12-15.

Gregerman, A. (1984). *Competitive advantage: Framing a strategy to support high growth firms.* Washington, DC: National Council for Urban Economic Development.

Growth Strategies Organization. (1992). *1992 report on GSO Survey of Economic Development Organizations* (Special ed.). Reston, VA: Author.

Hansen, R. W., & Munsinger, G. M. (1972). A prescriptive model for industrial development. *Land Economics, 48*, 76-81.

Hansen, S. B. (1989). Targeting in economic development: Comparative state perspectives. *Publius, 19*, 47-62.

Hunker, H. L. (1974). *Industrial development: Concepts and principles.* Lexington, MA: D. C. Heath-Lexington.

Jones, D., Manifold, D. J., & Spencer, J. S. (1980). The industrial screening matrix: A tool for industrial developers. In J. E. Pluta (Ed.), *Economic and business issues of the 1980s* (pp. 63-67). Austin: University of Texas, Business Research Bureau.

Klassen, L. K. (1967). *Methods of selecting industries for depressed areas.* Paris: Organisation for Economic Cooperation and Development.

Koebel, C. T., & Bailey, C. L. (1993). Putting targeted economic development to work: The role of targeted marketing in an economic development program. In P. B. Meyer (Ed.), *Comparative studies in local economic development* (pp. 17-29). Westport, CT: Greenwood.

Kotler, P., Haider, D., & Rein, L. (1993). *Marketing places.* New York: Free Press.

Kraushaar, R. (1986, April). *The planner as stock analyst: Identifying winners and losers in a regional context.* Paper presented at the annual meeting of the American Planning Association, Milwaukee, WI.

Kraushaar, R., & Feldman, M. (1987). Industrial restructuring and the limits of industry data: Examples from western New York. *Regional Studies, 23*, 49-62.

Krugman, P. R. (1983). Targeted industrial policies: Theory and evidence. In Federal Reserve Bank of Kansas City (Ed.), *Industrial change and public policy* (pp. 123-155). Kansas City, MO: Federal Reserve Bank of Kansas City.

Laird, W. E., & Rinehart, J. R. (1979). Economic theory and local industrial promotion: A reappraisal of usual assumptions. *AIDC Journal, 14*, 33-49.

Lee, K. C. (1987). Modeling target industries with differential weights. *Papers of the Regional Science Association, 62*, 125-135.

Levy, J. M. (1990). What local economic developers actually do: Location quotients versus press releases. *APA Journal, 56*, 153-160.

Lipman, B. J., & Miller, T. R. (1987). *Feasibility study to update, refine, enhance, or replace EDA's industrial location system (ILS)*. Washington, DC: Urban Institute.

Luke, J. S., Ventriss, C., Reed, B. J., & Reed, C. M. (1988). *Managing economic development*. San Francisco: Jossey-Bass.

Magaziner, I. C., & Reich, R. B. (1982). *Minding America's business*. New York: Vintage.

McGahey, R. (1990). Improving economic development strategies. *Journal of Policy Analysis and Management, 9*, 532-535.

McKee, W. L., & Froeschle, R. C. (1985). *Where the jobs are: Identification and analysis of local employment opportunities*. Kalamazoo: W. E. Upjohn Institute for Employment Research.

McLean, M., & Voytek, K. P. (1993). *Local economic analysis: Using analysis to guide local strategic planning*. Chicago: Planners Press.

Minshall, C. W., Miller, J. D., & White, A. S. (1976). *Development of a community level target industry identification program*. Prepared for the State of Missouri Division of Commerce and Industrial Development. Columbus, OH: Batelle Columbus Laboratories.

Moriarty, B. M. (1980). *Industrial location and community development*. Chapel Hill: University of North Carolina Press.

Pascal, A., & Gurwitz, A. (1983). *Picking winners: Industrial strategies for local economic development*. Santa Monica, CA: Rand Corporation.

Petshek, K. R. (1968). Can industrial development be systematically approached. *Land Economics, 44*, 255-268.

Porter, M. E. (1990, March-April). The competitive advantage of nations. *Harvard Business Review, 68*, 73-93.

Regan, E. V. (1988). *Government, Inc.: Creating accountability for economic development programs*. Washington, DC: Government Finance Research Center.

Ryans, J. K., & Shanklin, W. K. (1986). *Guide to marketing for economic development*. Columbus, OH: Publishing Horizons.

Siegel, B., Reamer, A., & Hochberg, M. (1987). Sectoral strategies: Targeting key industries. *Economic Development Commentary, 11*, 8-13.

Smith, D. M. (1981). *Industrial location: An economic geographic analysis*. New York: John Wiley.

Stevens, B., Treyz, G., Arvidson, E., Ehrlich, D., & Delmar, C. (1984). *Targeting measures and target industries for western Massachusetts*. Regional Science Research Institute.

Sweet, D. C. (1970a). An industrial development screening matrix. *Professional Geographer, 22*, 124-127.

Sweet, D. C. (1970b). The systematic approach to industrial development research. *AIDC Journal, 5*, 21-32.

Sweet, D. C. (1971). Identifying industrial potential. *AIDC Journal, 6*, 71-76.

Syversen, W. M. (1985, October). *Targeting industries for retention and expansion: The use of secondary data analysis*. Paper presented at the Conference on Community Economic Development Through the Retention and Expansion of Existing Business, Columbus, OH.

Thompson, W. R. (1987). Policy-based analysis for local economic development. *Economic Development Quarterly, 1*, 203-213.

Thompson, W. R., & Thompson, P. R. (1987). National industries and local occupational strengths: The cross-hairs of targeting. *Urban Studies, 24*, 547-560.

Thornton, L. (1984). Targeting industries for economic development. *Economic Development Review, 2*, 23-28.

Vaughan, R. J. (1976). *The urban impacts of federal policies: Vol. 2. Economic development.* Santa Monica, CA: Rand Corporation.

Vaughan, R. J., & Bearse, P. (1981). Federal economic development programs: A framework for design and evaluation. In R. Friedman & W. Schweke (Eds.), *Expanding the opportunity to produce* (pp. 307-329). Washington, DC: Corporation for Enterprise Development.

Vaughan, R. J., Pollard, R., & Dyer, B. (1985). *The wealth of states: The political economy of state development.* Washington, DC: Council of State Planning Agencies.

Wolman, H., & Voytek, K. P. (1990). State government as consultant for local economic development. *Economic Development Quarterly, 4,* 211-220.

COMMENTARY ON INDUSTRY TARGETING

Is industry targeting a viable local economic development strategy? The answer to this question depends on one's view of economic development's status as a profession. Economic development is an emerging profession in my view, and therefore many of its strategies and practices are still very experimental in nature. Judgments about the value and effectiveness of industry targeting should be made in this larger context. This suggests that practitioners, academic researchers, and policy makers may struggle for some time in determining the effectiveness of industry targeting and other economic development strategies.

The Ledebur and Voytek chapter discusses the evolution of local industry-targeting practices, their conceptual foundations, and examples of industry-targeting approaches. The use of targeting strategies has increased as economic developers turn to them to support both existing industry development and new business recruitment programs. Ledebur and Voytek discuss data and methodological problems that must be overcome if industry targeting is to garner greater respect from the academic research community. Many of these problems stem from a limited understanding of how local economies function within changing national and global economic contexts and, more important, how local policies and programs can affect local economic performance. I would agree with the authors' general assessment of these shortcomings, and I would urge practitioners to continue to refine these strategies in the future.

The real question remains: Is industry targeting a viable strategy? In other words, does it work in promoting local economic growth? Practitioners and researchers must solve two basic problems to make industry targeting more viable. The first is to improve research methods and data used to identify target industries and companies. The second is to improve local policy actions taken to develop target companies and industries.

The viability of industry targeting should be judged by the goals guiding its use. Industry-targeting strategies presume that communities and states want to intervene in local economic affairs to influence local economic performance. Economic development officials assume that local policy actions can affect local economic outcomes, although this is still an open question in many respects. Many communities and states use industry-targeting strategies to sharpen the *focus* of their development activities and in hopes of increasing their positive impact. Economic developers and community leaders assume that an improved focus on "targets of opportunity" will increase the probability that local economic development programs will produce desirable economic outcomes. Many development officials view industry-targeting analysis as a rational approach to the identification of valid development targets. Development officials are demanding that targeting studies provide more precise and reliable analytic results. In response, target-industry consultants and academic researchers are experimenting with new analytic methods and more reliable data sources to improve results. In the future, these studies must withstand even greater scrutiny on methodological grounds.

The identification of appropriate targets of opportunity is only half the battle. Communities and states must also identify and undertake effective action to develop industry targets—a task that may prove even more difficult than identifying the right targets. What actions will produce successful results from target-industry development efforts? Various business development, marketing, and business climate improvement strategies are currently used by economic development organizations to develop target industries and businesses. These include direct mailings to companies in target industries, business recruitment trips to visit out-of-town companies, investments in special local workforce and infrastructure resources required by target industries, and other activities that improve the area's competitive position relative to the target industry requirements. Local economic development competition has grown as an increasing number of local places strive to increase their share of jobs and business investment opportunities. These places use strategies such as industry targeting to differentiate themselves from their competitors. A future competitive edge will demand even greater creativity and effort from local economic development groups.

In conclusion, little is known about the actual impact and effectiveness of target-industry development efforts. In other words, it is unclear whether these strategies have their intended impact. Researchers may be quicker to make this point, but many practitioners would also agree that evaluation of these programs can help to strengthen their future effectiveness. Even with these problems, industry targeting is valuable to community and state officials in assessing future development opportunities and ensuring that these are consistent with an area's development values, goals, and resources. Industry-targeting strategies are more likely to be used by areas that favor proactive interventionist approaches to local economic development. Although the actual impact of these initiatives is unknown in most cases, these strategies appear to benefit communities and states by providing them with more systematic and rational approaches to economic development. Quite importantly, industry-targeting strategies can help area leaders to approach growth more selectively and thereby ensure that the growth to be encouraged is consistent with area strengths and goals.

COMMENTARY ON INDUSTRY TARGETING

Sabina E. Deitrick

Industrial targeting has grown in importance in economic development as patterns of uneven growth continue in the United States (Beeson, 1992) and as many areas that suffer from slow or stagnant job growth seek ways to identify potential "engines of growth" (George, 1983). With traditional bases of economic growth lagging in some regions and states, such as parts of the industrial Midwest, and even as high-growth regions, such as southern California, continue to reel from economic recession, industrial targeting has emerged as a form of strategic economic planning. It represents a more sophisticated technical method to identify potential growth areas in a regional economy compared to earlier development programs that relied on promoting export industries, improving supply-side conditions, or engaging in "smokestack" chasing.

Industrial targeting, however, does not set forth a policy agenda for economic development practitioners. Therein lies a problem common to technical solutions in economic development. Often policies undertaken after the industrial-targeting effort is complete look like more traditional economic development practice. On the other hand, different types of policies may be used to introduce or bolster more innovative approaches to economic development. Even so, engaging in a comprehensive industrial targeting effort does not guarantee that a policy agenda endorsing the approach will be implemented.

In this commentary, I will review the points made in Voytek and Ledebur's chapter on industrial targeting and continue with further points about the policy environment in which industrial targeting as an economic development strategy must operate.

■ Industrial Targeting: The Exercise

Voytek and Ledebur have laid out the important elements of any industrial targeting exercise. As they state at the outset, an industrial targeting exercise depends on the availability of good data. Data constraints present a central problem for conducting an effective industrial targeting exercise because data availability, or unavailability, can limit what can be attained from the targeting effort. For instance, when using secondary data, the analyst often relies on sectors defined by the Standard Industrial Classification (SIC) system. With these data, the selection of target industries is more reliable for manufacturing activities than for activities in other sectors because manufacturing data are better collected and organized under the SIC. Thus, important growth areas, such as producer services or environmental technologies that can bring money into the regional or local economy from outside, cannot be properly analyzed as potential target industries (Boyle, 1994).

For smaller areas, confidentiality in reporting limits what analysts can collect from secondary sources, particularly below the two-digit SIC level, thus limiting the scope of the targeting exercise. In addition, smaller economic development organizations, as well as larger ones, may have financial constraints that preclude the collection of survey data or other primary-source information necessary to develop a community resource inventory. As in many other comprehensive analyses, analysts must determine at the outset what data they can secure with the resources available to them before engaging in the industrial targeting exercise.

Voytek and Ledebur have rightly pointed out the importance of combining several methods in conducting an industrial targeting strategy. This "synthetic" approach has the advantage of not eliminating potential sectors from targeting, though they may not attain high scores on any one of several steps used in the targeting exercise. In a similar vein, the authors point out the problems of using national growth industries and standard benchmarks for local or regional industry targeting. They outline complementary approaches to targeting that use comparable communities in the analysis, such as in studying undersupplied industries or in performing the community resource assessment.

Assessing an area's strengths and industry growth potential against its strongest competitors can aid in identifying potential target industries. Despite additional costs, benchmarking exercises may unveil differences and information not attainable against national figures or standard benchmarks. In an industry-targeting exercise, comparing the area's competitive advantage against nearby or similar communities or regions can yield a different set of industries to target than those selected against national performance standards (Boyle, 1994). Likewise, the community resource profile can be enhanced. A recently conducted benchmarking exercise for the Pittsburgh region compared a number of locational factors among a set of comparable regions (Tripp Umbach & Associates, 1994). Pittsburgh's relatively high manufacturing unionization rate, seen as a detriment to potential firm attraction and firm retention, was equal to nearby Ohio's (Tripp Umbach & Associates, 1994, p. 7). Despite the similarity in rates, Ohio has successfully attracted new investment in the 1990s because organized labor has not been seen as a problem by prospective firms looking at Cincinnati and Columbus. The authors concluded that the perception of unionization in the Pittsburgh region was as much a problem as were real union rates.

Finally, the industrial targeting exercise set forth may omit from consideration other key industries that are part of a sectoral cluster. Much of the theoretical groundwork for the techniques outlined by Voytek and Ledebur is based on demand-side theories of growth. These may understate the importance of interdependencies found in a spatial association (Pred, 1976). Locally, the sectoral cluster allows analysis of complex interrelationships involving both private firms and public actors (Sternberg, 1991). Porter's (1990) work on national competitiveness and regional clusters points to a number of ways that policy can enhance the interdependencies among clusters of firms. Other techniques not addressed by Voytek and Ledebur, such as factor analysis, have been used by researchers to identify metropolitan specialization by groups of sectors (Lubov & Anton, 1994; O'Huallacháin, 1992).

■ Policy Environment

Although industrial targeting exercises may identify potential growth industries in a region, the programs needed to improve operating conditions, attract businesses, and upgrade workers' skills require resources and a means to implement them through the political and policy process. As the authors emphasize, the industrial targeting exercise cannot end with the report.

There may be conflict between political projects and targeting strategies. As Beauregard (1993) noted, "What is interesting about economic development practice is that it manages to combine repetition with novelty" (p. 279). Fiscally strapped state and local governments often grab the latest in economic development initiatives that deliver revenues and exposure more quickly than investments in smaller scale economic development based on industrial targeting efforts. Tourism-based strategies are a case in point. Though cities spend millions of dollars on tourist promotion activities—convention centers, sports arenas, and gambling operations, for instance—in lieu of other types of economic development policies, tourism rarely shows up on the consultants' list of industries to be targeted. In 1991, when U.S. states spent a total of $50 million on manufacturing extension services to increase the competitiveness of their smaller manufacturing firms, New Jersey spent the same amount to regulate its casinos, and state governments spent six times that figure—$300 million—to advertise state lotteries (Harris, 1994). There may be a mismatch between the technical targeting exercise and the activities that politicians and economic development professionals target. Why this occurs should be a question asked at the outset of any targeting exercise. Making the targeting exercise a useful policy—and political—document requires more than just good modeling and process. Unfortunately, as Voytek and Ledebur caution, a lack of action on the targeting exercise often is precisely the outcome.

Although Voytek and Ledebur do a good job of summarizing and analyzing the steps involved in setting up and conducting an industrial targeting study, their plea for integrating industrial targeting into an economic development strategy might be put more forcefully. As Mier and Fitzgerald (1991) noted, "The new literature on local economic development seems likely to continue the pattern of academe's limited success at connecting with professional practice, and we are plagued with the question 'why?' " (p. 275).

For whatever the merits of the industrial targeting exercise, however complete the process and however practical the solutions, without an effective strategy to integrate the results into policies such as business retention and expansion, modernization, and labor training and education, industry targeting will not be an effective strategy. More is needed to understand how industry targeting has been implemented into a successful economic development strategy and to evaluate its results.

REFERENCES

Beauregard, R. A. (1993). Constituting economic development: A theoretical perspective. In R. D. Bingham & R. Mier (Eds.), *Theories of local economic development: Perspectives from across the disciplines* (pp. 267-283). Newbury Park, CA: Sage.

Beeson, P. E. (1992). Agglomeration economies and productivity growth. In E. S. Mills & J. F. McDonald (Eds.), *Sources of metropolitan growth* (pp. 19-35). New Brunswick, NJ: Rutgers University, Center for Urban Policy Research.

Boyle, M. R. (1994). Economic development targeting in the nineties. *Economic Development Review, 12*, 13-22.

George, R. (1983). *Targeting high-growth industry.* Montreal: Institute for Research on Public Policy.

Harris, T. R. (1994, August). *Expansion of the gaming industry: Opportunities for cooperative extension.* Paper presented at the annual meeting of the American Agricultural Economic Association, San Diego.

Lubov, A., & Anton, P. A. (1994). Using factor analysis to identify industrial targets. *Economic Development Review, 12,* 74-78.

Mier, R., & Fitzgerald, J. (1991). Managing economic development. *Economic Development Quarterly, 3,* 268-279.

O'Huallacháin, B. (1992). Economic structure and growth of metropolitan areas. In E. S. Mills & J. F. McDonald (Eds.), *Sources of metropolitan growth* (pp. 51-85). New Brunswick, NJ: Rutgers University, Center for Urban Policy Research.

Porter, M. E. (1990). The competitive advantage of nations. *Harvard Business Review, 90,* 73-93.

Pred, A. (1976). The interurban transmission of growth in advanced economies: Empirical findings versus regional-planning assumptions. *Regional Studies, 10,* 151-171.

Sternberg, E. (1991). The sectoral cluster in economic development policy: Lessons from Rochester and Buffalo, New York. *Economic Development Quarterly, 5,* 342-356.

Tripp Umbach & Associates. (1994). *A comparative assessment of southwest Pennsylvania as a business location.* Pittsburgh: Conference on Real Estate of Southwestern Pennsylvania.

RESPONSE TO DONALD T. IANNONE AND SABINA E. DEITRICK

Kenneth Voytek
Larry Ledebur

Iannone and Deitrick pose a common test in answering the question addressed. Simply put, does industry targeting result in better economic development policies? Do policies change, and are they better? Does industry targeting change and improve actions to implement these policies?

We agree that this is the defining issue. For this reason, we have bemoaned the absence of a body of evaluation studies that can answer this question and, perhaps, point to how this type of industry performance information can be more effectively integrated into policy development and the formulation of economic development strategies. The overwhelming reluctance of local officials, elected or appointed, to seek or permit evaluation of economic development projects—their formulation, implementation, and outcomes—is a serious obstacle to efforts to improve economic development policies and practice.

While awaiting this evaluation information, we can draw some conclusions. The two commentators concur that industry targeting is a key information component necessary to drive effective strategic planning for economic development. These industry performance and outlook measures can be a guide to greater rationality in economic development policies and strategies. As we suggested in our chapter, one of the primary arguments in favor of industry targeting is its potential to guide local policies away from costly negative-, zero-, or minimal-outcome policies and programs. In an environment in which economic development policies are so often driven by conventional wisdom and promises of silver bullets, industry targeting can contribute to greater rationality in the pursuit of the grail of jobs, incomes, and tax bases.

Don Iannone raises a critical issue mentioned in our chapter but not sufficiently stressed. With our current level of knowledge, identifying effective action once industry targets are established may be much more difficult than identifying effective targets. Methodologies for industry targeting are growing increasingly sophisticated, although the process remains plagued by issues of data availability and reliability. We have not witnessed parallel progress in capacities to move from detailed information on industries and performance to the design and implementation of effective strategies and programs. At the policy level, all too often, elected officials and development practitioners remain captives of the conventional wisdom of economic development and, more specifically, industrial recruitment and essentially zero-sum interjurisdictional competition.

Sabina Deitrick raises an issue not specifically addressed in our chapter. She points to work in the Pittsburgh area that "benchmarks" the region against other perceived "regions in competition." Benchmarking can be an important outcome of industry targeting. It can identify "real" versus "perceptual" problems in the local economy and business climate, as well as comparative measures of industry performance indexed against regional or localized industry performance versus national averages.

Clearly, this information is important input into strategic planning and policy and program development. Perhaps no less important, this information does establish temporal "benchmarks" against which regional performance can be measured over time. It is this information—not simply growth measures, but progress against clearly defined goals and objectives responding to deficiencies and opportunities in the region— that may be more useful in accessing past policies and practices and charting new directions.

In the long run, the critical issue is not the industrial performance of one region as compared to another but the ability of a region to assess itself realistically, to set clear and achievable goals, to determine an effective course of action, and to monitor and measure its success or lack thereof relative to clear and valued "benchmarks."

8

Can City Hall Create Private Jobs?

LEON TAYLOR

The cognoscenti of economic development have long debated whether City Hall should bid for distant firms or cultivate local ones. Bidding prevails: Of 3,000 U.S. communities surveyed by *Site Selection* ("Survey," 1993), 9 in 10 offer incentives, most often property tax cuts or worker training. But many localities cooled toward auctions in the 1980s when, amid factory closings and double-digit unemployment, bids for jobs soared. General Motors received incentives valued at $700 million to put its Saturn factory in sleepy Spring Hill, Tennessee (Lyne, 1992). As bids went ballistic, disenchanted communities pulled out (Herbers, 1990). They tried instead to "incubate" firms—that is, offering a sterile office in which entrepreneurial wannabes could put up their feet until they worked up the nerve to strike out on their own and pay market rents like everybody else.

Incubators or incentives? That's the Great Debate—and most of it is humid oxygen. For whether a city courts a petrochemical giant from Germany or mothers a tool shop down the street, its goal is still to pile up capital—to persuade firms to add structures, equipment, and machines to the local economy. In either case, the city acts on the belief that giving each worker more capital will enrich him or her.

Unfortunately, over the long run, the city's supposition errs. Here's why. Suppose that each addition to the capital stock creates less income than any addition before it. Then it will also create less in savings that finance future growth. The day will dawn when the urban economy has so much capital that merely maintaining it claims all the energies that would otherwise have ensured future growth. City Hall will simply fill potholes in thousands of empty streets. Firms will simply replace old plants with new plants that make the same old products. The standard of

AUTHOR'S NOTE: I thank Rob Mier and Wim Wiewel for help in separating a pinch of wheat from the chaff of earlier drafts.

living will level off. In short, if cities merely pile up capital, they will someday stagnate.

Policies to develop urban economies fail because they ignore the engine of growth—technological change. To bribe an employer to hire 400 residents by handing him or her cash over the barrel or keys to a warehouse is to live from hand to mouth. Accumulating jobs today will not create jobs 10 years from today. Only ideas can do that: for instance, the idea of shrinking a computer until it fits a desk. City Hall should try to induce innovations, not capital; ideas, not jobs. The economic development agency should not be a smoky back room where henchmen for the mayor haggle with henchmen for the CEO. It should be a think tank, puzzling out how the city can grow in generations to come.

Ideas matter more than machines: That theme for the cities emerges, by analogy, from a debate among economists over how nations grow (Grossman & Helpman, 1994; Pack, 1994; Solow, 1994). The puzzle is to explain why poor countries are not catching up with the rich. It is not enough to say that the rich stay ahead by saving and investing more; for the usual model to work, the United States would have to be saving at a rate 30 to 100 times greater than the saving rate of the Philippines (Romer, 1994). Because that cannot happen, the model is missing something.

Its deficiency is in assuming that, per worker, a machine is worth far more in an economy with few machines than in an economy with many machines. In other words, the model assumes that a machine has more value to a worker in a poor country than to a worker in a rich country. As a consequence, poor countries should grow faster than rich ones because they get more output per worker from their average machine. But in reality, poor countries do not grow that much faster. To explain that cold fact, the usual model must suppose that rich countries are adding machines at a rate that is impossibly high.

One way out is to assume that a machine is worth almost as much in an economy with many machines as in an economy with few machines. If a rich country gets almost as much output per worker from its average machine as a poor country, then it will grow almost as fast. But that seems to defy common sense. Surely, a desktop computer is worth more to a worker if he or she has only one rather than two.

Suppose, however, that the worker learned to use a computer more effectively *because* she had two. For example, she could give one to a worker with none, and they could work side by side on models of watersheds. As one worker solved problems, she could share her knowledge with the other. Innovations would spread.

The diffusion of innovation, more than the mere accumulation of capital, spurs growth. The worker would gain nothing from loaning her second computer to a colleague who refused to share her findings. The importance of innovation is more than theory. Studying the U.S. economy from 1909 through 1949, Solow (1957) found that growth in capital per work hour accounted for only 19% of the growth in nonfarm output per work hour. The rest of the gain seemed due to improvements in equipment and in ways to produce. An overlapping study, from 1929 through 1982, broadly supported those conclusions (Denison, 1985). This study estimated that growth in capital per worker accounted for only 13% of the growth in output per worker. Most of the rest of the gain seemed due to progress in the education of workers as well as in science and technology.

To stimulate growth, urban policy makers must stimulate innovation—which, fortunately, is most likely to occur in cities. For the 35 largest cities of the United States, the number of patents filed per capita was 4.1 times the national average in 1860 and 1.6 times the national average in 1910 (Pred, 1966). In the late 1940s, 57% of the U.S. metropolitan population received 90% of the patents granted for chemicals, metals, and machinery, as well as engine products and processes (Thompson, 1962).

To see why invention occurs in the cities, consider first why it occurs. Some philosophers of science, prominently Denis Diderot, attributed invention to the accumulation of scientific knowledge. In this view, every giant stands on the shoulders of another, though perhaps unwittingly. Discovery is often serendipity: The anatomist Luigi Galvani found the electrical properties of metals while dissecting frogs. Invention, then, results from knowledge and luck. Thomas Edison tested 2,000 filaments to develop an electric lightbulb (Rescher, 1978).

This view of invention underlies the neoclassical theory of economic growth, or the "usual model." Suppose that invention roots in a mulch of knowledge that has nothing to do with economic motives. Then it makes sense to treat invention as external to an economic model of growth, as a parameter that economists cannot really explain. A model of urban growth would need to explain why scientific knowledge accumulates in cities rather than elsewhere.

The dark vision of a stationary state, in which the living standard levels off, also makes sense under a theory of exogenous invention. Max Planck, who won a Nobel Prize in physics, argued that the difficulty of more research in science increased with every advance. Indeed, the number of patents granted for inventions in the United States, per thousand scientists and engineers, fell steadily from 587 in 1900 to 49 in 1954 (Machlup, 1962). Meanwhile, the cost of obtaining a little more progress in science has risen. After World War II, total real spending in the United States on science, per scientist, rose 4% a year; the productivity of a scientist, measured by publications, fell slightly (Price, 1986). It is true that after 1960, spending rose more slowly; and in the late 1980s, it fell as the Cold War thawed (U.S. Bureau of the Census, 1993). Nevertheless, science costs more now than early in the 20th century. Then, a physicist could set up an experiment, with new equipment, for a few hundred dollars. "We have no money to spend," Ernest Rutherford told his students, "so we shall have to think" (Bowden, 1965, p. 851). Now high-energy physics costs more than $200,000 per scientist each year (Rescher, 1978). Some historians who believe that the wheels of science have been slowing since the Age of Reason predict that knowledge will stop growing. If they are right, and if people invent because of what they know, then urban policy makers need not worry about long-run economic growth. It will not happen.

But do people invent because of what they know? Invention is like a murder; it requires an opportunity and a motive. Certainly, knowledge provides opportunities. But does it motivate invention?

Not as a rule. The first steam engines that worked commercially were built by a blacksmith, Thomas Newcomen, and a military engineer, Thomas Savery. A barber who learned to write at age 50, Richard Awkwright, used the spinning machine to produce thread commercially, the first step toward converting the textile industry to machines. Evidently, Awkwright learned the technical secrets of textile production through conversations at the barber's chair (Mantoux, 1983).

If people do not invent because they know a lot, why do they invent? To find out, Schmookler (1966) studied the U.S. economy from 1800 to 1955, focusing on invention and change in agriculture, petroleum, paper, and railroad equipment. In each industry, what led directly to invention was not a scientific breakthrough; it was the need to solve well-defined technical problems.

People try to solve these problems to make money. Surveying the work of Alexander Graham Bell, Thomas Edison, Robert Fulton, Samuel Morse, and James Watt, the economist Frank Taussig (1989) concluded that "the *direction* in which the contriver turns his bent is immensely affected by the prospect of gain for himself" (p. 50). Neither did Eastman Kodak, B. F. Goodrich, General Electric, and AT&T set up their research laboratories to advance the social welfare. Innovation is the most powerful way to compete for profits: A hot new product, such as the personal computer, can remake an industry. In comparison, price cutting pales.

Profit seekers are more likely to innovate in cities than elsewhere because cities provide two motives for innovation: to cut high urban costs and to reap high urban revenues. Urban costs are high for two reasons. First, growth in production or population eventually congests an area of given size. Transport slows; health and physical capital decay through pollution and other hazards. In the 1660s, when plague and fire raced through crowded London, residents fled and created suburbs (Smallwood, 1965).

Second, high costs stem from the prices charged by input owners to exploit site values. At first, the firm profits from being near a pool of workers, suppliers, or buyers; a harbor; or a railroad terminal. But as the value of proximity becomes clear to all, the firm will have to yield some gains to those who control its site—the landowner and the city government. They will extract from the firm its profits from occupying the site.

Over time, the firm will develop technologies that cut location costs. Cities that support these innovations will attract firms and grow; cities that do not support them will stagnate.

Consider urban America early in the 19th century. In that era of unpaved roads, transporting bulk goods sometimes cost 100 times as much by horse or wagon as by ship or barge. Commerce clustered along navigable waterways. Urban workers had to walk to their jobs, so they lived within 3 miles—a bracing 1-hour trek—of the waterways. Landowners and governments that controlled bywater regions could extract location rent—the former through the lease and sale of property and the latter through taxation and regulation. Port cities flourished: New York, Boston, New Orleans, St. Louis (Glaab & Brown, 1983).

The stage was set for a revolution in transportation. By completing a canal between the Illinois River and Lake Michigan, as well as improving the harbor at the mouth of its namesake river, the upstart village of Chicago diverted corn and lumber traffic from St. Louis. Chicago saw its population rise from 5,000 in 1840 to 30,000 in 1850, when its leaders announced their intent to displace St. Louis as the great city of the region (Belcher, 1947).

The "New York of the West" paid little attention to the boastful town best known for producing corner lots, statistics, and wind. St. Louis was sitting pretty. Situated on the limestone cliffs of the Mississippi, downstream from where the Father of Waters received the Ohio, the Missouri, and the Illinois rivers, St. Louis was the natural point for moving cargo from the light steamers of the upper Mississippi to the heavy steamers of the broad stretch to New Orleans. St. Louis was thriving. Its levees received more than four times as much wheat and flour as Chicago; its population doubled by the decade. St. Louis was content with the steamboat. It saw no reason to bet all on the iron horse. "Nature having done so much for us has paralyzed our efforts," said a Missouri writer (Belcher, 1947, p. 52).

Chicago was not paralyzed. Its favorite son, Stephen A. Douglas, guided a land-grant bill through Congress to help build the Illinois Central (Johannsen, 1973). In 1852, the railroad reached the Windy City. Its population tripled in 10 years. As the terminus of 11 trunk roads and 20 branch and feeder lines, Chicago had become the leading railroad center in the West. Its strong terminal facilities and low rail rates attracted manufacturers throughout the region. From 1860 to 1910, industrial employment in Chicago increased from 5,400 workers to more than 325,000. From 1870 to 1900, manufacturing employment declined in southern Illinois and Indiana, as well as throughout Missouri (Pred, 1966).

Belatedly, St. Louis and Missouri fought back by supporting several railroads. But the spectacular success of Chicago left Eastern investors reluctant to put up money for Missouri rails; often the city and the state had to subscribe. Lacking money, the railroads were poorly built and poorly maintained. Four lines defaulted in 1860, and work halted on most of the others. Too little, too late: St. Louis never recovered (Belcher, 1947).

Just as traders left St. Louis for Chicago, firms will abandon the city that rests on its riches for the city that gambles on innovation. To revive its base, the once-rich city will have to support a new and more powerful innovation. Once it attracts firms, it (and landowners) will extract rent. The cycle will renew.

Each turn of history's wheel brings technological breakthroughs: the railroad, the telephone, the electric trolley, the power transmission line, the automobile, the aircraft, the air conditioner, the interstate. Partly through city rivalry for growth, such innovations spread rapidly. Consequently, the city in search of growth must act quickly, often in great uncertainty. If an innovation opens a window of opportunity for a city, rival cities will soon slam it shut.

The transport revolution again illustrates these ideas. Between 1800 and 1850, the cost of bulk transport overland in the United States fell by more than 90% (Tolley & Krumm, 1983). Steam-railroad lines spider-webbed across the country, their intersections attracting factories from city centers. For instance, from 1880 to 1900, large manufacturers moved from the center of Detroit to form a half-circle around it, with the Detroit River as the diameter and the rail lines as an arc engirdling the old built-up section of the city (Zunz, 1982).

To stimulate growth in areas abandoned by factories, the cities supported the streetcars by granting franchises and permitting consolidation of lines. The typical result was a system of tracks radiating from downtown, like spokes on a wheel. The influx of shoppers at the center of the wheel attracted the great department stores, such as John Wanamaker's in Philadelphia (Leach, 1993). In the downtowns reborn, land values soared. In the Chicago of 1910, one-half square mile of the Loop claimed almost 40% of the total assessed value of the land in the city (Hoyt, 1933). The high land rents of downtown, with the spread of the automobile, eventually spurred the rise of the suburban shopping mall.

The effects of location rent help explain why the rate of innovation increases with the rate of economic growth. Profits rise during recovery, providing rent seekers with bigger plums to pick. The landowner raises the rent on a lot in a booming district; the city hikes its taxes on a lustily growing industry. The firm must expect extractions to rise considerably, however, before it will shoulder the high setup costs of a major innovation. Thus, major leaps in technology are most likely in periods of high, sustained growth.

High urban costs are not the only stimuli to innovation. Firms also innovate to increase revenues by improving their products or expanding their market areas. Consider Nantes of the 1820s, in western France. Retired army officer Stanislas Baudry, who owned a flour mill powered by steam, used the excess hot water to open a public bath. Because it was more than a mile from the center of the city, Baudry provided 16-seat coaches, each drawn by two horses. He noted that most riders got off at intermediate points. Expanding the shuttle, he created the "omnibus," after the name over the door of the hatmaker Omnes (McKay, 1976). U.S. cities supported the omnibus by granting an exclusive franchise to the local operator, compensating him for risk (Schaeffer & Sclar, 1980).

In brief, the firm innovates in the city partly because the potential profits are highest there. Growth occurs, however, not necessarily where the invention is conceived but where it is applied. New Jersey's strength in the electrical industry owed much to the 1892 decision of Edison General Electric and Thomson-Houston to merge in order to share patents and raise prices. The new firm, General Electric, cut unit production costs by consolidating the manufacture of electrical lights at Edison Works in Harrison, New Jersey—and closing the inferior Thomson-Houston plant in Lynn, Massachusetts (Bright, 1949; Passer, 1953).

To attract such growing industries, the city can provide infrastructure. A statistical study of 264 metropolitan areas in the United States in 1977 found that the density of freeways in an area had a positive and permanent effect on the number of jobs in high-tech firms that were innovating or expanding (Markusen, Hall, & Glasmeier, 1986).

To attract a clustering industry, the city must complete its infrastructure before rival cities do. But should it not finish first, it might have to finish, anyway: Once begun, public works projects are hard to stop. Consequently, infrastructure competition often leads to excess capacity, as Los Angeles knows. In 1993, it opened the largest convention center on the Pacific. At $500 million, the renovation cost more (after inflation) than the 250-mile transport of water from the desert in 1913. The convention project is not carrying much water yet. In the last half of 1993, Los Angeles booked fewer than half as many conventions as its archrival, San Diego, which planned to double its own convention space. Fortunately, the Los Angeles center contains so much glass that when they take it down for lack of business, they can put up a see-through Washington Monument in its place (Mahtesian, 1994).

The city that competes by building might incur debt in the long run —debt that it does not need. In 1991, cities devoted 5% of their spending to interest payments, more than they spent on health and hospitals (U.S. Bureau of the Census, 1993). Still, some risks are worth taking. Consider a project that consumes 10% of what the city produces in 1 year. If the project raises the growth rate of the city's economy from 3% to 3.5% a year, then it will pay off in 20 years. The agency's heroic task is to identify projects that look as if they will pay off but that the city can stop if it later learns that they will not pay off.

One possibility is to provide venture capital to firms for research and development, with the proviso that they submit to periodic reviews and apply their innovations locally. Roughly along these lines, Connecticut has a New Product Development Corporation that has served as a model for Utah, New Mexico, and four other states. The corporation gives a grant to a firm to design, test, and prepare a product for market; the grant can cover up to 60% of these development costs. In return, the corporation receives a 5% royalty on sales, up to five times the amount of the grant (Fisher, 1990).

Certainly, the firm would rather get a grant than give up equity, or go in debt, during its early, struggling years. But should the government finance the development of products at all? If firms expect part of the net social benefits of innovation to show up in profits, then surely they will finance innovation themselves.

The problem is that financial markets are not perfect. Venture capital is chronically scarce in some areas: In 1982, more than half of all the firms financed by venture capital were in California and Massachusetts (Peltz & Weiss, 1984). The regional disparity in capital might be due to market structure and geography. Chinitz (1961) thought that in regions dominated by a few large firms, the banks got out of the habit of taking risks, and Adam Smith (1976) thought that firms in ports were the most adventurous because they could most easily tap large markets. Isolated regions with few firms, such as Appalachia, are most likely to lack venture capital.

If financial institutions could operate freely across state lines, then the largest banks might open Appalachian branches to finance promising ventures. But until governments remove barriers to interstate banking, the state is a logical candidate to provide venture capital: It is large enough to absorb a few risks, and its officials are close enough to the area to know which projects there are likely to pan out.

The state should provide a venture capital fund for no more than 10 years. This will encourage firms to innovate now, while they can tap the fund; and it will not thwart the later emergence of private financiers, should the fund demonstrate gains in providing venture capital. To disseminate the innovation and promote growth, the state might agree to reduce its royalty fee if the firm shared its patents with other local firms. As the lender of last resort, the state can expect most ventures to fail. Although Connecticut's corporation funded mainly established firms, over half of those funded by 1984 had covered less than 10% of their grant amounts by 1988. Any venture capital fund must hope for a few big strikes to pay for the many losses; over half of the corporation's royalty income came from five projects (Fisher, 1990).

Big strikes do occur. Consider what a private firm, American Research and Development Corporation, did for two engineers at the Massachusetts Institute of Technology in the 1950s. Ken Olsen and Harlan Anderson obtained $70,000 from American to start making the first minicomputers, on the second floor of an old textile factory near Boston. They named their enterprise the Digital Equipment Corporation to avoid frightening investors with the word *computer*. By 1972, no investor would have taken fright: American's investment in DEC was worth more than $400 million (Rosegrant & Lampe, 1992).

In sum, to attract innovative activity, the city or state might provide public works or venture capital. It might, however, succeed all too well. For once a thriving large firm is in place, it attracts buyers and suppliers; in turn, these attract related firms. No longer does the city define the industry; now the industry defines the city—as when the oil industry took off in Houston, or tourism in Miami, or auto making in Detroit. The city locks into the short run. All that it can then do is bribe a few firms to come add a dash of variety to its economy.

Rather than simply induce the occasional wave of growth, the city might be able to provide for continual growth by providing for continual innovation. Again, innovation requires motive and opportunity. The city can rarely do much about the profit motive, but it can create perpetual opportunities by squeezing many creators of ideas into a small space. After all, how quickly an innovation will occur depends in part on how many people are working on it—and on how many ideas they share.

Innovators prefer densely populated and educated areas for the same reason that novelists frequent crowded coffeehouses. Strolling among the patrons, novelists overhear gossip and gospel, quarrels and romance. From the wealth of material, juxtaposed in surprising ways, they weave a new plot. Similarly, the intellectual ferment of the city helps innovators gather pertinent data and organize them anew.

Creative juices flow in the coffeehouse because its patrons delight in ideas. The coffeehouse is not a roadhouse. Similarly, in the city, an increase in uneducated workers does not spur innovation—and indeed might discourage it by removing the need of firms to find substitutes for costly labor. Neither do workers themselves innovate in a system of mass production, of repetitive tasks, in which they leave their heads at the factory gates.

Innovators seek a coffeehouse—but will they find one? Why, in other words, would seekers and purveyors of ideas congregate? For the same reason that Adam Smith's fictional porter rode his horse to the city: "A village is by much too narrow a sphere for him" (Smith, 1976, p. 21). Only a city provides a market large enough for purveyors of ideas. In addition, they must come to the city to demonstrate the worth of their intellectual wares. Complex ideas take time to absorb and to size up; before recipients of the ideas will commit this time, they must have some indication that it will be time well spent. If they are perusing titles in the bookstore, they will check the reputation of the author, the publisher, the blurb writers. If the ideas are too new to have been published—and these, perhaps, would most interest innovators—then purveyors might have to demonstrate credibility through such body cues as eye-to-eye contact. "One can lie with the mouth," wrote Nietzsche (1973), "but with the accompanying grimace one nevertheless tells the truth" (p. 105). Several industries drawn to the downtown—lending institutions, corporate headquarters, government offices, and the editorial offices of newspapers—require meetings to verify honesty. They come to the center of the city because, with cutting of commuting costs, it is the cheapest place for people to meet.

If the city can somehow turn itself into a 24-hour coffeehouse, then it can spark continual innovation. To complete the conversion quickly, the city must attract a highly educated workforce rather than spending a generation to educate its own. To attract the highly educated, the city might develop parks and theaters; build highways and rapid-rail transit that, in saving time, particularly benefit those who value time highly; and enforce, in select areas, crime controls as well as environmental controls such as zoning.

The most basic way to attract the highly educated is to build a university. In Massachusetts, more than 90% of the electrical engineers received their highest degrees within the state. When construction of a loop highway, Route 128, opened scores of vacant acres within a 20-minute commute of MIT and Harvard University, hundreds of high-technology firms came to the corridor, including out-of-state giants like Sylvania, Honeywell, and Hewlett-Packard (Rosegrant & Lampe, 1992). A university, however, is not always the rock that attracts barnacles of high technology. In 1985, the Baltimore region had the lowest concentration of high-technology employment in the Baltimore-Washington corridor, although Johns Hopkins University was the largest recipient of federal research funds in 1989 (Feldman, 1994).

Rather than simply obtaining an intellectual enclave like a university, the city should cultivate a democracy of minds. Plato wanted to limit the city to 5,000 residents, the largest crowd that could hear an orator. Although the rapid dissemination of ideas is no longer constrained by a lungful of air, the principle remains the same: Every resident of the city should be able to understand every idea that shapes the city. A pet store employee should be able to contribute to the study of the city's downturns, and so should a PhD economist.

Start, then, with the public schools. One must doubt the prescription, standard in tax referenda, of larger doses of dollars. Eric Hanushek's (1986) survey of 147 statistical studies, focusing on U.S. public school systems from the late 1960s to the mid-1980s, concludes that a rise in spending per student on primary and secondary education barely affects test scores. Decreases in class size—and increases in teacher experience, pay, or education—also have little effect.

One can make too much of these results. Students can score high on a test without knowing much if the test is taught to them; indeed, teachers who lack education and experience are most likely to opt for this easy form of instruction. Still, note what happened over the 1970s and 1980s. Spending per pupil almost doubled after inflation, class sizes shrank by more than a fifth, and most teachers had picked up master's degrees (Chubb & Moe, 1990). Yet in 1991, one of every seven American youths had dropped out of high school; among Hispanic males, almost one of every two had dropped out (*Statistical Abstract*, 1993). Students were voting with their feet.

One clear conclusion, dating to Coleman (1966), is that family background strongly influences student achievement. Students in affluent, educated households score higher than other students, and fewer of them drop out. Education reform, then, should start at home.

The main problem is for families to convince youths that they can learn. Some psychologists' version of predestination—IQ theory—has become an article of faith for schools that practice tracking and urge youths to adjust to a life that few could want. The theory of "life adjustment" began its ascent in 1940, when a psychologist, Lewis Terman, concluded that 60% of American youths had IQs below 110. Led by vocational educators, theorists deduced that most youths were not smart enough for challenging courses, although they might be able to manage "industrial arts" (Hofstadter, 1963). That attitude persists: In 1990, two of every five teachers surveyed by the Carnegie Foundation for the Advancement of Teaching (1990) agreed that public schools could not really expect to graduate more than about 75% of all students.

The poor, learning in classrooms of their inadequacy, pass on their sense of resignation to their children. Jonathan Kozol (1985) quoted an illiterate parent:

I look at my 17-year-old son and my 12-year-old daughter and I want to help them with their homework, but I can't. My son was supposed to repeat the ninth grade for the third time this year. . . . He finally said he wanted to drop out. I see my handicap being passed on to my son. (p. 57)

Any program to make the city a factory of ideas must first dispel this despair. There is precedent. When the Civil War began, more than 90% of the black adults in the South could not read and write (Foner, 1988). But after the war, when an agent for the Freedmen's Bureau told 3,000 freed slaves in Mississippi that they were to have schools, "They fairly jumped and shouted in gladness" (p. 96). By 1870, blacks had spent more than $1 million on education. In Beaufort, North Carolina, 144 years after the town was settled, its first public schoolhouse ever was built by freedmen.

One way to improve education for youths today in poor urban households that have long been out of work is to give them greater freedom of choice: Convert entitlements, such as Medicaid, public housing, food stamps, and school lunches, to cash. Children will be happier and thus more willing to learn. In addition, the government can administer cash payments more cheaply than entitlements, so it can pass on savings to the household.

Converting entitlements to cash would go far toward closing the gap in family income between the best schools and the worst. Using data from the *Statistical Abstract of the United States* (U.S. Bureau of the Census, 1993), I have calculated that converting noncash benefits to cash in 1991 could have yielded $10,800 a year per poor household. Indeed, it is hard to think of another measure that could as quickly close a gap so large: Studying more than 500 high schools in the early 1980s, John Chubb and Terry Moe (1990) found that family income was 35% higher in the top fourth of schools (in terms of gains in achievement) than in the bottom fourth.

Setting aside income, how well a student does in school depends, of course, on how well prepared he or she is. Chubb and Moe found that if a student could somehow raise his or her 10th-grade test scores from the lowest fourth of all sophomores to the top fourth, then he or she would later gain the equivalent of two thirds of a school year.

Students prepare if they want to learn. To make them want to learn, the city can educate their parents for free. Its schools can offer—in evenings and weekends—small, vivid, and interactive classes in family relations, domestic violence, race and gender relations, personal finance and health, music, drama, criminology, current affairs, and neighborhood history. Such subjects have proven popular among uneducated households through TV dramas, films, and music. Transferring some of this popularity to the school can make it the focal point of the neighborhood, as the church was in the 19th century. Designing classes for the neighborhood will reconcile the resident to the local school, perhaps for the first time since the Progressives tried to stop corrupt contracting in big-city schools by replacing ward control with one board elected at large (Tyack, 1974).

The goal of classes for the neighborhood is not to train adults—or even primarily to inform them—but to encourage them to acquire a taste for ideas in hopes that they will pass it on to their children. A class that sketches the long decline of the neighborhood can then discuss whether history is made by individuals, institutions, or fate. By sating an intellectual hunger, the class will fulfill an emotional one as well: Sympathetic, stimulating discussion alleviates the self-absorption and depression of parents that make it hard for the children to keep their minds on their books. "Families don't talk to their children," said a principal in Hartford, Connecticut (Fiske, 1991). "We have kids who arrive at kindergarten unable to put together a sentence" (p. 226).

The main challenge is to design a class that will attract parents: With 98% of American households owning TVs, P.S. #195 will have to compete with Hollywood (U.S. Bureau of the Census, 1993). This will demand creative minds, but one can draw on the excess supply in the entertainment industry.

Neighborhood education can develop grassroots leaders whose vision is not just to stuff downtown with convention centers, stadiums, and casinos. "Leaders do the right thing," said Warren Bennis (1989). "Managers do things right" (p. 18). Robert Moses was a leader. At the Student Nonviolent Coordinating Committee in 1964, the shy high school teacher pursued ideals, not power, attracting 900 youths from comfortable lives to a Mississippi summer that would end in six murders (Dittmer, 1994; Weisbrot, 1991). A mayor like that could stir the city to risk income for the grandchildren. But where can the city find a Moses? Not in the pulpits: The churches have lost believers and the power to persuade. Not in the ivy-cloaked classrooms: The great universities train youths for office nights, not freedom summers. The city's best hope is to raise its own leaders by first creating a shared sense of morality, by pursuing programs in which residents teach one another. The city can design classes for the neighborhood; let parents help govern and plan policy for schools, as in "Comer schools" like Columbia Park Elementary in Landover, Maryland (Fiske, 1991); and revive the Vista program, with residents as volunteers. Teaching is an act of sharing. It is a small sacrifice that makes one willing, by degrees, to make larger sacrifices.

Can City Hall create private jobs? Sometimes. Possibly always. Urban growth is a prize in a lottery that can bankrupt losers. But there is a way, over time, to fix the lottery: The city can bribe fewer firms to come—and teach more adults to read and think. I have calculated on the back of an envelope that for $700 million, Tennessee could have offered small neighborhood classes to 1.37 million poor adults.

Above all, educate: That is a radical agenda for City Hall, which devotes only $1 of every $10 spent to education and libraries. Unfortunately, city leaders will find it hard to surrender the thrill of the chase, the neck-and-neck race for 400 headlines and 40 jobs, for the quiet incubation of ideas. Most likely, they will pleasantly preside over decline. C. P. Snow (1993) would compare them to the doges of the Venetian Republic, which descended over three centuries from controlling the Mediterranean to losing its own lagoon:

> They had become rich, as we did, by accident. They had acquired immense political skill, just as we have. A good many of them were tough-minded, realistic, patriotic men. They knew, just as clearly as we know, that the current of history had begun to flow against them. Many of them gave their minds to working out ways to keep going. It would have meant breaking the pattern into which they had crystallized. They were fond of the pattern, just as we are fond of ours. They never found the will to break it. (p. 40)

REFERENCES

Belcher, W. (1947). *The economic rivalry between St. Louis and Chicago 1850-1880*. New York: Columbia University Press.

Bennis, W. (1989). *Why leaders can't lead*. San Francisco: Jossey-Bass.

Bowden, L. (1965). Expectations for science. *New Scientist, 27*, 849-853.

Bright, A. (1949). *The electric-lamp industry*. New York: Macmillan.

Carnegie Foundation for the Advancement of Teaching. (1990). *The condition of teaching 1990*. Princeton, NJ: Carnegie Foundation.

Chinitz, B. (1961). Contrasts in agglomeration: New York and Pittsburgh. *American Economic Review, 51*, 279-289.

Chubb, J., & Moe, T. (1990). *Politics, markets and America's schools*. Washington, DC: Brookings Institution.

Coleman, J. (1966). *Equality of educational opportunity*. Washington, DC: Government Printing Office.

Denison, E. (1985). *Trends in American economic growth, 1929-1982*. Washington, DC: Brookings Institution.

Dittmer, J. (1994). *Local people: The struggle for civil rights in Mississippi*. Urbana: University of Illinois Press.

Feldman, M. (1994). The university and economic development: The case of Johns Hopkins University and Baltimore. *Economic Development Quarterly, 8*, 67-76.

Fisher, P. (1990). Connecticut's new product development corporation. In R. Bingham, E. Hill, & S. White (Eds.), *Financing economic development: An institutional response*. Newbury Park, CA: Sage.

Fiske, E. (1991). *Smart schools, smart kids: Why do some schools work?* New York: Simon & Schuster.

Foner, E. (1988). *Reconstruction: America's unfinished revolution, 1863-1877.* New York: Harper & Row.

Glaab, C., & Brown, A. (1983). *A history of urban America.* New York: Macmillan.

Grossman, G., & Helpman, E. (1994). Endogenous innovation in the theory of growth. *Journal of Economic Perspectives, 8,* 23-44.

Hanushek, E. (1986). Economics of schooling: Production and efficiency in the public schools. *Journal of Economic Literature, 24,* 1141-1177.

Herbers, J. (1990). A third wave of economic development. *Governing, 3*(9), 43-50.

Hofstadter, R. (1963). *Anti-intellectualism in American life.* New York: Knopf.

Hoyt, H. (1933). *One hundred years of land values in Chicago.* Chicago: University of Chicago Press.

Johannsen, R. (1973). *Stephen A. Douglas.* New York: Oxford University Press.

Kozol, J. (1985). *Illiterate America.* New York: Plume.

Leach, W. (1993). *Land of desire: Merchants, power and the rise of a new American culture.* New York: Vintage.

Lyne, J. (1992). State budgets bleed, but incentives still flow. *Site Selection, 37,* 868.

Machlup, F. (1962). *The production and distribution of knowledge in the United States.* Princeton, NJ: Princeton University Press.

Mahtesian, C. (1994). Escalation in the convention center war. *Governing 7*(10), 19-21.

Mantoux, P. (1983). *The industrial revolution in the eighteenth century.* Chicago: University of Chicago Press.

Markusen, A., Hall, P., & Glasmeier, A. (1986). *High tech America.* Boston: Allen & Unwin.

McKay, J. (1976). *Tramways and trolleys: The rise of urban mass transport in Europe.* Princeton, NJ: Princeton University Press.

Nietzsche, F. (1973). *Beyond good and evil.* New York: Penguin.

Pack, H. (1994). Endogenous growth theory: Intellectual appeal and empirical shortcomings. *Journal of Economic Perspectives, 8,* 55-72.

Passer, H. (1953). *The electrical manufacturers, 1875-1900.* Cambridge, MA: Harvard University Press.

Peltz, M., & Weiss, M. (1984). State and local government roles in industrial innovation. *Journal of the American Planning Association, 50,* 270-279.

Pred, A. (1966). *The spatial dynamics of United States urban industrial growth, 1800-1914.* Cambridge, MA: MIT Press.

Price, D. (1986). *Little science, big science—and beyond.* New York: Columbia University Press.

Rescher, N. (1978). *Scientific progress: A philosophical essay on the economics of research in natural science.* Pittsburgh, PA: University of Pittsburgh Press.

Romer, P. (1994). The origins of endogenous growth. *Journal of Economic Perspectives, 8,* 3-22.

Rosegrant, S., & Lampe, D. (1992). *Route 128: Lessons from Boston's high-tech community.* New York: Basic Books.

Schaeffer, K., & Sclar, E. (1980). *Access for all: Transportation and urban growth.* New York: Columbia University Press.

Schmookler, J. (1966). *Inventions and economic growth.* Cambridge, MA: Harvard University Press.

Smallwood, F. (1965). *Greater London: The politics of metropolitan reform.* Indianapolis: Bobbs-Merrill.

Smith, A. (1976). *An inquiry into the nature and causes of the wealth of nations.* Chicago: University of Chicago Press.

Snow, C. (1993). *The two cultures.* New York: Cambridge University Press.

Solow, R. (1957). Technical change and the aggregate production function. *Review of Economics and Statistics, 39,* 312-320.

Solow, R. (1994). Perspectives on growth theory. *Journal of Economic Perspectives, 8,* 45-54.

Survey: 91% of U.S. communities offer development incentives. (1993). *Site Selection, 33,* 294.

Taussig, F. (1989). *Inventors and money-makers.* New Brunswick, NJ: Transaction.

Thompson, W. (1962). Locational differences in inventive effort and their determinants. In Universities-National Bureau Committee for Economic Research (Ed.), *The rate and direction of inventive activity: Economic and social factors.* Princeton, NJ: Princeton University Press.

Tolley, G., & Krumm, R. (1983). On the regional labor-supply relationship. In R. Grieson (Ed.), *The urban economy and housing.* Lexington, MA: D. C. Heath.

Tyack, D. (1974). *The one best system: A history of American urban education.* Cambridge, MA: Harvard University Press.

U.S. Bureau of the Census. (1993). *Statistical abstract of the United States.* Washington, DC: Government Printing Office.

Weisbrot, R. (1991). *Freedom bound: A history of America's civil rights movement.* New York: Plume.

Zunz, O. (1982). *The changing face of inequality: Urbanization, industrial development, and immigrants in Detroit, 1880-1920.* Chicago: University of Chicago Press.

COMMENTS ON "CAN CITY HALL CREATE PRIVATE JOBS?"

William M. Bowen

This chapter advances the hypothesis that supply-side investment in human capital development causes regional job growth. Although there is likely to be a grain of truth to it, City Hall may create jobs in other ways not directly related to investments in human capital development. Moreover, supply-side human capital investments may not always be the most suitable job creation tactic for certain cities.

Urban and regional scientists posit several conceptually distinct possible causes of job creation (Leven, 1985). The most prominent comes from economic base theory. This theory suggests, in essence, that local basic jobs are created by increased demand from *outside* the region for goods and services produced *inside* the region. Accordingly, tax subsidies and financial incentives to attract, retain, and expand locally based export firms may create jobs by increasing the size and/or number of local export-oriented firms. Other possible causes of job creation suggested by conventional theory include (a) substituting local production for goods and services previously imported from outside the local jurisdiction (import substitution), (b) relaxing the regulatory burden in a local region to enhance the likelihood of a spatial price equilibrium, and (c) increasing the locale's aggregate production function by improving the marginal efficiency of private (nonhuman) capital investment. The latter may be accomplished through programs and processes designed to change strategically the region's industrial mix, to stimulate agglomeration economies, or to enhance the quality of the local infrastructure or environment.

Although supply-side investments in human capital development may have merit in certain situations, the rather bleak record of efforts to implement speculative theory in complex societal situations suggests that one might not want to commit to them on such a basis alone. A speculative theory may be considered as a tentative expression of what people have seen as a regular pattern in a multitude of facts. Though a good one is valuable insofar as it facilitates our ability to make predictions, without systematic reference to actual events even the best cannot, by itself, provide us with knowledge of reality. This is because speculative theory always requires the limitations inherent in its assumptions. Its ability to provide us with knowledge of reality depends on continual, systematic, and critical reference to empirical realities, in search of evidence regarding its assumptions. Absent such reference, a theory in effect arbitrarily halts at a particular set of assumptions.

Empirical research and evaluation are needed to answer the job creation question. This is because empirical research and evaluation place the highest possible premium on the amount of information and the relevance and fidelity of the system used to connect theoretical speculations with facts or observations regarding job creation. Moreover, in the case of the job creation question, this advantage is of paramount importance because the dynamic realities of locational differences in the geography of price within the free enterprise system are unfathomably complex. There are many different forms of intervention: tax abatements, tax increment financing, leasing arrangements, incubators, provision of venture capital, training, infrastructure, and parking, to name a few. Any time City Hall intervenes at a certain level with one or more of these, it does so at a particular time in a particular locale within the context of a particular set of demographic and workforce characteristics, a given capital-to-labor ratio in each of the local industries, and its own characteristic industrial mix and set of expected returns on investment in land, labor, and capital. These regionally specific factors are all combined within the context of the realities of growing human knowledge, newly emerging technologies, an increasing scale of economic activity, historically unprecedented modes of production, and entirely new corporate structures, all of which are evolving and combining to massively reformulate local conditions. Because there are practical limitations on our ability to continually reevaluate and revise our speculative theories against these constantly changing realities, all of them must make assumptions about what can be known with certainty. Empirical research and evaluation methods provide the most systematic, rational means available with which to test whether these assumptions comport with reality.

Though theoretical speculation and anecdote may have value, Bartik's (1991, pp. 81-108) extensive review of the myriad empirical studies—as well as his own empirical research—on the issue led him to conclude that state and local policies can indeed have a significant positive effect on local job growth. It is therefore a mistake to dismiss on the basis of speculative theory and anecdote alone the possibility that state and local tax and financial incentive policies may have a significant long-term beneficial effect on local job growth. When speculative theory conflicts with sound empirical research, it is time to revise the theory.

Even if the existing empirical research were not to thus bring the speculative theory to task, the supply-side human capital investment hypothesis would have theoretical defects. The most obvious stems from the fact that human capital is highly mobile. Because of this high mobility, a City Hall that implements this tactic in a relatively low-wage region is likely to be a loser. Those in whom the investments in human capital are made can be expected to migrate to relatively high-wage regions where they can get a greater return on the investment. Evaluation of the practical significance of this defect in any certain region would—again—require empirical research.

Empirical research and evaluation rather than theoretical speculation and anecdote are key to answering the job creation question because they alone enable policy makers to distinguish knowledgeably between types of programs that are successful and types that are not, to implement the successful ones, to avoid implementing the ones that fail, to learn from successes, and thereby to enhance the decision processes in City Hall where attempts are made to create local jobs. The main tasks in such evaluation are (a) to identify the specific goals and objectives of the program, (b) to develop conceptually quantitative and qualitative indicators of the degree to which the goals are accomplished, and (c) to select and implement appropriate research methods to operationalize the indicators and to measure the program's effects (James, 1991). It would be highly desirable to follow this approach at the state level as well (Bingham & Bowen, 1994).

On balance, it is clear that a healthy skepticism regarding the details of particular official claims is in order. First, the real issue is income, not jobs. If a job is created but does not provide enough income for the recipient to pay his or her bills, then neither the recipient nor the region will experience an improved quality of life. Second, official claims regarding the numbers of jobs created by particular programs undoubtedly lack credibility in certain cases (Hill, 1994). However, at the same time, it is fallacy to doubt the veracity of all of City Hall's claims to have created jobs on the basis of observed overstatements about having done so. What is true in particular cases is not necessarily true in general. Third, good speculative theory may at times facilitate prediction, but if it is to be convincing and of practical value in certain situations, it must be continually tested empirically and revised in light of changing realities. And in the case of the job creation question, it must consider regional differences. Finally, though the chapter's spirit of advocacy regarding regional investments in human capital development may have merit in certain situations, much further empirical research is needed before any general conclusions are reached. Certainly, the force of the existing empirical research suggests that other conventional forms of intervention may also have some lasting beneficial effect. For all anybody really knows, City Hall may well be able to create jobs in any number of ways. Ladd (1994) is probably dead on target when she says that "supply side incentives alone appear to be a costly and not very effective means of generating new jobs" (p. 208).

REFERENCES

Bartik, T. J. (1991). *Who benefits from state and local economic development policies?* Kalamazoo, MI: W. E. Upjohn Institute for Employment Research.

Bingham, R. D., & Bowen, W. M. (1994). The performance of state economic development programs: An impact evaluation. *Policy Studies Journal, 22,* 501-513.

Hill, E. W. (1994). *Tax abatement: War within a state.* Unpublished manuscript, Cleveland State University, Levin College of Urban Affairs.

James, F. J. (1991). The evaluation of enterprise zone programs. In R. E. Green (Ed.), *Enterprise zones: New directions in economic development* (pp. 225-240). Newbury Park, CA: Sage.

Ladd, H. F. (1994). Spatially targeted economic development strategies: Do they work? *Cityscape: A Journal of Policy Development and Research, 1,* 193-211.

Leven, C. L. (1985). Regional development analysis and policy. *Journal of Regional Science, 25,* 569-592.

REPLY TO WILLIAM M. BOWEN

Leon Taylor

In his fine response, William Bowen asserts that human capital mobility would obviate the attempt of a city, in a low-wage region, to boost its economy by investing in human capital. But in a study of 48 states between 1973 and 1980, state and local educational spending—as a share of state income—had important positive effects on total job growth as well as on job growth in two of six industries studied, retail trade and finance (Wasylenko & McGuire, 1985). Helms (1985) studied 48 states from 1965 to 1979 and accounted for factor mobility. The short-run rise in personal income (with standard error) due to an increase of $1 per $1,000 of income in property tax financing of local schools was .068% (.035); for financing by other taxes of higher education, the income increase was .109% (.052).

Perhaps educational spending garners area economic gains partly because poor households can be immobile. Dunn (1979) surveyed 200 textile workers who lost their jobs when the town's one large industrial employer closed. Multivariate regressions indicated that the average worker said he would take a cut of one seventh in weekly income to stay in town rather than move to a similar job. Studying poor renter households in Pittsburgh and Phoenix, Venti and Wise (1984) found that the average household would forego income equal to a 14% rise in utility rather than move. Transaction costs affected decisions to move much more in Pittsburgh than in Phoenix; maybe community value matters. Still, the more educated devoted more of their income to rent. Dunn did not control for education. As for the highly educated, multivariate regressions suggest that the number of engineers had a strong and statistically significant effect on the number of births of one-establishment firms among standard metropolitan statistical areas (SMSAs) in the late 1960s and early 1970s, particularly in communication-transmitting equipment (Carlton, 1979).

I do not know of much statistical evidence that tax breaks negotiated with individual firms boost jobs or income, net of the opportunity costs of those breaks.

Before modeling, one tells a story to create hypotheses in context. To tell a story is not to espouse policy; that requires empirical work.

REFERENCES

Carlton, D. W. (1979). Why new firms locate where they do: An econometric model. In W. Wheaton (Ed.), *Interregional movements and regional growth*. Washington, DC: Urban Institute.

Dunn, L. F. (1979). Measuring the value of community. *Journal of Urban Economics, 6*, 371-382.

Helms, L. J. (1985). The effect of state and local taxes on economic growth: A time series-cross section approach. *Review of Economics and Statistics, 67*, 574-582.

Venti, S. F., & Wise, D. A. (1984). Moving and housing expenditure: Transaction costs and disequilibrium. *Journal of Public Economics, 23*, 207-243.

Wasylenko, M., & McGuire, T. (1985). Jobs and taxes: The effect of business climate on states' employment growth rates. *National Tax Journal, 38*, 497-511.

9   Do Leadership Structures Matter in Local Economic Development?

LAURA A. REESE

■ Who Controls Local Economic Development and to What Ends?

Local economic development issues are often framed in the language of economics—tax incentives, business location theory, venture capital, revenue shortfalls, bond ratings, and the like. However, the question of whether leadership structures make a difference in local economic development, or in fact in any policy area related to urban affairs, is essentially political. To assess whether leadership "matters" is to address questions such as:

- Can local leaders affect development?
- How autonomous are leaders in determining local policy?
- Do local government actions affect economic development outcomes?
- Who benefits from local development policies?
- What are the goals of local development?
- How effective are local development policies in achieving goals?
- How are issues of public monies and private outcomes resolved?

All of these questions are inherently "political." They revolve around issues of control, agency, quality, and benefit, or in the words of Easton (1953), "the authoritative allocation of values" (p. 129).

Arguing that the root decisions about local economic development are inherently political begs the question regarding the relationship between "leadership" and "structure" in local government. Formal governing structures—mayor or manager, or various council arrangements and electoral systems—do not always determine leadership. Indeed, "leadership" in development may be provided by actors inside and outside the formal

224

organizational structure. In this sense, Stone's (1989) distinction between government and governing is well taken:

> All governmental authority in the United States is greatly limited—limited by the Constitution, limited perhaps even more by the nation's political tradition, and limited structurally by the autonomy of privately owned business enterprise. The exercise of public authority is thus never a simple matter; it is almost always enhanced by extraformal considerations. (p. 3)

These "extraformal" arrangements, made among business, community, and elected and appointed governmental leaders, provide the political and monetary resources necessary for the government to govern and ultimately to shape economic development policy.

Given this perspective, this chapter explores the independence of local actors in economic development, assesses the impact of differences in formal structure, and suggests points of decision making within the governing regime. It concludes with some speculation about the importance of leadership in securing positive economic development outcomes through the nexus of symbolism, political entrepreneurship, and organizational culture.

■ The Extent of "City Limits": Does Local Leadership Matter?

For leadership structures and arrangements to "matter" in local economic development, it must first be concluded that local units and leaders can, in fact, affect their environment. This matter has been the subject of considerable debate in recent years. One argument emphasizes the external constraints on cities, suggesting that public officials basically react to external pressures (see, e.g., Kantor, 1987; Katznelson, 1976; Molotch & Logan, 1985; Peterson, 1981). Such external forces include federal and state governmental policies, business location and production decisions, capital and labor flows, and even actions of other local governments. Assuming a common goal of expanded economic growth, economic development decision makers will presumably "shoot anything that flies, claim anything that falls" (Rubin, 1988). In an effort to respond to local economic interests, local governments pursue expansionary policies in a "growth machine" process (Molotch & Logan, 1985). In short, this dynamic results in a policy system in which cities attempt all techniques with which they are familiar or that have been employed by their competition in an effort to obtain and retain the vaunted economic progress (Bingham, 1976; Bowman, 1988; Pelissero & Fasenfest, 1989).

Another argument stresses the importance of local initiatives, essentially suggesting that "politics does matter" and that local officials "are the architects of their own responses to the structural constraints and changing conditions in which city politics is embedded" (Stone, 1987, p. 14). City officials are viewed not as passive reactors but as independent, though bounded, actors influencing local development (Jones & Bachelor, 1993; Stone, 1987). These independent decisions are based on a rational analysis of financial conditions, goals, and risks (Pagano & Bowman, 1989), although they are perhaps subject to bureaucratic and procedural limits (Jones, Greenberg, & Drew, 1980; Lineberry, 1977; Pagano & Bowman, 1989; Reese, 1993; Rubin, 1988, 1990). In short, "considerable slack exists in the social cables that bind the local polity and the local economy" (Jones & Bachelor, 1993, p. 253), leaving local leadership enough room to be agents of their own destinies even in cases of severe economic constraint.

The debate between city limits and city power has both macro and micro elements, with problems of language and units of analysis. The discussion of power highlights the macro discourse: Given the very real limits placed on cities, what is the extent of their power in relation to the environment, specifically business interests in the case of economic development? The roots of this debate can be traced to early community power studies that featured the opposing elitist and pluralist arguments of Hunter (1953) and Domhoff (1978) and Dahl (1961) and Polsby (1980) respectively.

Historically, pluralists posited that political conflict was organized by a variety of relatively equal interest groups competing before an independent government answerable to voter demands. Elitists argued that a resource-laden upper class drove decisions of elected bodies in their own interests. Both positions, however, reflect a variety of conceptual and empirical deficiencies. Elitism has always suffered from its paranoid flavor and its lack of attention to electoral processes and divergence in interest among elites. Pluralism, on the other hand, has never satisfactorily dealt with issues of limited representation within and resource variation among groups, and it strains the concept of government neutrality. Different worldviews were also influenced by methodological issues such as the locale and policy arena examined and the technique employed to identify "powerful" participants.

The "systemic power" contribution of Stone (1980) did much to further the understanding of community power structures. Stone reconciled elitist and pluralist approaches by acknowledging that business leaders have divergent interests and that governmental leaders have independent powers. However, he argued, the demands of interest groups and citizen coalitions and the need for monetary and political resources from the business community create a situation in which attention to corporate needs is more a function of environmental conditions than of overt control.

Currently, regime theory, arguing that coalitions of government officials and business interests form to carry out governmental policy or "social production," reigns supreme (Elkin, 1987; Stone, 1989). Peterson's (1981) critique notwithstanding, the prevailing doctrine suggests that politics does matter and that local governments can affect the larger environment.

The micro discourse raises a slightly different question, one whose answer can be either affirmative or negative within the doctrine of "politics matters." Specifically, can individuals affect the economic environment? In other words, does leadership matter? These issues were raised in early studies of "great leaders" and policy entrepreneurs (Eyestone, 1978; Mollenkopf, 1983). Subsequent literature has pointed to the influence of individual leaders on their environment and has identified the independent effects of leadership under conditions of fiscal stress and revenue raising (Clark & Ferguson, 1983; Levine, Rubin, & Wolohojian, 1981).

Local government officials are not completely constrained by business elites and can have some impact on their environment. Indeed, "different leaders with differing abilities will exploit the situation differently, and different outcomes will be produced" (Jones & Bachelor, 1993, p. 15). Thus, economic development policies, and ultimately the local economy, can be affected by individual local leaders. But although the literature suggests that individuals can have an impact, less guidance is provided regarding how individual leadership is exercised. It is to this issue that the following sections will speak.

■ The Impact of Structure and Decision-Making Arrangements

The Case for Unreformed Systems

Although Stone's (1989) point that government does not necessarily imply governing is well taken, research to date has indicated that formal structure can have an independent impact on the types of economic development policies and mechanisms proposed and adopted in a locality. The debate regarding policy outcomes of reformed versus unreformed political systems sets the tone for later research on the effects of structure on economic development, including the conflicting results. The preponderance of early research suggested that unreformed systems—mayoral, ward based, and partisan—were more responsive to organized group interests (Aiken, 1970; Grimes, Bonjean, Lyon, & Lineberry, 1976; Karnig, 1975; Lineberry & Fowler, 1967). On the other hand, some evidence was also presented that city managers, reacting to implicit career incentives, also increasingly responded to group pressures (Greenstone & Peterson, 1968; Northrop & Dutton, 1978).

Recent research on the specific relationship between governmental structure and economic development policy reflects similar disagreement. Unreformed systems, however, appear to be more active in the economic development arena and more responsive, particularly to business interests. For example, higher levels of politicization (Clingermeyer & Feiock, 1987) and electoral competition (Reese, 1991) appear to lead to increased activity in economic development, particularly in the use of tax incentives.

The role of a strong mayor in such systems seems to be of particular importance (Jones & Bachelor, 1993; Stone, 1989; Swanstrom, 1985). Cities with mayors evidence higher usage of policies that seek to address social equity concerns or guarantee a return on "investment" for the locality. These include linkage programs tying incentives to housing, for example, and performance guarantees that mandate hiring of local workers or guarantees on length of operations in the community (Clavel & Kleniewski, 1990; Drier, 1989; Elkins, 1993); marketing and land development incentives (Reese, 1992); and a wider array of different incentives (Feiock, 1989; Reese, 1992). This appears to reflect both electoral pressures and a greater institutional capacity to act.

Further, strong mayoral systems appear to be able to marshall resources for major development projects, keep order within the system, and bargain effectively with business interests (Jones & Bachelor, 1993; Schumaker, 1986). Indeed, Schneider and Teske (1993) found that a full-time mayor provides the necessary visibility and power base to attract persons willing to act as "progrowth entrepreneurs," although city managers can also fill that role.

That unreformed systems are more active in economic development is not surprising. The ward system and elected executive give interests—both business and citizen—greater control over both the council and the mayor. A more competitive electoral environment also stimulates government officials to be more attuned to constituent desires. These increased group pressures translate into more economic development policy activity for several reasons. First, with the exception of the growth-limitation movement in cities with the luxury of growth, citizens in most cities expect their elected officials to produce jobs and increase the tax base (Rubin & Zorn, 1985). Citizen and business perceptions of whether a city is "business friendly" can be strongly affected by policy actions on the part of officials (Goldstein, 1985; Ledebur & Hamilton, 1986; Schmenner, 1982). Further, although unreformed systems provide more pressure outlets for citizens, they do the same for business interests, prompting incentives as electoral rewards. Finally, the independent electoral base provides a mayor with greater leverage and authority in interacting with business interests and in ensuring that projects are carried through to completion. There is some disagreement in the literature, however. Mayoral systems have also been found to affect only tax abatements and urban development block grants (Feiock & Clingermeyer, 1986) or to have almost no effects (DeSantis, 1991). And reformed systems do not appear to differ significantly from unreformed systems in taxing and spending behavior (Welch & Bledsoe, 1988).

Organizational Arrangements

Other structural aspects besides mayoral leadership also appear to be important in economic development policy making. The organizational arrangement of the economic development function and the extent of bureaucratic professionalization seem significant. The existence of a free-standing economic development department and higher levels of professionalization (size and training) appear to stimulate economic development policy activity (Green & Fleischmann, 1989; Jones & Bachelor, 1993; Pelissero & Fasenfest, 1989; Rubin & Zorn, 1985) and to increase the likelihood that the community will have an economic development plan (Green & Fleischmann, 1989). Further, greater professionalization appears to reduce reliance on techniques with questionable efficacy, such as tax abatements (Reese, 1991), while stimulating the use of a wider variety of development incentives (Reese, 1992).

Finally, mechanisms for citizen input in the economic development decision-making process also appear to affect the techniques employed. Greater citizen input through advisory commissions, citizen surveys, and neighborhood commissions leads to more activity in economic development. This appears to be particularly true with the use of marketing and financial incentives (Reese, 1992) and in governmental activity in response to fiscal stress (Sharp & Bath, 1993). This supports suggestions in the literature that voters expect local officials to create an attractive environment for business development, thus explaining the effects of greater politicization in the system (Rubin & Zorn, 1985).

Politics Versus Bureaucracy

It seems clear that governmental structure and administrative arrangements make a difference in the types of economic development policies pursued. Another debate revolves around the relative impact of two institutional actors: elected officials and bureaucratic professionals. If one were looking for true leadership in economic development, where would it most likely be found—in the mayor's or city manager's office or in the office of the director of economic development? In a world where

Wilson's (1887) politics/administration dichotomy prevailed, the answer would be clear. The locus of policy making would rest with elected officials, whereas the job of implementation would fall to bureaucrats. Much of the literature to date seems to suggest that this is the case. Descriptions of economic development policy making in major cities from Detroit to Atlanta highlight the role that elected officials play in determining economic development policy (Jones & Bachelor, 1993; Stone, 1989). These cases, however, have several elements in common. They occur in large communities and involve a "big event": a major new auto plant, a riverfront market or aquarium, or a significant attraction such as the World's Fair. Such situations are often characterized by "peak bargaining," in which mayors are central in negotiations with the leaders of a small number of other major interests or "command posts" and "expensive capital projects and issues of current high salience" are involved (Jones & Bachelor, 1993, p. 53).

However, most economic development decisions are not "big events." Indeed, Jones and Bachelor (1993) suggested that "economic development is a prime candidate for peak bargaining, but only when major projects are contemplated. (Much of policy making in the economic-development sphere is as routinized as it is in other areas of policy)" (p. 241). Economic development policy activity is a combination of a few "big events" and many routinized decisions and day-to-day actions. And such routines, often controlled by professionals rather than elected officials, have been found to affect arenas from foreign policy (Allison, 1971) to urban service delivery (Jones et al., 1980).

Research has suggested that professional actors such as economic development directors are the most influential participants in economic development decision making (Reese, 1993). Indeed, a recent survey of economic development professionals, mayors, and city managers revealed that 69%, 33%, and 46%, respectively, agreed that economic development decisions were largely left to "the professionals" (Reese, 1993). One third of mayors also described themselves as most influential, suggesting perceptions of an equal division of power among mayors.

To the extent that professional actors provide leadership, if only for routine decisions, the nature of decision making in this sector becomes critical. If peak bargaining, coalition building, and regime maintenance characterize leadership in the electoral sphere, decision rules and standard operating procedures reign supreme in the administrative sphere. Personal attitudes, professional training, past experiences, and the actions of others appear to affect policies and techniques selected and promoted by professional actors (Pagano & Bowman, 1989; Reese, 1993; Rubin, 1988, 1990). Professional leadership appears to be influenced by turbulence in the environment, producing goal displacement—substituting what is being accomplished for what should be accomplished—and "shooting anything that flies"—using all possible activities and hoping that something will hit the target (Reese, 1993).

In this sense, "leadership" is replaced by "group-think." "The formal and informal arrangements that characterize the regime encourage a particular definition of a city's problems and tend to limit search processes for solutions to those problems" (Jones & Bachelor, 1993, p. 249). Thus, what has been done in the past strongly influences what is being done in the present. To the extent that such decision-rule behavior characterizes both bureaucratic decision making for routine events *and* regime decision making for "big events" (referred to as *solution sets* by Jones and Bachelor, 1993), any analysis of leadership must deal with this reality. This issue will be revisited later.

■ Leadership Structure and Development Outcomes

The research just discussed illustrates an unfortunate trend. Much research has been devoted to the relationship between structure and policy selection, but little has focused on linking structure to economic development outcomes. Very little is known about which structural arrangements are most successful in attracting and maintaining economic growth, and what research exists is largely suggestive rather than substantive. It has been suggested that mayors have an advantage over city managers in their ability to bargain from a position of power. This reflects their citywide electoral base and the ability to develop and maintain the necessary regime coalitions to ensure stability and coherence in the process (Jones & Bachelor, 1993; Schumaker, 1986). Research also indicates that business location decisions may be affected by the extent to which elected officials can ensure regularized, predictable, and speedy interaction between government and business (Reese, in press; Waters, 1990). Beyond this, however, the literature is essentially silent on the relationship between structure or leadership and economic development outcomes.

This limitation is endemic to the policy evaluation literature as a whole. To say that leadership "matters" leads directly to that *other* question, "Matters to what?" The "what" can be any number of things: policy choice, policy goals, resource allocation, outputs, outcomes, or the distribution of benefits. The first of these, policy choice, has been fairly well covered in the literature noted above. A relationship between leadership structure and resource allocation could also be implied in unreformed systems. Because of their greater use of tax abatements, marketing activities, and land development incentives, mayoral or unreformed systems should reflect higher spending levels for economic development. Although this has not been fully examined in the literature, my analysis of International City/County Management Association survey data indicates that cities with manager/council systems do tend to have smaller budgets for development. But the relationship is not statistically significant and may well be an artifact of community size or other factors.

In a similar vein, analysis also indicates that mayor and manager cities are significantly different in the types of goals pursued in economic development. Specifically, more reformed cities are likely to give a higher priority to the attraction of new business and downtown development as policy goals. On the other hand, the only difference in perceived goal attainment is in neighborhood commercial development, in which respondents in manager cities feel significantly more successful.

None of this speaks to actual goal attainment, however, as is not surprising given the literature generally. For practitioners and academics alike, local development policy is formulated and discussed with little agreement about or consideration of how the locale is defined, what the terms *economic* and *development* really mean, and to which of the community's competing interests policy is to be responsive (Cox, 1993). Methodologies for assessing goal attainment or outcome are so problematic that although programmatic efforts are well examined, their impact and effectiveness remain unclear. Clarke and Gaile (1992) noted that "attempting to assess the effects of local economic development strategies is a quagmire of good intentions and bad measures" (p. 193). Further, disagreement on whether the ultimate goals should be traditional considerations such as performance in job creation and tax base growth or other values such as equity or quality of life (i.e., the social economy) complicates the issue (Reese & Fasenfest, 1993).

Mier and Giloth (1993) stressed that evaluative criteria for leadership involve knowledge of past and alternative present options and circumstances; communication style; use of policy examples, "metaphors," or stories about development experiences; and use of data and advanced technology to convey information. Thus, the definition of "effective" economic development becomes the ability of public officials to promote "cooperative leadership" that stresses inclusion over authority. In short, although there is very little literature providing empirical evaluation of economic development techniques, and none explicitly relating structural or leadership issues to programmatic "success," recent voices have called for more explicit attention to definitions of effectiveness that include social equity and representation considerations.

■ Leadership, Symbolism, and Organizational Change

Unresolved Issues

Several critical issues regarding the relationships between structure, leadership, and economic development remain unresolved. As previously noted, significant deficits in the literature include the definition and assessment of outcome or policy "success"; the impacts, if any, of structural or leadership differences on "success"; and the power relationships between political and bureaucratic actors. Further, several shortcomings can be identified even for areas that are well covered. These tend to revolve around methodological issues and are summarized below.

Measures of "structure" too often reflect measurement convenience rather than substantive distinctions. Operationalizing government structure as mayoral or manager systems, for example, ignores important differences in the political influence exercised by various managers or mayors. In what is putatively a reformed system, a manager with 20 years' experience in a community may nevertheless be as much a political "boss" as any strong mayor and may have substantial electoral support, though unexpressed at the polls. On the other hand, a mayor facing a council composed of members of the opposition party may have limited ability to secure passage of a policy agenda.

Indicators used to describe policy choice are also based primarily on convenience. Variables such as fiscal health, economic growth, bureaucratic professionalism, electoral competition, and politicization have all been measured in different ways. This leads to uncertainty about conflicting findings: Are they the result of actual empirical differences or an artifact of measurement technique?

Findings regarding economic development policy selection or usage present a picture of surface variety but little information regarding effort. In other words, most studies identify the techniques used in cities but provide little sense of how extensively or with what frequency each is employed. There may be significant differences between a city that has used a tax abatement once and one that routinely relies on them.

Research on structure and leadership in economic development has tended to treat all development policies and mechanisms as if they were the same. Indeed, regime theory posits the same governing arrangements across all policy areas. However, it is well known that the different findings of elitist and pluralist schools had much to do with the fact that different policies were examined; research on apples and oranges found two different fruits. The literature on structure presented here suggests that mayor and manager cities differ in their use of some economic development incentives but not others. Leadership issues may well have differential impacts depending on the type of development technique explored; demand, supply, and redistributive policies have very different natures and should be examined separately.

The literature has also tended to focus on the influence of individual actors, with less attention devoted to the interaction of participants. The "policy determinants" literature, due in part to the requirements of the correlation and regression analysis typically employed, treats each structural element as a discrete entity. And although regime analysis by definition focuses on the interaction of various interests—governmental, business, citizen—the treatment tends to be too broad to include the more intricate relationships within groups, specifically between elected and appointed governmental officials.

Given these failings, what would constitute a viable and comprehensive methodology for evaluating the impact of leadership structures on economic development? Bartik (1994), in a call for economic development policy evaluation, suggested the following criteria for analysis: Be of high quality, focus on programs/policies directed toward "intensive intervention," focus on a wide range of programs and possible outcomes over time, and be conducted by professionally credible organizations or actors (pp. 101-102). This suggests that what is needed is a large-scale meta-analysis that would provide comparative, quantifiable information on a number of efforts across the country. Mier and Bingham (1993), on the other hand, called for the use of metaphors and stories to understand the economic development ramifications of leadership and effect. Such efforts would produce in-depth, intuitively understandable examples and lessons based on practice and experience in a small number of locations.

This, then, frames the spectrum of methodological options in economic development research: large-scale, high-tech evaluations that would provide a wealth of comparative information with little depth or sensitivity to locale, or experience-rich, detailed stories with limited generalizability. Both options suffer from the defects of their advantages. What is needed, obviously, is a synthesis of these extremes. The Fiscal Austerity and Urban Innovation Project (FAUI) methodology comes to mind as a viable amalgam, though focused on an international rather than a state level. Researchers could obtain both quantitative macro and qualitative micro data on local leadership issues as they relate to economic development within a state. This would allow the accumulation of case study information and the development of metaphors that would be readily understood by practitioners and academics alike. The compilation of such state analyses would then allow for the large-scale analysis envisioned by Bartik (1994). In this way, practitioners would benefit from policy and outcome data across the country while also including the detailed information necessary for emulation or avoidance.

■ Do Leadership Structures Matter?
Some Speculation

Several conclusions have been drawn thus far. Local leadership matters within the very real constraints posed by the economic system and other internal and external actors. Although unreformed structures seem to strengthen the hand of elected officials in negotiation and interaction with business, there is ample room for individual leadership. In the words of Jones and Bachelor (1993), "Because exceptional, non-routine actions on the part of leaders can affect the general circumstances, the course of events can be changed by acts of leadership" (p. 15). Thus, both structural and individual factors determine the extent of control possible and the leadership exercised in economic development.

But how does an individual have an impact in a locality run by a regime with policies implemented by professional bureaucrats? The literature has highlighted the operative roles of each of these institutions. Governing regimes, typically initiated by elected officials and composed of primarily business or other "resourced" interests (at least for corporate regimes), identify problems, define solutions, and marshall resources to carry out necessary social production functions. However, according to Jones and Bachelor (1993), each regime conducts this process within the confines of "solution sets": systems of problem definitions and preferred solutions developed and employed by the regime in the past and changed only incrementally over time. Thus, in the collective memory and conscious-ness of the regime, economic development goals become defined in predictable ways, with fairly uniform techniques or incentives applied. Such solution sets regularize patterns of interaction among regime par-ticipants and reduce the need for constant control of the process by elites.

At the same time, to the extent that decision making is left to profes-sionals—as is particularly the case for routine matters—further static forces operate. When the regime is functioning as it should, and conflict over development issues is controlled, elected officials are even more prone to leave decisions to the "advice of experts" (Baumgartner, 1989). The literature has suggested certain decision-making procedures here as well. Professionals, particularly in the face of environmental uncertainty, tend to use decision rules, the standard operating procedures, or profes-sionally based criteria for decision making. These too are heavily influ-enced not only by professional training but by past action.

The solution sets employed by regimes and the decision rules used by professionals create a dual-edged sword for leaders trying to exert inde-pendence. This is particularly true under conditions of uncertainty be-cause their dynamics often flow in opposing directions. Indeed, this defines the difference between solution sets and decision rules.

For the regime, environmental uncertainty or change is likely to open the door for a "regime shift." New actors may enter the process, new solution sets are defined, and a greater role for individual leadership appears possible. As Baumgartner (1989) suggested, "Shifting situations leave more room for individual leadership than do well defined and objectively understood ones" (p. 114). Along similar lines, Jones and Bachelor (1993) noted "the possibility of radical regime change based on new ideas that activate new participants" as a result of new actors entering the system or stress in the environment (p. 15). Such periods are ripe for leadership as prior problem definitions and solutions are challenged.

Even a casual observation of contemporary politics in the city of Detroit, for example, suggests such dynamics. Though it is possible for mayors to change without affecting the regime system, in Detroit, which has mayor-centered machine politics (DiGeatano, 1988), a change in executive after 20 years might well stimulate a major regime shift. Though environmental forces may necessitate Mayor Dennis Archer's maintenance of former Mayor Young's corporate regime, the transition may also herald new possibilities for leadership.

However, a countervailing force remains at work within the professional bureaucracy. Turbulence and uncertainty may promote regime shift and a concomitant redefinition of the solution set, thus opening the door for individual leadership. The same forces, however, are likely to increase bureaucratic reliance on standard operating procedures. Because decision rules serve to reduce uncertainty and provide guidelines for operating in a changing and often hostile environment (at least as perceived by bureaucrats), just as the door opens for leadership within the regime, the professional bureaucracy is trying desperately to keep it shut. The challenge to leadership is then compounded: providing guidance and motivation within the regime while also affecting entrenched bureaucratic behaviors. Thus, while Mayor Archer may be poised to guide the city of Detroit on new paths to development, the economic development professionals may be least willing or able to follow.

■ Symbolic Politics Revisited

What do the 1990s and beyond hold in store for leadership in economic development? Will "cooperative leadership" models lead to an economic development agenda that emphasizes social equity, minority representation, community development, citizen participation, and cooperation among units of government? Or will continuing economic stress, political distaste for cities and their residents, and increasing concentration of capital lead to ever-increasing levels of reliance on standardized solutions and a "leadership" that embraces corporatism out of abject need? The outcome that occurs in most central cities will depend not only on the vagaries of economic fortune but also on political decision making. For example, state policies regulating local economic development incentives, thus limiting intercity competition, and ultimately a federal urban and/or industrial policy could tip the scales toward more desirable outcomes. In the absence of this, local leadership will bear the burden of trying to affect future scenarios, a situation that necessitates a regeneration of local actors, organizations, and interests.

If it is true that "politics is talk" (Bell, 1975) and that symbols are the main form of communication (Elder & Cobb, 1983), then Baumgartner (1989) is correct in arguing that "leadership is often a matter of rhetoric" (p. 115). This is particularly true for leadership within the context of economic development. Indeed, Jones and Bachelor (1993) noted that the return on a city's investment in economic development is in the "symbolic perception of the city as a success story, a place in which things get done efficiently and effectively. This adds to the credibility of officials in their dealings with other corporations, and presumably it adds to the well-being of the city in the long run" (p. 243). The challenge for leadership in economic development, then, is to direct policy making and build consensus using the symbols through which issues are framed.

Thus, Mayor Archer, or any elected executive, must find the means to affect organizational change both within the regime and within its constituent bureaucracies. To date, the use of symbols by the new mayor in Detroit has been extensive: longer workdays, more public visibility, greater concern with neighborhood services, frequent public forums, more interaction with city council, visible interaction with other units of government in the area, even an emphasis on family. Bureaucratic personnel have been instructed to respond to all citizen contacts promptly and with consideration. A symbolic new day has dawned as the new mayor seeks to exert his influence on both routine and nonroutine decisions.

Whether this leads to a change in regime and alters the prevailing bureaucratic culture in Detroit remains to be seen. However, "one can always envision a different pattern of interactions leading to different solution-sets and thereby different policy actions. So long as this can happen, a somewhat different urban political world is possible, and there remains a real and important role for public leadership" (Jones & Bachelor, 1993, p. xii). Leaders can manipulate the organization of decision-making institutions—the structure of public/private authorities, citizen avenues for input, legislative environment—to reduce conflict "either by negotiating actual changes in the initiatives or through symbolic manipulation of perceptions" (Sharp & Bath, 1993, p. 223). Reduction in conflict over economic development can then pave the way for addressing both business and community needs: the central goal of cooperative leadership. At the confluence of external boundaries, regime characteristics, governmental structure, and bureaucratic imperatives, individual leadership can make a difference through symbolism, environmental opportunities, and the skillful use of public resources.

REFERENCES

Aiken, M. (1970). The distribution of community power: Structural bases and social consequences. In M. Aiken & P. Mott (Eds.), *The structure of community power* (pp. 487-525). New York: Random House.

Allison, G. T. (1971). *Essence of decision: Explaining the Cuban missile crisis.* Boston: Little, Brown.

Bartik, T. J. (1994). Better evaluation is needed for economic development programs to thrive. *Economic Development Quarterly, 8,* 99-106.

Baumgartner, F. R. (1989). Strategies of political leadership in diverse settings. In B. D. Jones (Ed.), *Leadership and politics* (pp. 114-134). Lawrence: University Press of Kansas.

Bell, D. (1975). *Power, influence and authority.* New York: Oxford University Press.

Bingham, R. A. (1976). *The adoption of innovation by local government.* Lexington, MA: D. C. Heath.

Bowman, A. M. (1988). Competition for economic development among southeastern cities. *Urban Affairs Quarterly, 4,* 511-527.

Clark, T. N., & Ferguson, L. C. (1983). *City money.* New York: Columbia University Press.

Clarke, S. E., & Gaile, G. L. (1992). The next wave: Postfederal local economic development strategies. *Economic Development Quarterly, 6,* 187-198.

Clavel, P., & Kleniewski, N. (1990). Space for progressive local policy: Examples from the United States and the United Kingdom. In J. R. Logan & T. Swanstrom (Eds.), *Beyond city limits: Urban policy and economic restructuring in comparative perspective* (pp. 199-234). Philadelphia: Temple University Press.

Clingermeyer, J., & Feiock, R. (1987, September). *Municipal representation, interest group activity and economic development policy choice.* Paper presented at the annual meeting of the American Political Science Association, Chicago.

Cox, K. (1993). The concept of local economic development policy: Some fundamental questions. In D. Fasenfest (Ed.), *Community economic development: Policy formation in the U.S. and U.K.* (pp. 45-64). New York: St. Martin's.

Dahl, R. A. (1961). *Who governs?* New Haven, CT: Yale University Press.

DeSantis, V. (1991, April). *The effect of political structures on local economic development.* Paper presented at the annual meeting of the Midwest Political Science Association, Chicago.

DiGeatano, A. (1988, March). *Machine politics in the post-industrial era.* Paper presented at the annual meeting of the Urban Affairs Association, St. Louis, MO.

Domhoff, G. W. (1978). *Who rules America now?* Englewood Cliffs, NJ: Prentice Hall.

Drier, P. (1989). Economic growth and economic justice in Boston: Populist housing and jobs policies. In G. Squires (Ed.), *Unequal partnerships: The political economy of urban redevelopment in postwar America* (pp. 35-58). New Brunswick, NJ: Rutgers University Press.

Easton, D. (1953). *The political system.* New York: Knopf.

Elder, C. D., & Cobb, R. W. (1983). *The political use of symbols.* New York: Longman.

Elkin, S. (1987). *City and regime in the American republic.* Chicago: University of Chicago Press.

Elkins, D. (1993, April). *Testing competing explanations for the adoption of managed growth (type II) policies.* Paper presented at the annual meeting of the Midwest Political Science Association, Chicago.

Eyestone, R. (1978). *From social issues to public policy.* New York: John Wiley.

Feiock, R. C. (1989). The adoption of economic development policies by state and local governments: A review. *Economic Development Quarterly, 3,* 266-270.

Feiock, R. C., & Clingermeyer, J. (1986). Municipal representation, executive power, and economic development policy activity. *Policy Studies Journal, 15,* 211-229.

Goldstein, M. L. (1985). Choosing the right site. *Industry Week, 15,* 57-60.

Green, G. P., & Fleischmann, A. (1989). Analyzing local strategies for promoting economic development. *Policy Studies Journal, 17,* 557-573.

Greenstone, J. D., & Peterson, P. E. (1968). Reformers, machines and the war on poverty. In J. Q. Wilson (Ed.), *City politics and public policy* (pp. 267-292). New York: John Wiley.

Grimes, M. D., Bonjean, C. M., Lyon, J. L., & Lineberry, R. L. (1976). Community structure and leadership arrangements: A multi-dimensional analysis. *American Sociological Review, 41,* 706-725.

Hunter, F. (1953). *Community power structure: A study of decision makers.* Chapel Hill: University of North Carolina Press.

Jones, B. D., & Bachelor, L. W. (1993). *The sustaining hand.* Lawrence: University Press of Kansas.

Jones, B. D., Greenberg, S., & Drew, J. (1980). *Service delivery in the city: Citizen demand and bureaucratic rules.* New York: Longman.

Kantor, P. (1987). The dependent city: The changing political economy of urban development in the United States. *Urban Affairs Quarterly, 22,* 493-520.

Karnig, A. K. (1975). Private-regarding policy, civil rights groups, and the mediating impact of municipal reforms. *American Journal of Political Science, 19,* 91-106.

Katznelson, I. (1976). The crisis of the capitalistic city: Urban politics and social control. In W. D. Hawley (Ed.), *Theoretical perspectives on urban politics* (pp. 214-290). Englewood Cliffs, NJ: Prentice Hall.

Ledebur, L. C., & Hamilton, W. H. (1986). The great tax-break sweepstakes. *State Legislatures,* pp. 12-15.

Levine, C. H., Rubin, I., & Wolohojian, G. C. (1981). *Politics of retrenchment.* Beverly Hills, CA: Sage.

Lineberry, R. L. (1977). *Equality and urban politics.* Beverly Hills, CA: Sage.

Lineberry, R. L., & Fowler, E. P. (1967). Reformism and public policies in American cities. *American Political Science Review, 61,* 701-716.

Mier, R., & Bingham, R. D. (1993). Metaphors of economic development. In R. D. Bingham & R. Mier (Eds.), *Theories of local economic development* (pp. 284-304). Newbury Park, CA: Sage.

Mier, R., & Giloth, R. P. (1993). Cooperative leadership for community problem solving. In R. Mier (Ed.), *Social justice and local development policy* (pp. 165-181). Newbury Park, CA: Sage.

Mollenkopf, J. (1983). *The contested city.* Princeton, NJ: Princeton University Press.

Molotch, H., & Logan, J. R. (1985). Urban dependencies: New forms of use and exchange in U.S. cities. *Urban Affairs Quarterly, 21,* 143-170.

Northrop, A., & Dutton, N. (1978). Municipal reform and group influence. *American Journal of Political Science, 22,* 691-711.

Pagano, M. A., & Bowman, A. M. (1989, September). *Risk assumption and aversion: City government investment in development.* Paper presented at the annual meeting of the American Political Science Association, Atlanta.

Pelissero, J. P., & Fasenfest, D. (1989). Suburban economic development policy. *Economic Development Quarterly, 3,* 301-311.

Peterson, P. E. (1981). *City limits.* Chicago: University of Chicago Press.

Polsby, N. (1980). *Community power and political theory* (2nd ed.). New Haven, CT: Yale University Press.

Reese, L. A. (1991). Municipal fiscal health and tax abatement policy. *Economic Development Quarterly, 5,* 24-32.

Reese, L. A. (1992). Explaining the extent of local economic development activity: Evidence from Canadian cities. *Government and Policy, 10,* 105-120.

Reese, L. A. (1993). Decision rules in local economic development. *Urban Affairs Quarterly, 28,* 501-513.

Reese, L. A. (in press). Modeling economic development decision-making: The case of tax abatements. *Policy Studies Review.*

Reese, L. A., & Fasenfest, D. (1993, April). *Measuring the effects of local economic development policy: Considering social outcomes.* Paper presented at the annual meeting of the Urban Affairs Association, Indianapolis.

Rubin, B. M., & Zorn, C. K. (1985). Sensible state and local economic development. *Public Administration Review, 45,* 333-340.

Rubin, H. J. (1988). Shoot anything that flies, claim anything that falls: Conversations with economic development practitioners. *Economic Development Quarterly, 3,* 236-251.

Rubin, H. J. (1990). Working in a turbulent environment: Perspectives of economic development practitioners. *Economic Development Quarterly, 4,* 113-127.

Schmenner, R. W. (1982). *Making business location decisions.* Englewood Cliffs, NJ: Prentice Hall.

Schneider, M., & Teske, P. (1993). The progrowth entrepreneur in local government. *Urban Affairs Quarterly, 29,* 316-327.

Schumaker, P. (1986). Economic development policy and community conflict: A comparative issues approach. In T. Clark (Ed.), *Research in urban policy* (pp. 25-46). Greenwich, CT: JAI.

Sharp, E. B., & Bath, M. G. (1993). Citizenship and economic development. In R. D. Bingham & R. Mier (Eds.), *Theories of local economic development* (pp. 213-231). Newbury Park, CA: Sage.

Stone, C. (1980). Systemic power and community decision-making: A restatement of stratificationist theory. *American Political Science Review, 74,* 978-990.

Stone, C. (1987). The study of the politics of urban development. In C. N. Stone & H. T. Sanders (Eds.), *The politics of urban development* (pp. 3-22). Lawrence: University of Kansas Press.

Stone, C. (1989). *Regime politics: Governing Atlanta, 1946-1988.* Lawrence: University of Kansas Press.

Swanstrom, T. S. (1985). *The crisis of growth politics.* Philadelphia: Temple University Press.

Waters, G. A. (1990, January). *Insights from the economic developer.* Paper presented at the annual meeting of the Michigan City Management Association, Grand Rapids, MI.

Welch, S., & Bledsoe, T. (1988). *Urban reform and its consequences.* Chicago: University of Chicago Press.

Wilson, W. (1887). The study of administration. *Political Science Quarterly, 2,* 197-222.

LEADERSHIP AND ECONOMIC
DEVELOPMENT POLICY: A COMMENTARY

Robert A. Beauregard

No single research methodology can fully answer the question of whether, and if so how, leadership influences economic development. As Laura Reese rightly points out, our choices range from large-sample studies that use quantitative data to document the patterning of relationships across political jurisdictions to case studies that expose in detail the personalities and processes involved. Each, and the variations in between, including longitudinal studies, answers a different question about leadership.

Consequently, we should focus our energies on how to conceptualize the issue. Recognizing this, Reese spends the greater part of her chapter making distinctions. These distinctions constitute a framework of concepts and relationships for generating research questions.

To investigate the relationship between leadership (the independent concept) and economic development (the dependent concept), we must be as precise as possible as to the meanings of each. On the dependent side of the equation, economic development policies have to be distinguished from their outcomes: that is, their material and symbolic impacts. Policies also need to be further divided. Reese talks about peak events (e.g., pursuit of a professional sports team) versus routinized programs and also notes the substantive diversity of economic development policies. Because most states and localities run multiple economic development programs concurrently and because program effects are interactive, the mix of policies becomes significant.

On the independent-variable side of the equation, the key distinction is between leadership and "structures." The former refers to specific positions—elected officials, bureaucrats, policy experts, corporate elites—and the personalities who occupy them. The latter refers to the institutionalized responsibilities and resources expressed in the type of government (e.g., city manager) and in government-constituency relationships (e.g., regime coalitions or public-private partnerships).

This leadership-structure distinction is central to contemporary questions about leadership and economic development. The current debate is primarily between those who view governments as severely constrained by the institutional strictures of a capitalist democracy and consequently as unable to pursue redistributive policies and those who believe that citizens can influence economic development through political structures that are responsive to their (and not capital's) needs and desires. Thus, research into the role of leadership rubs against the grain of a structural and economistic (but also institutional) argument. Note, too, that politics is the elusive independent variable, not leadership, the agency of political action.

Given this perspective, let me suggest ways to frame the research. First, we should leave to evaluation research the assessment of the efficacy of policy. Policy outcomes are susceptible to extralocal forces beyond the influence of local actions, and although leadership should indicate what outcomes are expected, determining the effectiveness of a policy is a separate task. Because public officials are responsible for public bureaucracies, however, implementation has to be addressed.

Second, we need to add to our framework the socioeconomic conditions to which policy is a rational response. The major function of leadership is to orchestrate a policy response that makes sense in terms of the perceived threats and opportunities facing the locality. This includes an accounting of the jurisdiction's institutional strengths and weaknesses.

Third, leadership itself needs to be problematized. Its meaning is not obvious. Reese talks about leadership structures (which in this debate seems an oxymoron) and also notes the importance of specific personalities. In turn, she recognizes that professionals as well as elected officials might lead economic development policy making, but she seems less open to the possibility that leaders could come from the business sector or the larger community.

Moreover, leaders can lead in a variety of ways. We expect them to define issues and provide direction, explain situations, create opportunities for others to exercise initiative, mobilize resources and groups, organize responses, and (most important) perform: that is, achieve their proclaimed goals. Not all of these activities are equally important to us, and few leaders are capable on all of these dimensions. Of particular concern is the extent to which the contemporary fascination with "image" distorts our understanding of leadership performance.

These distinctions suggest a specific question that can be used to investigate the relationship between leadership and economic development (it is only one of a number of questions that could be posed): Within institutional constraints, can political leadership mobilize resources and accommodate constituents so as to (a) generate an appropriate match between economic development policy and socioeconomic conditions and (b) ensure that the policies are properly implemented? Is this different from what would have occurred in the absence of that leadership?

Research into this question, first, controls for institutional structures and resource constraints. Subsequently, it verifies that leaders actually bring together people, organizations, and resources and focus their attention on a specific set of economic development policies that are congruent within current socioeconomic conditions, local institutional capacities, and the position of the locality in the larger political economy.

To conclude that leadership makes a difference, we must also show that the resultant correspondence of policies to conditions would not have occurred *but for* the intercession of these leaders. If we cannot show this, then we cannot claim that leadership matters.

Leaders might have a distorted understanding of the conditions to be addressed or might simply opt for the most feasible or currently most popular policies. In either case, their influence is wasted. Leaders will draw on their authority not only to massage public perception but also to give direction to governmental actions. If this direction leads to progrowth policies, it is no less leadership; the existence and efficacy of leadership are not contingent on whether redistributive policies are pursued.

Finally, the research has to address implementation. Leadership is thwarted to the extent that bureaucratic structures fail to manage policies properly. Making sure this does not happen is the responsibility of elected officials.

Clearly, even this relatively precise question poses theoretical problems and formidable research challenges. A research design that tests the proposition that leadership matters must distinguish leadership not only from structure but also from politics and must then compare the actions of leaders with the policies that ensue when leaders are absent. A design lacking these components will not answer the leadership question.

RESPONSE TO COMMENTARY

Laura A. Reese

The research agenda described by Robert Beauregard is important, ambitious, and indeed somewhat daunting. This is not to suggest that it should not be pursued, of course. Examining the connection between development policy and socioeconomic conditions as a measure of leadership raises a number of interesting issues. What policies are appropriate to what conditions? Given that the "conditions to be addressed" are politically defined and lack objective agreement, how is a researcher to create a measure to assess this? To the extent that researchers currently do not agree or are not certain about what policies lead to what results, this task clearly pushes the envelope of current research methodologies and theories. The call for study of what would have happened in the absence of an act of leadership raises similar challenges. Is there such a condition as an "absence of leadership?" Is this different from "lack of effective leadership?" How does one, within the context of a research design, create a counterfactual situation representing a lack of leadership? Again, from a methodological standpoint, these are ambitious questions.

Finally, regarding future research agendas, I would not be as quick to leave the assessment of the efficacy of policy to "evaluation research." It seems to me that whether certain types of leadership styles or leadership configurations or regimes lead to more effective policies is a topic pivotal to research on leadership in economic development. Including business-sector or community leadership in these analyses, as Beauregard notes, is essential. I did not mean to exclude them. The interaction of leaders from different roles—governmental, community, elected, appointed, private sector, and so on—is another area ripe for research. Do different actors have different leadership styles? Do different configurations of leadership produce different policies?

I do have some concerns regarding the distinction Beauregard makes between leadership and structure. Although the distinction itself is important in clarifying both research objectives and real-world dynamics, I would argue that the terms do not differ between specific positions and institutionalized responsibilities, as he suggests. Instead, a more useful conceptual distinction might be as follows: *Structure* refers to the formal government organization as outlined in a city charter (mayor versus manager, at-large versus ward elections, and so on), whereas *leadership* refers to the actions, activities, personalities, and interactions between individuals and groups that ultimately affect policies and programs. Indeed, much research in organizational theory identifies structure as the positions individuals hold and the authority that comes with them. Leadership, on the other hand, is the exercise of influence. Hence, the test of effective leadership is the ability to marshal all the resources and forces necessary to meet organizational goals. Or, to use a metaphor common in that literature, we see the leader as the "conductor" bringing the musicians together to create the symphony.

Further, I am not convinced that "politics" is the ultimate independent variable of interest, as opposed to leadership as defined above. Nor is it clear how Beauregard is using the term. For example, is he suggesting that politics causes leadership? Is politics an intervening variable that needs to be taken into account when assessing the types or success of leadership? In other words, is it an environmental factor—the *challenge* that leaders have to address? Or, because leadership is the ability of the leader to influence the policy process, is leadership politics?

These issues are vital in thinking about future research. Defining and clarifying variables is essential to understanding leadership and its presumed policy effects. In this, Beauregard and I are in complete agreement. And, as the dialogue here highlights, the task of variable definition and operationalization is far from over.

10 Can Economic Development Programs Be Evaluated?

TIMOTHY J. BARTIK
RICHARD D. BINGHAM

The question addressed in this chapter seems simple: Can economic development programs be evaluated? But the answer is not simple because of the nature of evaluation. To determine a program's effectiveness requires a sophisticated evaluation because it requires the evaluator to distinguish changes due to the program from changes due to nonprogram factors. The evaluator must focus on the outcomes caused by the program rather than the program's procedures.

Evaluations can be divided into two categories: process, or formative, evaluations and outcome, impact, or summative evaluations. Process evaluations focus on how a program is delivered. Impact evaluations focus on the program's results. Although process evaluations are important, the focus of this chapter is on program outcomes—thus the concern with impact evaluations. However, both types of evaluations need to be defined.

■ Types of Evaluations

Practitioners often disagree with professional evaluators, particularly academic evaluators, about what is a satisfactory evaluation. For some practitioners, a response to a question asked of a recipient of a state or local economic development program concerning the number of jobs "saved" or "created" by government assistance constitutes "proof" that the program is worthwhile. For the evaluator, however, such responses may be suspect.

Some of the disagreements over evaluation methodologies between practitioners and academics stem from conflicting opinions about what constitutes evaluation. One of the major textbooks in the field (Rossi & Freeman, 1985) defined *evaluation* as "the systematic application of social research procedures in assessing the conceptualization and design, implementation, and utility of social intervention programs" (p. 19). As the authors elaborated, "Evaluation research involves the use of social research methodologies to judge and to improve the *planning, monitoring, effectiveness,* and *efficiency* [italics added] of health, education, welfare, and other human services programs" (p. 19). A fairly broad definition as to what constitutes evaluation is reasonable. Thus, the concepts of planning, monitoring, effectiveness, and efficiency have a place in the evaluation of economic development programs.

It is helpful to look at evaluation as a continuum moving from the simplest form of evaluation, monitoring daily tasks, to the more complex, assessing impact on the problem. Such a continuum is illustrated in Figure 10.1. Each of the six points illustrated on the continuum represents a different level of evaluation, with each level building on the previous one. These evaluation activities take place in roughly the same sequence as the implementation of the program. First, tasks are monitored; then activities are assessed, outcomes are enumerated, and the effectiveness of programs is measured; and finally, a judgment is made as to whether the problem has been reduced (Trisko & League, 1978). The lower-level functions are best at providing information about how a program can be improved. The higher-level evaluation techniques determine if a program works (e.g., actually creates jobs). This partially explains the economic developers' proclivity for process evaluations (discussed later).

The lowest level of evaluation, monitoring daily tasks, simply examines the internal workings of a program. The monitoring activity examines questions about the management of the organization: Are contractual obligations being met? Are staff members working where and when they should? Is the program administratively sound? Are daily tasks carried out efficiently? Are staff adequately trained for their jobs? Monitoring activities are designed to uncover and deal with management problems. Work analysis, resource expenditure studies, management audits, procedural overhauls, and financial audits are considered monitoring-level evaluation activities.

The next level of evaluation is assessing program activities. Here, the characteristics of current activities are identified. In assessing program activities, the following types of questions are asked: What activities are taking place? Who is the target of activity (businesses, cities, etc.), and what are the target's problems or needs? How well is the program implemented? Could it be done more efficiently? Are clients satisfied? Does the program have a favorable image? Thus, this level of evaluation assesses numbers and types of program services, the recipients of those services, program efficiency, and other similar characteristics of the program.

The preceding levels of evaluation are process or formative evaluations. The next level, enumerating outcomes, is the first of the outcome

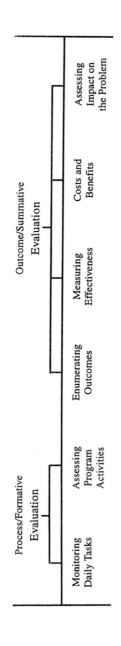

Figure 10.1. Continuum of Evaluation

or summative evaluations. Enumerating outcomes allows one to determine whether program objectives—the immediate short-term outcomes—have been achieved. Typical questions at this level might be: What is the result of the activities described in the process evaluation? What happened to the target population? How is it different from before? Have unanticipated outcomes occurred, and are they desirable? Have program objectives been achieved? To ascertain what the results of the program are, some over-time measures are needed. How are the program recipients different from the way they were before?

Merely enumerating outcomes, however, does not allow one to attribute changes that occurred simultaneously with the program's existence to the impact of the programs. This is essentially the problem solved by measuring program effectiveness. Effectiveness measurement tells one whether program goals have been accomplished. The questions of causality are specifically addressed: What would have happened in the absence of the program? Does the program work? What are the other factors that may have contributed to changes in the recipients? To answer these questions, a cause-and-effect relationship must be established between the program and the outcome. Did the tax abatement "cause" an increase in employment in the target company?

However, simply because a program is shown to be substantatively effective—that is, it actually does create jobs—does not mean that the program should have ever been implemented. Cost-benefit analysis allows one to determine if the program benefits outweigh the program costs. Often they do not. For example, it is questionable whether the public costs of subsidies for the construction of several large auto plants will ever be recovered (Elder & Lind, 1987; Jones & Bachelor, 1984). Thus, cost-benefit analysis simply asks: Do costs of the program outweigh the benefits of the program?

Finally, because a program has been shown to be both substantively effective and cost-effective does not mean that there is an improvement in the problem situation. The reality is that most programs do not have the necessary resources to have a measurable impact on the problem. Assessing the impact requires answers to the following types of questions: What changes are evident in the problem? Has the problem been reduced as a result of the program? What new knowledge has been generated for society about the problem or the ways to solve it? From a practical standpoint, it is difficult to answer these questions even if the program has had adequate resources. This level of evaluation moves from the program to policy—and true policy analysis is enormously expensive to carry out. Also, so many external variables exist that it is virtually impossible to determine real impact on social problems resulting from policy intervention.

These six levels of evaluation, although they have no strict boundaries, provide a framework for assessing the quality of evaluations that have been conducted on economic development programs. As was shown earlier, to effectively "prove" that a program accomplishes its goals, the evaluation must be at the highest two levels of evaluation—measuring effectiveness or assessing impact. Simply enumerating outcomes is not sufficient.

■ The Problem of Outcome Evaluation

The question "Can economic development programs be evaluated?" is hard to answer because it is difficult (but necessary) to determine what would have happened to the program participants if the program did not exist. The evaluator wants to compare what actually happened with what "would have happened if the world had been exactly the same as it was except that the program had not been implemented" (Hatry, Winnie, & Fisk, 1981, p. 25). But because it is impossible to determine exactly what "would have happened if," the answer is to use procedures that give an approximate idea of what would have happened.

Threats to Validity

One reason it is hard to determine what "would have happened if" is that changes taking place in the world may affect the program participants so as to make it look as if the program worked when it actually did not work or, conversely, to make it look as if the program did not work when it actually did. For example, historical events occurring during the program may cause changes in program participants that could appear to be the result of the program. In the economic development arena, one common historical event that can lead to such misinterpretations is the business cycle. Suppose a city gives a firm a property tax abatement that coincides with the bottoming out of a national recession. During the following 3 years, the firm completes its physical expansion, does well in the market, and increases its employment by 50%. The difficulty for the evaluator is determining how much of the employment increase is due to the abatement, how much is due to the national economic recovery, and how much is due to the firm's performance.

Maturation of a firm or an industry can cause a similar problem for the evaluator. The growth pattern of a new firm or industry approximates an S-shaped curve. During the initial years, the firm grows slowly as it innovates and perfects its product. This period is followed by a time of rapid growth when the firm's product holds a competitive advantage in the market and the firm grows rapidly. Then, as other companies enter the market, the firm's market share (and employment) level off and may even begin to decline. The evaluation problem is that if a firm uses economic development assistance just before beginning its "natural" period of rapid growth, it will be difficult to distinguish that natural growth in employment from the growth resulting from the economic development assistance.

Another reason it is hard to determine what "would have happened if" is that program participants may differ from nonparticipants because of how the program selects participants or how participants select themselves. Almost all economic development programs are selective in that only a minority of firms participate, and the firms that participate are not chosen randomly. The firms that select themselves for participation or are selected by program managers might have performed differently from nonparticipants without the program. Evaluations of the program will be biased if they attribute these performance differences to the program.

In theory, this "selection bias" in evaluations of economic development programs could make the programs look either better than they are or worse than they are. In practice, for most economic development programs, this "selection bias" will make the programs look better than they are. Firms applying for economic development assistance will frequently be firms that are aggressively seeking to expand. The desire of these firms to expand is why they applied for a property tax abatement, a training grant, help with modernization, or help with exporting. Even without economic development assistance, such aggressive firms would be more likely to expand, train, modernize, and export.

For some economic development programs, selection bias may make the programs look worse than they are. For example, enterprise zone programs select for assistance all firms located in neighborhoods that are hostile to business, with high crime, poor infrastructure, inadequate labor skills, and low consumer demand. Firms in such neighborhoods would be likely to do much more poorly than firms in average neighborhoods without the program. Even a successful enterprise zone program will find it difficult to stimulate firms in these depressed neighborhoods to outperform firms in other neighborhoods.

In the lexicon of program evaluation, the problems caused by history, maturation, or selection are referred to as "threats to validity." Evaluators typically use different evaluation designs to attempt to overcome the threats to validity. Three common designs are illustrated in Figure 10.2.

Models to Assess Outcomes

The design shown in Figure 10.2 (A), the one-group pretest-posttest design, is one of the most commonly used evaluation designs. It is not a powerful design and is subject to most of the traditional validity problems. Here, the group receiving the program is measured before the initiation of the program (A_1) and again after the program is completed (A_2). The difference scores are then examined, and any improvement $(A_2 - A_1)$ is usually attributed to the impact of the program. The major drawback of this design is that changes in the participating firms may be caused by other events and not the program. The longer the time lapse between the preprogram and postprogram measurements, the more likely it is that other variables besides the program affected the postprogram measurement.

The design in Figure 10.2 (B), the pretest-posttest comparison-group design, is a substantial improvement over the simple one-group design because the evaluator attempts to create a comparison group that is as similar as possible to the program recipients. Both groups are measured before and after the program, and their differences are compared $[(A_2 - A_1) - (B_2 - B_1)]$. (Note: The percentage change might be used rather than the absolute amount of change.) The use of a comparison group can alleviate many of the threats to validity, as both groups are subject to the same external events. The validity of the design depends on how closely the comparison group resembles the program recipient group.

The pretest-posttest control-group design (Figure 10.2 [C]) is the most powerful and "truly scientific" evaluation design presented here. The only significant difference between this design and the comparison-group design is that firms are assigned to the program and control groups randomly. The firms in the program group receive program assistance, and those in the control group do not. The key is random assignment. If the number of subjects is sufficiently large, random assignment implies that the characteristics of subjects in both groups are likely to be quite similar prior to the initiation of the program. This initial similarity, and the fact that both groups will experience the same historical events, mature at the same rate, and so on, reasonably ensures that any difference between the two groups on the postprogram measure will be the result of the program.[1]

Other Approaches to Measuring Outcomes

These different formal evaluation designs all use "objective" evidence from the program itself to infer the extent to which the program "causes" various outcomes. For economic development programs, there are other, more informal approaches to determine the program's effects. One approach is to survey firms that are program participants and ask them whether the program caused various changes in the firm's actions. This "subjective" approach seems unscientific, but natural scientists seldom are able to ask their experimental subjects (rats, electrons, whatever) about the causation issue.

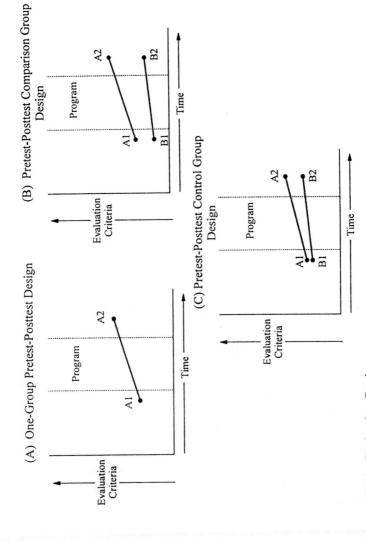

Figure 10.2. Evaluation Designs

One problem with surveying firms about causation is that firms may have difficulty giving precise and accurate responses. Even if the program surveys the relevant decision maker in the firm, that decision maker may find it difficult to give a precise quantitative assessment of the effect of the program on the probability of making an investment decision. A second problem with surveying firms is that some firms may have incentives to exaggerate the program's effects on the firm's behavior. The economic development program may provide assistance only if a firm says the assistance is essential to the decision, in which case admitting that the assistance was not essential might raise legal issues. The firm's managers may also believe that future assistance is more likely if they give a favorable view of the program.

Despite these problems, surveying firms about causation may yield useful evidence in cases in which firms do not have strong incentives to lie. If firms at least try to tell the truth, the responses to surveys about causation may provide a rough guide to the effectiveness of the economic development program. Comparing responses from several programs with similar target groups of firms and similar goals may allow some assessment of which program is more effective.

Another approach to estimating the effects of an economic development program is to simulate plausible firm reactions to the program's change in the incentives facing the firm. This simulation of the change in firm behavior would be based on some explicit or implicit model of how firms behave. Such a model would presumably be based on some other source of data. For example, suppose an economic development program lowers wages per hour by a wage subsidy or lowers effective wages by increasing labor productivity. The resulting average effect on firm employment might be inferred from previous studies of how firms' labor-demand decisions respond to changes in wages. The accuracy of such simulations depends on the extent to which the simulation model captures the main relevant features of the economic development program. In the present example, the simulation of the employment effects of a wage subsidy would be less accurate if firms also viewed the wage subsidy as symbolizing a friendly local "business climate," an aspect of the program that would be difficult to model.

The preceding discussion has implicitly assumed that the evaluator of the economic development program already knows the outcomes to be measured. But determining the outcomes to be measured is a difficult art. It is important to distinguish between "ultimate outcomes," the outcomes one would really like to measure, and "proximate outcomes," the outcomes that actually can be measured. For example, suppose, in evaluating an export assistance program, that the program's effect on a community's total employment is viewed as the "ultimate outcome." But this might be too hard to measure given the small size of the program, so an evaluation might want to look at the program's effects on jobs in the assisted businesses. If this was too hard to measure, the evaluation might look at the program's effects on exports in the assisted businesses, and if that was too hard to measure, it might examine the program's effects on a firm's efforts to export.

Another important distinction is between outcomes at the local level and outcomes on some broader geographic scale, such as the national level. A program that subsidizes new business investment in Mississippi is likely to have quite different effects from Mississippi's perspective than from a national perspective because much of any increased investment in Mississippi will come at the expense of reduced business investment in other states. Which geographic perspective to appropriate depends on who will be using the evaluation and for what purposes.

An even more complicated issue is determining what should be the ultimate outcomes by which one might judge economic development programs and how to value these ultimate outcomes. Suppose new jobs is viewed as an appropriate ultimate outcome. Then an immediate question is how to place a value on the program's job creation, to allow a comparison with the costs of the program. The value of this job creation is likely to depend on who gets the jobs from the program and what people would have been doing with their time if they had not obtained the jobs created by the program. Thus, the process of trying to determine the real value of a program's ultimate outcomes will usually lead to some redefinition of these outcomes.

Most economists and other professional public policy analysts would argue that economic development programs should ideally be subjected to cost-benefit analysis:

> Cost-benefit analysis is an intuitively easy process to conceptualize and appreciate. One gathers all of the costs of providing a good or service and weighs those costs against the dollar value of all the subsequent benefits provided by the good or service. If the benefits outweigh the costs, the good or service should be continued; if the costs of providing the service exceed the benefits obtained, the service should be terminated. (Bingham & Felbinger, 1989, p. 207)

Thus, with cost-benefit analysis, both the costs and benefits of a program are expressed in dollar terms. For example, in the case of a tax abatement, the cost of the program would include the value of the taxes abated. Benefits would include taxes generated for the locality by the salaries of the new employees, multiplier effect of salaries paid to employees living in the jurisdiction, and so forth.

Although cost-benefit analysis is conceptually the correct way to approach economic development evaluation or any other public policy evaluation, it can be difficult to apply in practice. For example, it is difficult to know exactly what dollar value is to be attached to a job that goes to a person who otherwise would be committing crimes versus a job to a person who otherwise would be "unemployed" but taking care of children at home. But even if such questions are difficult to answer conclusively, asking them at least directs the evaluator's attention to important issues. In the present example, asking the question of the value of a job directs the evaluator's attention to the important issue of determining who gets the jobs from a particular economic development program.

■ A Critical Review of Some Examples of Different Types of Economic Development Evaluations

It is impossible to provide a comprehensive review of all economic development evaluations in this chapter, so this section will critique examples of different types of economic development evaluations. Types of economic development evaluations considered here include process evaluations, before-and-after evaluations, survey evaluations, evaluations based on models of firm behavior, comparison-group evaluations, randomized-control-group evaluations, and evaluations that use models of community impact.

Process Evaluations

Many state economic development programs have been the subject of process evaluations by state audit agencies or other such "good-government" groups. These process evaluations typically focus on dull but important issues such as the need for better planning, increased targeting of resources, and improved monitoring of program activities.

A good example of a high-quality process evaluation is the recent evaluation of Virginia's economic development programs by the Virginia Joint Legislative Audit and Review Commission (1991). Among other things, this report recommended that Virginia should try to reduce duplication of services between small-business development centers and technology transfer programs. It recommended greater targeting of programs: Industrial training assistance should be targeted more toward higher-wage industries, retention calls should be targeted toward firms in key local industries that the local economy is having trouble retaining, and marketing efforts should be targeted toward areas of the United States and the world from which Virginia has successfully recruited companies in the past. The report recommended improved program monitoring: better monitoring of the activities of marketing staff, better efforts to measure the payback of industrial training programs, and more standardized measures of the activities of small-business assistance programs.

These recommendations seem sensible. But the report does not provide much evidence on whether Virginia's economic development programs make a difference to the assisted firms or to Virginia's economy.

Enumerating Outcomes: Before-and-After Evaluations

Another common evaluation approach is to enumerate outcomes and to assume that any new activities of assisted firms must be attributable to the economic development program. For example, a recent study by the Pennsylvania Economy League (1992), an independent government "watchdog" group, on the Northeast Tier Ben Franklin Technology Center concluded that "through December 1990, a total of 3817 jobs have been created and a total of 4420 jobs have been retained as a direct result of [the Northeast Tier's] Ben Franklin Center administered projects" (p. 17). These job numbers were derived by assuming that any job created or retained in a project with Northeast Tier assistance was due to the program.

More sophisticated outcome enumeration studies at some point admit some doubts about causation. Knowledge Systems and Research Inc., in a 1991 study of New York State's Entrepreneurial Assistance Program, claimed at one point that

> an estimate of business creation impact is provided by the survey of clients 2-3 years after initial assistance was received by EAP where the reported percentage of new businesses among start-up clients is 30.8%. . . . Other measures of entrepreneurial assistance impacts for FY 1989 are: business expansions—68 (24.3%); businesses saved—60 (21.5%); jobs created (other than business operators)—36. (pp. 4-7 to 4-8)

Later on in the report, however, the consultant admitted that "it cannot be determined in the scope of this evaluation how many of these jobs have been or will be created but for EAP assistance" (p. 4-11). In other words, what were previously called "impacts" of the program may not really be "impacts."

Enumerating Outcomes: Survey Evaluations

Many evaluations of economic development programs have asked clients of the programs to assess their impact. The reliability and usefulness of this survey evidence varies a great deal from one study to another, depending on the type of economic development program and how the survey results are used. Mt. Auburn Associates and the Corporation for Enterprise Development, in a 1990 study of New York State's business financing programs, surveyed 186 firms that had received loans either from the state Job Development Authority, the Urban Development Corporation, or the Science and Technology Foundation. The report used surveys to examine the effects of the financing in several ways:

> For 41% of the surveyed firms, availability of state financing was the deciding factor in making the investment in the community. Without state financing, these firms would have canceled the investment, moved to another community, or gone out of business. For another 40% of the surveyed firms, state financing affected the scale or timing of the investment decision. Eighteen percent of the surveyed firms indicated they would not have altered their investment decision had the state financing not been available. (p. 17)

> We asked the companies surveyed to provide their employment level as of November 30, 1989, and to answer hypothetically what their employment *would have been on the same date in the absence of the financing*. The difference between actual and hypothetical employment levels counts those jobs retained and created *attributable to the financing*. . . . Since the question related to employment in the absence of the financing is hypothetical, the answers are not precise. Yet, only the firm's owner or management fully understands the factors responsible for employment growth or decline. The company itself is the only source of information to address the issue of "attribution." . . .
> Based on this job impact measure, the level of job retention and creation was 6,876 jobs. . . . The difference between this figure and the job growth of 7,137 . . . results from firms reporting that some jobs created could not be attributed to the financing. The job impact measure recognizes the job retention due to financing and specifies if the financing was responsible for job creation. (pp. 23-25)

Although this survey approach is preferable to giving the program credit for everything that happens at firms that receive financing, there may be some incentives for firms to misrepresent the effects of the financing. For many financing programs, laws or program rules require that loans go to firms only in cases in which the loan is necessary for the project to proceed. If a firm said that a project would have proceeded without the state financing, and the firm's response became public, serious political or legal problems could arise.

Another study that used survey evidence to assess program effectiveness is the previously mentioned study of New York State's Entrepreneurial Assistance Program (EAP). EAP actually consisted of two separate programs—support centers and development centers—with support centers apparently providing more formal training of potential entrepreneurs. In response to the question "To what degree did the center help you achieve your goal?" a majority of the clients of both types of centers felt the center helped at least somewhat, but "support center clients were most positive, with development center clients less convinced that the program was helpful" (Knowledge Systems and Research, Inc., 1991, p. 4-9).

This relative ranking of these two types of centers is convincing. Center clients were not required to claim that the service was essential to receive service. If the entrepreneurial assistance was not helpful, what incentive would there be for clients to lie and claim that the service was helpful? If the service was not helpful, clients would not perceive any reason to maneuver to keep the programs alive. In contrast, for programs that provide cash assistance, there is always an incentive for firms to keep the programs alive and available, because cash assistance is always helpful to the firm even if the assistance does not change the firm's behavior.

Even if EAP clients give truthful survey responses, it is unclear how a client's saying that EAP helped in achieving the client's goals translates into a precise quantitative effect on the probability that the client will start a small business or expand a small business. But because two programs with similar goals and clientele are considered, the relative ratings of the two programs may give a rough qualitative idea of the relative effectiveness of the two programs. The comparison suggests that formal training sessions are a useful component of an entrepreneurial training program.

A third example of an evaluation using survey evidence is a large-scale 1992 study, conducted by Mt. Auburn Associates, of Ohio's Edison Technology Centers program. This study included both surveys of firms involved with the Edison Centers and focus groups with some of these firms. Overall, 40% of Edison member firms indicated strong effects of the Edison programs on process, skills, customers, cooperative relationships, or business practices. Perhaps as important, the reported impact varied a great deal by center:

> On a firm by firm basis, EBTC [Edison BioTechnology Center] appears to be the Center with the most impact among its members. It had the highest percentage in several impact categories—60% said EBTC had a strong impact on at least one aspect of business operations, 33% said there were Ohio employment impacts, and 74% said that if EBTC did not exist present and/or future performance would suffer. . . . There is very high enthusiasm among the member firms. . . .

- CAMP [Cleveland Advanced Manufacturing Program] works intensively with far more Ohio-based firms than any other Center. . . . A higher percentage of members said that CAMP aided in manufacturing process improvements and in business practice improvements than was so for any other Center. . . . Enthusiasm among interviewees and focus group members for CAMP services is very high.
- EABC [Edison Animal Biotechnology Center] has only worked with three Ohio firms over the course of its existence. Its impact has been minimal.
- Measures of IAMS' [Institute of Advanced Manufacturing Sciences] current performance must take into account that the current incarnation of IAMS is less than two years old. . . . A lower percentage of members said that IAMS had a strong impact on firm operations, or any positive impact, than for any other Center. . . . Focus group members tended to have less enthusiasm for belonging to the Center than was so at other Centers.

The analysis in the individual chapters and a review of the survey tables indicates that the economic effectiveness of the four other Centers tends to be between those discussed above. (Mt. Auburn Associates, 1992, pp. 12-6 to 12-7)

These comparative rankings of the different Edison Centers are convincing for the reasons previously stated: It is unclear why center members would have an incentive to say that the services had an impact if they were useless, and the differences in survey results across centers are likely to correspond to a rough qualitative ranking. In addition, because the survey results are backed up by focus groups, they seem more convincing. It requires a more extended effort to lie throughout a focus-group discussion. Also, the focus-group discussions provide some insight into why different centers are more or less effective.

Enumerating Outcomes: Evaluations
Based on Models of Firm Behavior

Another approach to enumerating outcomes is to use some model of firm behavior that simulates whether a program made some difference to a firm's behavior. One example of the use of such a model is a study by the Illinois State Office of the Auditor General (1989) of the economic development programs of the Illinois Department of Commerce and Community Affairs (DCCA). This report blasted DCCA for poor management of its loan and grant programs to firms. DCCA was criticized for poor documentation of the programs and an unclear decision-making process for making grants. These criticisms appear to be well founded.

But the report also analyzed the effectiveness of DCCA's loan and grants on the basis of a model of when a state subsidy will alter a firm's location or expansion decision, and this part of the report is more questionable.

The Auditor General's report made two main adjustments to decide when a DCCA loan or grant was responsible for a firm's location or expansion decision. First, the Auditor General made an adjustment on the basis of the firm's rate of return before and after the DCCA loan or grant:

> If a firm had a rate of return at the time of the subsidy that was either negative (the firm was losing money) or below industry averages, we concluded that the firm needed assistance. If subsequent to the subsidy the firm's rate of return became positive and comparable to the industry average, we concluded the subsidy had been effective. If these conditions were absent, we concluded the subsidy was unwarranted or ineffective or both. (p. 98)

On the basis of an audit of a sample of 35 DCCA projects, the Auditor General concluded that in 12 projects DCCA's assistance was ineffective, making up 21% of DCCA's claimed jobs for all 35 projects.

Second, the Auditor General adjusted DCCA's job creation numbers on the basis of the percentage of capital provided by DCCA: "In our calculations we prorated the number of jobs created on the basis of the proportion of capital provided. If DCCA contributed 10 percent of the capital, then 10 percent of the jobs created were attributed to DCCA's activity" (p. 7). Because on average DCCA supplied about 13% of the capital for its projects, this eliminated about 87% of DCCA's claimed job creation. These two adjustments, plus other adjustments made by the Auditor General, reduced DCCA's claimed creation of 7,501 jobs in this sample of 35 projects to only 608 jobs.

The Auditor General's model of the effectiveness of economic development loans or grants is questionable. The two adjustments do roughly correspond to factors that influence the effectiveness of economic development subsidies, but the Auditor General's specific job creation numbers could easily be wildly inaccurate. With respect to the first adjustment, even if a firm is making above normal rates of return, a subsidy may increase profits in Illinois over profits in some alternative location, changing that location decision. An effect on location or expansion decisions is more likely if the subsidy has a larger effect on rates of return, but there is unlikely to be any fixed cutoff point determining when a subsidy will make a difference. With respect to the second adjustment, one would not think that the subsidy's effect on the probability of a location or expansion decision would be exactly proportional to the percentage of capital supplied, although perhaps the more capital supplied, the more likely the subsidy is to affect the decision. If the loan or guarantee alters the firm's profits enough so that location or expansion in Illinois is preferred to the alternative of not locating or expanding in Illinois, whereas before it was not, then 100% of the jobs involved are attributable to the subsidy, regardless of the percentage of capital supplied.

Another program evaluation that uses an implicit model of business behavior is a recent "self-evaluation" done of Michigan's Capital Access Program (CAP; Rohde, Cash, & Ammarman, 1990). Under CAP, a business borrowing money and the bank each pay a fee of 1.5% to 3.5% of the loan, with the bank's fee typically passed on to the borrower. These fees, and a state contribution of 150% of the combined bank/borrower fees, are paid into a separate reserve fund for each bank participating in the program. This reserve fund is available to cover losses in loans made under the program; a bank's losses beyond its own reserve fund are the bank's responsibility. CAP is designed to encourage banks to take greater risks with small-business loans, but not too much risk. Most banks aim for less than a 1% loss rate on small-business loans. Under CAP, loss rates of 10% or so can easily be tolerated and still be profitable to the bank.

CAP is designed so that in a competitive banking market, rational business borrowers and rational banks would not want to participate in CAP unless the loan was above normal risk because of the extra costs associated with CAP loans. Rohde and the other staff running CAP conclude in their report that the above-normal loss rate of CAP shows that the program is working as intended:

> There is strong evidence from a variety of sources that the Capital Access Program is causing banks to make loans that they otherwise would not make.
>
> The most compelling evidence comes from the . . . loss rate under the program. The loss rate under the program has been running probably at least 7 times . . . a normal bank loss rate. If banks, on average, when using the program take 7 or more times the risk they normally take without the program, it is a reasonable conclusion that the program is making a difference in causing them to take this added risk. (Rohde et al., 1990, p. 46)

This is strong evidence, but it really comes from a model of how banks and borrowers behave in a competitive banking market rather than any direct empirical proof that loans are being made under the program that otherwise would not have been made.

Some of Bartik's (1991, 1992) work on economic development has essentially argued that "a cost is a cost" and that financial subsidies to firms should have similar effects regardless of whether provided through regular taxes or through special tax subsidies. For example, Bartik (1992) argued that the annual cost per job created through reductions in general state and local business taxes is somewhere between $2,000 and $11,000 per year. In the absence of evidence to the contrary, there is some plausibility that special tax and other financial subsidies to firms will have similar average effects. For example, if a branch plant with 1,000 workers is given a financial subsidy equivalent to $1 million annually—$1,000 per worker per year—it could be argued that the subsidy is unlikely on average to affect the probability of the plant choosing that location by more than 50%. General business tax reductions seem to cost at least $2,000 per year per job created, so it seems unlikely that on average a special tax subsidy for a typical firm will be much more effective than that. Of course, in any particular case, the subsidy either tips the decision or it does not, but we are unlikely ever to know whether the subsidy made a difference for certain.

There are two problems with this line of reasoning. First, economic development financial subsidies may have more than a strictly economic value to the firm—they may symbolize a good business climate. Second, some of the evidence on effects of different types of changes in costs seems contradictory. For example, the average estimated effect in most business location studies of a 10% reduction in wages is not much greater than the effect of a 10% reduction in state and local business taxes, even though wages are a much greater proportion of costs than state and local business taxes. Bartik (1991, pp. 49-52) suggested some reasons that estimated effects of wages on business location decisions may be biased toward zero, but this pattern still raises some doubts about the "model" that all cost changes have similar effects on business location.

Evaluations Using a Comparison
Group Without Random Selection

Another way to measure the effects of economic development programs is to compare the performance of program participants with a comparison group of nonparticipants. By definition, this comparison group will not be randomly chosen, but there will be some effort to make the group comparable to the program participants.

One example of a nonrandom comparison evaluation is the study by Price Waterhouse (1992) of the U.S. Small Business Administration's (SBA's) 7(a) Guaranteed Business Loan Program. This study compared recipients of SBA guaranteed loans in fiscal year 1985 with a comparison group of firms drawn from SBA's Master Establishment List, which is derived from Dun and Bradstreet records and yellow pages listings. In selecting the comparison group, Price Waterhouse tried to match recipients to nonrecipients by two-digit industry, number of employees, and 10

SBA regions. Price Waterhouse did a phone survey of recipients and the comparison group in 1991, asking about 1984 and 1989 revenues and employment. The study found that recipients were more likely to be in business in 1989 than nonrecipients and had more growth from 1984 through 1989 in revenues and employment. But SBA loan guarantee recipients were also more likely than comparison group firms to be start-up firms, and many firms in the comparison group said they did not obtain an SBA loan because they did not need one. These facts imply that perhaps SBA loan guarantee recipients would have been more aggressive in expanding even if this SBA program had never existed. Price Waterhouse is careful to note that we cannot tell what would have happened to SBA loan guarantee recipients without the loan.

Another study with a comparison group is a study by Public Policy Associates and Brandon Roberts and Associates (1992) of Oregon's small-business service programs. This study included surveys of assisted small businesses and a general survey of Oregon small businesses. The report notes that users of state small-business services did better on increases in sales, profitability, and jobs during the 1989-to-1992 period. This finding is not emphasized in the report. This deemphasis seems appropriate because, again, users of services may have been more aggressive about expanding even without the services.

A somewhat different example of a comparison-group evaluation is Leslie Papke's (1991) study of the effects of Indiana's enterprise zone program. Her study differs from most other studies of economic development programs because it seeks to look directly at overall performance of some geographic area (in this case, enterprise zone neighborhoods) compared to other geographic areas. Ideally, all economic development programs would be evaluated on the basis of direct evidence on how they affect their target geographic area, whether that is a state, metropolitan area, city, or neighborhood. But such communitywide impact evaluations are not possible for most economic development programs because the programs are so small relative to their geographic area of interest that any estimated impact is likely to be spurious. Enterprise zone programs at least claim to be large enough relative to their chosen neighborhoods to have a detectable overall community economic impact.

Papke found that after zone designation, zone areas experience a reduction in unemployment claims compared to nonzone areas. One problem with her study is that she never described in any detail how she selected the comparison group of nonzone areas in Indiana. Hence, it is unclear whether the nonzone areas are truly comparable to the zone areas.

Another innovative comparison-group evaluation of economic development programs was done by Holzer, Block, Cheatham, and Knott (1993) on Michigan's former program of industrial training grants for manufacturing firms undergoing modernization. The evaluation compared firms that received training grants in 1988 and 1989 with firms that applied for such grants, but too late in the state's fiscal year to receive a grant. Apparently the state's modernization training grant program generally funded all eligible firms until they ran out of that year's budget. Holzer and his colleagues surveyed both groups of firms in 1990. They found that firms that received grants did more training afterwards than nongrantees and that their product scrappage rates were reduced more.

The Holzer evaluation could be criticized using the argument that firms that apply too late for state grants are less capable than firms that apply on time. But the two groups of firms did not seem to differ significantly in observed characteristics. In addition, one could argue that firms have some incentive to claim that the grants led to reductions in scrappage rates. But the incentive to lie is relatively low because firms that received training grants were ineligible for future grants.

Evaluations Using a Randomly Assigned Control Group

Another evaluation approach is to compare the performance of program participants to a randomly assigned control group of firms. The random assignment ensures that the control group will be similar to program participants on all relevant characteristics, both observed and unobserved.

The only economic development evaluation that has used random assignment is an Abt study (Benus, Wood, & Grover, 1994) of two experiments, sponsored by the Department of Labor, of entrepreneurship training for unemployment insurance (UI) recipients in Washington State and Massachusetts. The experiments first ran orientation sessions explaining the entrepreneurship training program; the 2% to 4% of UI recipients who expressed interest in such training were then randomly assigned to a treatment group and a control group. Both experiments showed significant increases in self-employment in the treatment group compared to the control group, with no significant increase in the probability of business failure. In Massachusetts, 47% of the treatment group ended up with some self-employment experience compared to 29% of the controls, whereas in Washington State, 52% of the treatment group entered self-employment compared to 27% of the controls. As this experiment shows, many individuals who received no entrepreneurial assistance still started up their own business. This emphasizes the point that one cannot attribute all business start-ups associated with an economic development program to the program itself (and clearly illustrates the threat to validity posed by self-selection).

Evaluation Models of Community Impact

Even if an evaluation study has determined the effects of an economic development program on firms participating in the program, an important remaining issue is what impact the change in behavior of the program participants has on the overall local community. As mentioned above, because most economic development programs are so small relative to the community, any impact will be difficult to detect, but the impact may be large relative to the size of the program.

To analyze how changes in participant firms in turn affect the community, an economic impact model must trace how the initial employment and other impacts on participant firms lead to multiplier effects on the activities of other firms, as well as effects on unemployment and labor force participation and on wage rates. A fiscal impact model that relates these economic impacts to effects on local taxes and public service costs is also needed.

A wide variety of models already seek to do economic impact and fiscal impact analyses. One such model that has been specifically adapted to economic development programs is the University of Illinois-Chicago's adaptation of George Treyz's (1993) well-known REMI economic impact model. Wiewel, Persky, and other UI-Chicago researchers have adapted the REMI model to trace how a given economic development project will have fiscal impacts on Chicago and economic impacts on Chicago residents (Persky, Felsenstein, & Wiewel, 1993; Wiewel, Persky, & Felsenstein, 1994).

One important adjustment in community impact models is that the community effects of assistance to export-base firms—firms that export their product outside the regional economy—are likely to be greater than the community effects of assistance to locally oriented firms. Economic development programs that help export-base firms are unlikely to have negative effects on the sales of other local firms. In contrast, economic development programs that help some firms that mainly sell to the local market are likely to have some negative effects on other firms that also sell to the local market. Helping locally oriented firms may have some positive effects on the local economy because some of the expansion in these firms' sales may come at the expense of imports to the local economy, not other local firms. But these positive effects on the local economy will probably be less than the positive effects of assistance to export-base firms.

Overall Assessment of Current Evaluations

This review of economic development evaluations may be misleading because examples of several types of evaluations were provided, and this could give the impression that each type of evaluation has been done with roughly similar frequency. The reality is that the vast majority of existing evaluations are process evaluations. A process evaluation is the standard type of evaluation done by a state audit agency or a good-government group. Also, a growing number of economic development evaluations use qualitative ratings from surveying the program's client firms. Only a few evaluations of economic development programs use comparison groups. In most of these evaluations, the chosen comparison group does not seem comparable. Only the Holzer et al. (1993) study of Michigan's economic development training program uses a comparison group taken from unsuccessful program applicants, which at least controls for the greater likelihood that the more aggressive firms will apply. Only the Abt study of entrepreneurial training programs (Benus et al., 1994) uses a randomly assigned control group.

The emphasis on process evaluations and surveys enumerating outcomes is unfortunate because evaluations with a good comparison group or a control group are likely to do the best job of measuring the true quantitative effects of economic development programs. Surveys of client firms provide some qualitative evidence on these programs' effectiveness. This qualitative evidence may be particularly valuable when one is comparing similar programs. But surveys of clients alone are not a reliable means of quantifying exactly what difference the economic development program has made to a firm's performance. Such surveys cannot really address the crucial issue of whether the quantitative effects of economic development programs are worth their costs.

This is an era that demands that government programs show that their benefits are greater than their costs. An important but puzzling question is why, in this atmosphere of accountability, there have not been more well-done comparison- or control-group studies of economic development programs. That issue is considered in the next section.

■ Why Aren't There a Greater Number of More Sophisticated Economic Development Evaluations?

If only evaluations with comparable groups can really tell us whether economic development programs are successful, why aren't there more such evaluations? There are six reasons for this situation. All of these reasons amount to saying that whatever the social benefits and costs of more sophisticated evaluations, such evaluations are not perceived as offering sufficient benefits to justify their costs from the viewpoint of those groups that would usually have to authorize, pay for, or conduct such evaluations.

The first reason for the lack of sophisticated evaluations is that such evaluations are difficult to do, compared either to no evaluation at all or to evaluations focused on process or on subjective responses to survey questions. Evaluations with a comparable group by definition require careful procedures to select this comparison group. Such evaluations also require collection of extensive quantitative data over a period of time from both the firms participating in the economic development evaluation and the comparison group. To allow for a comparable group and for the collection of baseline data, the evaluation should ideally be built into the program design.

These data collection and design efforts not only may be expensive in direct budgetary costs but also may require extensive administrative time and may be disruptive to the staff setting up the program. It is much easier to do no evaluation at all, or simply to interview some program staff or program clients later on and ask them what they think of the program.

A reason that *randomized*-control-group evaluations are unpopular is that such evaluations require explicitly denying services to specific firms. Of course, economic development services are implicitly denied to many firms because so little in funds is provided to economic development programs. In addition, many experiments with medical treatments or welfare programs do use randomized-control-group assignment explicitly to deny services to some individuals. But somehow the explicit denial of services seems more bothersome when those being denied services are firms, perhaps because this violates our sense of how a free-market system should work.

A third reason for the paucity of comparison/control-group evaluations is that more rigorous evaluations will have a disproportionate part of their benefits go to groups other than those paying for the evaluation. Hard quantitative evidence on the effectiveness of a particular approach to economic development will have benefits to all state and local areas, not just to the state or local area that has the program and is funding the evaluation.

A fourth reason for the unpopularity of rigorous evaluations is that such evaluations too often seem to avoid telling program administrators how they can improve their program. Outcome evaluations are frequently written as if the program is a "black box," and they shy away from trying to determine why a program was or was not successful. Knowing whether the program was or was not successful is of course the most important issue, but the program administrators want to know how to improve the program. A process evaluation would seem to offer some clues as to how to improve the program, even if the evaluation by itself does not document what the program really accomplished. Process and outcome evaluations should not be thought of as mutually exclusive alternatives, but as complementary.

A fifth reason for the lack of sophisticated evaluations is that many of the groups that typically do economic development evaluations, such as state audit agencies and some consulting firms, do not seem to be oriented toward evaluation studies that use sophisticated methods. State audit agencies frequently do not have staff who are trained in how to do studies that correct for selection bias due to a nonrandomly selected comparison group. Nor do state agencies typically have staff with much familiarity in how to set up an experiment with a randomly assigned control group.

A final, and perhaps most crucial, reason that more rigorous evaluations are rare is that program administrators fear the political consequences of a negative evaluation. If a program is not evaluated, one can always claim success. A process evaluation or a survey evaluation is subjective enough so that it may be easier to manipulate the evaluation process or reinterpret the results to make the program look better. But if a study shows that firms participating in the economic development program, compared to a truly comparable group, show no difference in performance, then it is difficult to argue that the program works.

These fears of the possible negative political consequences of evaluation are realistic given the American political tradition of suspicion of government. Christopher Jencks (1992-1993), a well-known sociologist at Northwestern University, recently commented on the peculiar American attitude toward government programs:

> In every other society with which I am familiar, the political system assumes that when a program is not working you try to improve it. Only in the U.S., where we doubt that government can ever do anything right, do we assume that if a program stumbles in its first couple of years we ought to terminate it. No wonder we have so few successful programs. It's like deciding that if babies get sick they should be thrown away. (p. 12)

As a result, economic development program managers, or indeed managers of any government program in the United States, are perhaps justified in following the advice of economists Gary Burtless and Robert Haveman (1984): "If you advocate a particular policy reform or innovation, do not press to have it tested" (p. 128).

The real issue in encouraging more sophisticated evaluations of economic development programs is how to change the perceptions of economic development program managers, state legislators, and others who would have to authorize such evaluations about the benefits and costs of more rigorous evaluations. Some thoughts about how to do this will be offered in the conclusion.

■ A Suggested Approach to Economic Development Evaluation

Three types of economic development evaluations are needed:

- *Community evaluations* that directly estimate how some economic development initiative affects an overall community
- *Firm evaluations* that directly estimate how an economic development program affects individual firms
- *Community impact evaluations* that calculate the impact of some assumed change in firm behavior on a community

Community Evaluations

As was mentioned, directly estimating the effect of some economic development program on a community is only feasible when the program is "large" relative to the size of the community. Enterprise zones are one case in which such evaluations seem feasible. In addition, it seems feasible to make such an evaluation in the case of large-scale public-works investment, such as highways, in rural counties (see, e.g., Rephann, 1992).

Community evaluations should ideally use random assignment. Such random assignment is feasible when only a small number of deserving communities can be helped with the program. For example, the Clinton administration's recently enacted enterprise zone law is providing assistance to 9 "empowerment zones" and 95 "enterprise communities." Although the selection of these "zones" and "communities" was not based on random assignment, it would have been possible to do random assignment, for many more communities applied for designation than could be accommodated under the program. If random assignment had been used, a simple comparison of changes in "treatment" and "control" communities would have allowed an unbiased assessment of the effects of the program.

If random assignment is not used, then community evaluations should compare assisted communities with other eligible communities that applied for the program but were not assisted. Using other eligible applicant communities increases the similarity between the "treatment" communities and "comparison" communities. Differences will still remain. To control for such differences, researchers will have to make a careful effort to identify other variables that might predict the economic performance of treatment and comparison communities. Past economic performance is one good control variable.

Firm Evaluations

For economic development programs that are too small plausibly to have a detectable direct effect on the community, evaluators should focus on the effects of the program on the performance of assisted firms. Most economic development programs fall in this category.

Such evaluations of effects on individual firms should ideally use random assignment to divide firms into treatment and control groups. Random assignment means that a simple comparison of the performance of the treatment-group and control-group firms should give an unbiased estimate of the effects of the program, as treatment- and control-group firms must on average be similar.

Random assignment is most feasible in cases in which there is significant excess demand by many similar firms for a program, the program provides similar assistance to all assisted firms, and the timeliness of the assistance is not the key issue. This situation is likely to occur in cases in which a program provides cash (e.g., loans or grants) or near-cash assistance (e.g., dollars for training) to many small and medium-sized firms: Many small firms are likely to want the assistance even if they have to wait a bit, the assistance provided to firms is similar, and the program is likely to need some type of application deadline and decision procedure to determine who gets the assistance. There may also be excess demand by similar firms for "classroom"-style educational and information programs, for example, in entrepreneurial assistance training programs.

Excess demand by similar firms for similar services is much less likely in industrial extension service programs: These programs provide customized assistance to individual firms and strive to provide the assistance as quickly as possible. Asking industrial extension customers to wait for random assignment is likely to hurt the reputation and effectiveness of the program. Excess demand by similar firms is also less likely for programs that provide cash or near-cash assistance to large firms. Each large-firm client and location or expansion decision is likely to be unique. Because each large-firm decision is likely to have sizable economic impacts, few government officials are going to want to risk a particular case's being assigned to the control group.

Even if there is excess demand, random assignment may not be politically or administratively feasible. A second-best approach is to use, as a comparison group, rejected firms that applied for assistance under the program. This procedure at least controls for the greater aggressiveness and growth orientation of applicant firms compared to nonapplicant firms. The analysis comparing assisted firms with rejected applicants should include variables controlling for other factors affecting the economic performance of firms, particularly if such factors were part of the process by which the program selected which applicants to fund.

This caveat is extremely important. Because of the enormous number of products manufactured, distributed, and sold in the United States, random assignment of establishments into experimental and control groups will almost never produce similar product mixes (unless the number of establishments is extremely large). This is important because industries differ widely in terms of their product and profit life cycles (see Markusen, 1985) and because of a given industry's place on the business cycle (leading versus lagging indicators).

It is also important because for many jurisdictions and programs, the number of firms receiving assistance is quite small. Thus, even if it were possible to assign firms randomly into treatment and control groups, it is unlikely that the two groups would be similar.

In both of the above instances, the only realistic alternative is to develop a matched-pair comparison group so that each firm receiving the program is carefully matched with a similar firm not receiving the program. Extreme care must be taken to ensure that each comparison-group firm is closely matched to its equivalent control-group firm on all relevant variables.

Thus, a matched-pair comparison group might be a more-than-satisfactory alternative. There are a number of ways to identify such firms. They can be identified by using secondary sources—for example, by combining the use of ES202 data, industrial directories, and annual reports. The simplest and most effective way, however, is simply to ask the firms receiving the program. Survey experience with these firms has shown that they can easily identify similar firms.

If either the program does not have excess demand by similar firms for similar services or the evaluator is unable to get data on rejected applicants, then other firms may also be used as a comparison group. Because firms that did not apply to the program are likely to be less aggressive and growth oriented than applicant firms, the researcher should try to include in the empirical analysis variables that control for a firm's growth orientation—for example, prior growth of the firm. The analysis will also be helped if the researcher can find variables that help predict whether a firm is assisted by the program but are plausibly not otherwise strongly correlated with the firm's economic performance. For example, how close a firm is to the program's offices could plausibly affect the likelihood of a firm's receiving assistance but might not be strongly related to the firm's performance otherwise. If one can identify variables that predict program participation but do not directly affect performance, then these variables can be used as "instrumental variables" to control for the nonrandom selection of firms for participation in the program. This is not the place to explain in detail how such instrumental-variable selection-bias correction procedures work—this topic is covered in many econometrics texts.

In cases in which no comparison-group data are available—the program is targeted at large firms, or no funds are available for collecting data on unassisted firms—surveys of assisted firms may be useful. Surveys collecting qualitative information from assisted firms may also be a useful supplement to more quantitative information available from control-group surveys. Surveys asking assisted firms to assess the impact of the program are likely to be most believable if the surveys are anonymous and if the program is designed so that firms do not have any legal or political pressure to claim that the program was effective. Qualitative surveys will be more useful if several similar programs can be compared using similar surveys. National standards for survey questions would be helpful. These national standards would allow state and local policy makers to compare their economic development program's performance with similar programs in other areas.

Community Impact Evaluations

If one can know or assume some effect of an economic development program on individual firms, it still is of interest to estimate or simulate what impact that change in individual firm behavior will have on the community's fiscal and economic situation. Even if this community impact is not large enough to be detectable, it could be large compared to the size of the economic development program.

A good economic and fiscal impact study should be sensitive to at least the following five issues:

1. *Export-base versus non-export-base companies.* As was mentioned earlier, economic development assistance to non-export-base companies, companies that primarily sell locally, is likely to have some adverse effects on the market share and performance of other locally oriented firms. These adverse competitive effects on unassisted local firms are less likely for economic development assistance to export base firms.

2. *Population in-migration.* Increases in local employment caused by economic development programs are likely to lead to in-migration. Research reviewed by Bartik (1991, 1993) indicates that for every five jobs created in a metropolitan economy, four jobs are likely to go to in-migrants in the long run (5 years or so). This increase in population reduces the employment benefits of economic development programs to the original local residents and is likely to raise the fiscal costs of economic development programs.

3. *Adverse effects due to increased costs and other congestion effects.* As an area grows, land and housing prices increase, forcing up wages. In addition, growth may cause greater traffic congestion and environmental problems. All these greater costs are likely to discourage employment growth in some firms. Hence, the initial increase in local employment growth due to some economic development program is likely to be offset to some extent in the long run due to these higher costs.

4. *Public-service costs.* Additional business activity and population will create the need for additional public expenditure. Too many analyses of economic development seem to pretend that growth only increases the tax base without increasing spending needs.

5. *Public-capital costs.* Some of the most expensive public expenditures required by growth are likely to be the capital costs associated with building new infrastructure or retrofitting old infrastructure to accommodate growth. These increased costs for additional population and business will often be more expensive, on a per-person or per-employee basis, than the capital costs for the current population and employment base. Retrofitting infrastructure is often expensive. In addition, the existing infrastructure may have already been paid for by past generations and may have been partially funded by federal grants.

■ Conclusion

Can economic development programs be evaluated? Our answer is, yes they can—and furthermore, some decent evaluations have already been done. Good quantitative evaluations of economic development programs require the use of a comparable group of unassisted communities or firms. Such quantitative studies can be usefully supplemented by well-designed qualitative surveys of assisted firms, done so that results can be compared across similar programs.

But although some reasonably good evaluations have been done, such evaluations are far too rare. Good evaluations are rare because the policy makers who would have to approve such evaluations perceive too much political risk and/or cost in economic development evaluations.

To encourage more good evaluations, the perceived benefits and costs of economic development evaluations need to be changed. This can be done in three ways.

The first force for change would be *federal funding and standards.* Federal funding of evaluations, along with federal funding of demonstration programs, would be a useful carrot for encouraging state and local governments to do more evaluation. In addition, federally funded economic development programs should have the "stick" of requiring rigorous evaluation.

To ensure that evaluations are rigorous and comparable across different state and local areas, the federal government should sponsor a cooperative effort with state and local economic development organizations to develop standards for high-quality economic development evaluations. Standards are particularly important for qualitative surveys, whose value is enhanced by having evidence from similar surveys of similar programs. The results of such rigorous evaluations should be disseminated, with federal funds, through publications, conferences, and electronic networks.

Which federal agency should be charged with promoting economic development evaluations is difficult to decide. One logical agency to lead this effort would be the Economic Development Administration (EDA), in the U.S. Department of Commerce. Given EDA's limited funding, it might be more effective for EDA to focus on improving the economic development programs of state and local areas rather than on carrying out its own programs.

The second force for change would be *greater professionalization of economic development agencies and legislative review committees.* Over time, the economic development profession has become more professionalized. This trend needs to be continued, and the idea of high-quality evaluation as an essential part of carrying out high-quality economic development programs should be encouraged.

The third force for change would be the *demonstration effect of good evaluations.* If some good evaluations are done, this should lead to more good evaluations. It is difficult to get people to do something that has not been done before. Once state and local economic development policy makers have seen that a high-quality evaluation of economic development programs can help improve the program's performance and political viability, the interest in economic development evaluations should increase. Although much of the discussion here has tended to emphasize outcome evaluations (because they are so infrequent) in economic development, good practices would suggest that process and outcome evaluations be accomplished simultaneously. After all, a program can hardly be shown to accomplish what it intends if it is not properly implemented.

Developing a tradition of high-quality evaluations of economic development programs is likely to take some time. Economic development evaluation is where job training evaluation was 20 years ago—with a few good evaluations, more low-quality evaluations, and too few evaluations overall. Over the next 20 years, enough high-quality evaluations of economic development programs should be done so that development professionals really know what works and what does not work in economic development.

NOTE

1. Although a number of authors in the evaluation field use the terms *control group* and *comparison group* interchangeably, they are not equivalent. Control groups are formed by the process of randomization. Comparison groups are groups that are matched to be comparable in important respects to the experimental group. In this chapter, the distinction between control groups and comparison groups is strictly maintained.

REFERENCES

Bartik, T. J. (1991). *Who benefits from state and local economic development policies?* Kalamazoo, MI: W. E. Upjohn Institute for Employment Research.

Bartik, T. J. (1992, February). The effects of state and local taxes on economic development: A review of recent research. *Economic Development Quarterly, 26,* 102-110.

Bartik, T. J. (1993, September). Who benefits from local job growth, migrants or the original residents? *Regional Studies, 27,* 297-311.

Benus, J. M., Wood, M., & Grover, N. (1994, January). *A comparative analysis of the Washington and Massachusetts UI Self-Employment Demonstrations.* Unpublished report prepared for the U.S. Department of Labor, Employment and Training Administration, Unemployment Insurance Service under Contract No. 99-8-0803-98-047-01.

Bingham, R. D., & Felbinger, C. L. (1989). *Evaluation in practice: A methodological approach.* New York: Longman.

Burtless, G., & Haveman, R. H. (1984). Policy lessons from three labor market experiments. In R. Thayne Robson (Ed.), *Employment and training R&D* (p. 128). Kalamazoo, MI: W. E. Upjohn Institute for Employment Research.

Elder, A. H., & Lind, N. S. (1987, February). The implications of uncertainty in economic development: The case of Diamond Star Motors. *Economic Development Quarterly, 1,* 30-40.

Hatry, H. P., Winnie, R. E., & Fisk, D. M. (1981). *Practical program evaluation for state and local governments* (2nd ed.). Washington, DC: Urban Institute.

Holzer, H. J., Block, R. N., Cheatham, M., & Knott, J. H. (1993, July). Are training subsidies for firms effective? The Michigan experience. *Industrial and Labor Relations Review, 46,* 625-636.

Illinois State Office of the Auditor General. (1989, July). *Management and program audit of the Department of Commerce and Community Affairs economic development programs* (Report prepared for the Legislative Audit Commission). Springfield, IL: Author.

Jencks, C. (1992-1993, Winter). *Focus* (University of Wisconsin-Madison, Institute for Research on Poverty), *13,* 12.

Jones, B. D., & Bachelor, L. W. (1984). Local policy discretion and the corporate surplus. In R. D. Bingham & J. P. Blair (Eds.), *Urban economic development* (pp. 245-267). Beverly Hills, CA: Sage.

Knowledge Systems and Research Inc. (1991, August). *Evaluation of twelve New York State technical assistance and financing programs.* Unpublished report prepared for New York State Department of Economic Development and New York State Urban Development Corporation.

Markusen, A. R. (1985). *Profit cycles, oligopoly, and regional development.* Cambridge, MIT Press.

Mt. Auburn Associates. (1992, December). *An evaluation of Ohio's Thomas Edison Technology Centers.* Unpublished final report submitted to the Ohio Department of Development.

Mt. Auburn Associates & the Corporation for Enterprise Development. (1990, June). *The economic impact of business financing by New York state agencies.* Unpublished report prepared for the Department of Economic Development, Urban Development Corporation, Job Development Authority, and Science and Technology Foundation.

Papke, L. E. (1991, December). *Tax policy and urban development: Evidence from an enterprise zone program* (Working Paper No. 3945). New York: National Bureau of Economic Research.

Pennsylvania Economy League, Inc. (1992, May). *Impact of Northeast Tier, Ben Franklin Technology Center on its region.* Unpublished report prepared for Northeast Tier, Ben Franklin Technology Center.

Persky, J., Felsenstein, D., & Wiewel, W. (1993, July). *A methodology for measuring the benefits and costs of economic development programs: Technical documentation.* Chicago: University of Illinois at Chicago, Center for Urban Economic Development.

Price Waterhouse. (1992, March). *Evaluation of the 7(a) Guaranteed Business Loan Program.* Unpublished final report prepared for the U.S. Small Business Administration under Contract Number SBA-5033-FAD-90.

Public Policy Associates and Brandon Roberts and Associates. (1992, October). *Oregon Small Business Services Evaluation.* Unpublished report.

Rephann, T. J. (1992). *Highways and rural economic development: Evidence from quasi-experimental control group studies* (Research Paper No. 9210). Morgantown: West Virginia University Regional Research Institute.

Rohde, S., Cash, J., & Ammarman, K. (1990, March). *Study of the Capital Access Program.* Unpublished working paper, Michigan Strategic Fund, Lansing, MI.

Rossi, P. H., & Freeman, H. E. (1985). *Evaluation: A systematic approach* (3rd ed.). Beverly Hills, CA: Sage.

Treyz, G. I. (1993). *Regional economic modeling.* Boston: Kluwer.

Trisko, K. S., & League, V. C. (1978). *Developing successful programs.* Oakland, CA: Awareness House.

Virginia Joint Legislative Audit and Review Commission. (1991). *Review of economic development in Virginia* (Virginia General Assembly, House Document No. 39). Richmond, VA: Author.

Wiewel, W., Persky, J., & Felsenstein, D. (1994). Are subsidies worth it? *Economic Development Commentary, 18*(3), 17-22.

COMMENTARY ON "CAN ECONOMIC DEVELOPMENT PROGRAMS BE EVALUATED?"

Robert Giloth

Bartik and Bingham's critical review of evaluation research and economic development programs is an important and timely contribution. In a field that typically values action above reflection, they answer the question "Can economic development programs be evaluated?" by usefully contrasting different evaluation designs, summarizing real-world evaluation studies, and, most important, emphasizing the salience of "outcomes" studies and the use of comparison and control groups to isolate the "real" effects of economic development interventions and hence to overcome common "threats to validity" that haunt economic development analysis. Moreover, they recognize the relative underdevelopment of evaluation and knowledge utilization in the economic development arena, and they formulate recommendations for overcoming evaluation barriers, challenging economic development evaluators to be more creative, expansive, and systematic in their work.

Though Bartik and Bingham suggest that evaluation research for economic development is 20 years behind that for employment training, many researchers and policy makers believe that employment training research, despite its measurement and research design prowess, has failed to deepen our understanding about why training programs do and do not work (Osterman, 1988). Evaluation methods are better, but program effects are modest at best. We do not know what works. A part of the problem may be how we think about what is "useful knowledge." Accordingly, the fundamental limitations of this chapter are the authors' premature dismissal of "process" evaluation and their overly narrow definition of the economic development interventions that should be evaluated. The consequences of these limitations are illustrated by the following questions:

- Are economic development programs sufficiently well designed?
- Do we understand all the factors that affect the operation of economic development programs, and hence their outcomes?

AUTHOR'S NOTE: This work has been partially supported by the Annie E. Casey Foundation. I am solely responsible for its contents, and the contents do not necessarily represent the views of the Annie E. Casey Foundation.

- How does economic development affect community well-being and contribute to poverty alleviation?
- Is the traditional role of the "remote and neutral" evaluator viable in the context of reinventing social and economic policy and administration?

One way to begin discussing these concerns is to reflect briefly on my experience in directing a research and development division in the city of Chicago's Department of Economic Development (DED) in the mid-1980s. Much of our time was consumed in trying to understand the job impacts of the city of Chicago's Industrial Revenue Bond (IRB) program at the same time that Congress was passing IRB restrictions, lower interest rates were making IRBs noncompetitive, and there was community outcry about the lack of IRB job-creation performance. Although IRBs are considered an outmoded "supply-side" intervention as defined by Eisinger and others, the IRB experience is useful to examine because of its lessons about program purpose, design, implementation, and impacts and because IRB programs share a number of characteristics with current business finance tools.

Over the course of 8 years, DED conducted six monitoring surveys of IRB job impacts. In addition, five independent evaluation studies (based in part on the DED data) used different evaluation designs—economic modeling, case studies, and comparison groups. The DED monitoring studies showed, in general, that the IRB firms created few net new jobs, failed dismally to reach job projections, and in a number of cases had closed their doors within several years of receiving the incentives. The independent evaluation studies documented a number of seemingly contradictory outcomes, including that there was discrimination in who got the jobs, that IRBs were necessary investments for firm growth, and that there was selection bias in which firms received IRBs—namely, a bias toward large firms from declining manufacturing sectors that had been growing before they obtained IRBs (Giloth, 1991, 1992).

These IRB monitoring and evaluation studies themselves shaped administrative and policy responses. The DED finance division, which was responsible for processing IRB applications, felt threatened and questioned whether the jobs indicator was useful and accurate. DED rhetoric about the public purpose of IRBs changed from emphasizing job creation to job retention. The city council pushed for tighter IRB performance criteria. Meanwhile, the mayor lobbied Congress for a continuation of the IRB program, arguing that the plight and disinvestment of cities mandated the continuation of programs with "positive" urban impacts such as IRBs. Eventually, the state of Illinois captured what was left of Chicago's IRB program; they had a quicker, more "efficient" IRB pipeline and fewer political hurdles. Evaluations of their loan portfolio, not surprisingly, mirrored the Chicago results (Giloth, 1992; Illinois Development Finance Authority, 1988).

This story reveals that IRBs were poorly designed, targeted, and implemented. But this is not the complete story; the tale of Chicago IRBs also identifies more general problems about economic development incentives that should be acknowledged:

- Evaluation studies produce an array of conflicting results about impacts.
- The "black box" of program administration is not an object of evaluation inquiry.
- Programs are not well designed. A part of the problem is that we do not always know what constitutes a good economic development intervention.
- A simple demarcation between process and outcome evaluation does not exist: Case studies can reveal how incentives produce specific impacts; and comparison groups can shed light on program selection processes, embedded in program design and administration, as well as outcomes.
- Typical economic development indicators, such as jobs, taxes, multipliers, and wealth creation, are frequently imprecise and extremely difficult to measure.
- Economic development, more than many public policy interventions, is highly sensitive to political context. That is, the design and operation of economic development programs are susceptible to influence by public and private actors.

Each of these issues directly or indirectly affects the quality of evaluation studies, and together they suggest that Bartik and Bingham's framework for answering whether economic development programs can be evaluated is too narrow.

Two types of contemporary economic development strategies underscore the inadequacy of Bartik and Bingham's scope of inquiry. The first type includes an array of economic development interventions—sectoral strategies and microenterprise—that are long term, with multiple outcomes, and that integrate traditional and nontraditional economic development components such as schools and human services. Developing a biotechnology industry, for example, involves R&D, multiple partners, incubation, venture capital, infrastructure, commercialization, and labor force development. This is a long and complicated process, involving new forms of partnerships and new institutions. It is even more complicated when there is the public expectation that a biotechnology strategy will have employment payoffs for disadvantaged populations. Conventional, outcome-oriented evaluation tools are not particularly helpful in the short run in gauging how this "development process" is doing.

Another example of this strategy involves microenterprises. Microenterprise programs, based in large part on Third World development models, build systems of support to nurture increased self-employment in an economy in which self-employment makes up a rather small percentage of the labor force. There has been much hope placed on microenterprise to contribute to urban and rural poverty alleviation, tapping latent entrepreneurial talent and neglected economic niches. Early evaluation studies of microenterprise, however, questioned the number of business start-ups that were actually being created by low-income participants (Guy, Doolittle, & Fink, 1991). This result had a chilling effect on the microenterprise movement. What had been missed or downplayed by this evaluation approach, according to the practitioners, was that microenterprise programs had important positive outcomes in addition to business start-ups—self-esteem, self-sufficiency, part-time income, investment in human capital, and the building of social networks. There was no single indicator of success (Friedman, 1991; Keeley, 1991).

Reflection on this first type of economic development points to the inadequacy of strictly outcome-oriented evaluations. What is needed to help make sense of unfolding program designs and institutional processes is evaluation that promotes learning, feedback, and self-reflection and that enables emerging interventions to evolve, to refine themselves, and to be evaluated in part by meaningful intermediate outcomes that program operators help to formulate. Bartik and Bingham recognize that program designers are quite skeptical about evaluation, but they do not explore why in any depth. Nor does this reluctance prompt them to think about reconceptualizing their own evaluation paradigm. Traditional evaluation is not useful in many contexts.

New paradigms for evaluation are being explored in the fields of children and family services, comprehensive community rebuilding, services integration, and microenterprise that may have more relevance for interventions such as empowerment zones/enterprise communities than traditional program evaluation does (Connell, Kubisch, Schorr, Weiss, & Carol, 1995; Fishman & Phillips, 1993; Self-Employment Learning Project, 1991; Usher, 1994; Williams, 1991). These forms of evaluation blur the lines between technical assistance and evaluation, attempt to help program administrators become ongoing evaluators, and are concerned with the "evaluability" of the intervention—whether it is ready and willing to be evaluated from the outcome perspective (Usher, 1994).

The second type of economic development strategy that received short shrift from Bartik and Bingham is the so-called "big bang" projects: stadiums, convention centers, world's fairs, major plan relocations, airports, and waterfront development. These projects pose a myriad of evaluation challenges that include ex ante versus ex post evaluations, long-term market impact projections, questions of market substitution and displacement, distribution and equity concerns, and the geography of costs and benefits. Here, the overall focus is on strengthening the economic competitiveness of places within an increasingly global urban hierarchy. Most of the analytic energy associated with evaluation is typically spent on the front end to obtain the approval or defeat of projects; significantly less time is spent on long-term impact evaluations. Successes are obvious, and who wants to talk about failed Rouse projects?

The evaluation issues surrounding "big bang" projects are troubling. First, the impacts of big projects are inextricably tied to markets, project design, and civic focus—the ability of public and private actors to advocate for development. They are civic events, icons of public/private partnerships in action. Evaluation, therefore, inevitably must deal with institutions, their roles in making or breaking projects. A worrisome fact is that the "bidding war" mentality still prevails in many cities and states regarding these types of projects despite the knowledge that they produce minimal or modest impacts in exchange for frequently massive public subsidies and negative "spillover effects" (Schweke, Rist, & Dabson, 1994). Certainly, the research methods used in these studies could be improved, but there remains the fundamental question of knowledge utilization: Bartik and Bingham's framing question should perhaps be changed to "Why bother evaluating economic development when no one cares?"

The second troubling problem with big-project evaluation concerns what is being evaluated—the project itself, a cluster of urban attributes that make up the new downtown, the city, or even the region? A paradox of contemporary urban development is that "urban success" stories often do little for their populations who are most in need and, hence, many would argue, for the long-run viability of cities and regions (Wolman, Barnes, & Neithercut, 1992). Might there have been better investment of resources to achieve these specific ends? Answering that question requires insight into what would have happened without the big project and whether there were better investment alternatives. Is this what economic evaluation should concern itself with?

For too long, evaluation research of economic development programs has not received critical attention. Monitoring and advocacy-based studies have been poor substitutes for well-designed evaluation studies. Bartik and Bingham address this gap by emphasizing the importance of comparison and control groups to assess the outcomes of programs. They offer a sample of research designs and studies to show how it can be done.

My basic concern has been that an outcomes approach to evaluation, though essential as an antidote to often blind and self-justifying policy making by local economic development officials, only makes sense if and when programs are clearly designed and implemented. These preconditions are not easily met when we expand the definition of economic development to include a broader range of "emergent" strategies and projects, in particular those that have an antipoverty intent or that seek to link social, community, and economic development. A challenge for the future of evaluation and economic development is to develop a better sense of the connections between process and outcomes evaluation, an appreciation and methodology of self-evaluation, and a more considered approach for accounting for the role of power and politics in the design, operation, and performance of economic development interventions. This would be a useful complement to the plea for more control and comparison groups.

REFERENCES

Connell, J. P., Kubisch, A. C., Schorr, L., Weiss, B., & Carol, H. (1995). *New approaches to evaluating community initiatives: Concepts, methods, and contexts.* Washington, DC: Aspen Institute.

Fishman, N., & Phillips, M. (1993, June). *A review of comprehensive, collaborative persistent poverty initiatives.* Unpublished manuscript, Northwestern University, Center for Urban Affairs, Evanston, IL.

Friedman, R. E. (1991, June). *Self employment: The frontier of asset based opportunity strategies.* Unpublished manuscript, Corporation for Enterprise Development, San Francisco.

Giloth, R. (1991). Designing economic development incentives: The case of industrial development bonds in Chicago: 1977-1987. *Policy Studies Review, 10,* 161-171.

Giloth, R. P. (1992). Stalking local economic development benefits: A review of evaluation issues. *Economic Development Quarterly, 6,* 80-90.

Guy, C., Doolittle, F., & Fink, B. L. (1991, August). *Self employment for welfare recipients: Implementation of the SEID Program.* New York: Manpower Development Research Corporation.

Illinois Development Finance Authority. (1988). *Industrial Revenue Bond and Direct Loan Program survey results.* Chicago: Author.

Keeley, K. S. (1991, June). *Economic opportunity: Another agenda.* Unpublished manuscript, WomenVenture, Minneapolis.

Osterman, P. (1988). Rethinking the American training system. *Social Policy, 19*(1), 28-35.

Schweke, W., Rist, C. A., & Dabson, B. (1994). *Bidding for business: Are cities and states selling themselves short?* Washington, DC: Corporation for Enterprise Development.

Self-Employment Learning Project. (1991). *Self-Employment Learning Project: Assessment framework.* Washington, DC: Aspen Institute.

Usher, C. L. (1994, April). *Improving evaluability through self-evaluation.* Unpublished manuscript, University of North Carolina at Chapel Hill, School of Social Work.

Williams, H. S. (1991). Learning vs. evaluation. *Innovating, 1*(4), 12-30.

Wolman, H., Barnes, S., & Neithercut, M. (1992). *Urban economic performance and income distribution: Do the poor benefit from urban economic success?* Detroit: Wayne State University, College of Urban, Labor, and Metropolitan Affairs, Center for Urban Studies.

COMMENTARY: EVALUATION YES, BUT ON WHOSE BEHALF?

David Fasenfest

Bartik and Bingham raise a central question for any development planner, practitioner, or politician, one that addresses the core of what constitutes a successful economic development program. Simply stated, it is: "If we cannot measure the effectiveness of a program or policy, how can we hope to know whether we are having a positive impact on a community?" Furthermore, the authors argue that such evaluations not only must be done but can be done, and they show us how.

The first part of the chapter is an extensive exposition of what sorts of evaluations are possible, which methods are preferable for assessing economic development, and the kinds of problems one might encounter in conducting these evaluations. After proposing some alternative models of evaluation, Bartik and Bingham very carefully and critically review examples of the various evaluation strategies to demonstrate the outcomes and pitfalls of each. Finally, the authors tell us that only through comparative group assessments can we really get a handle on whether policies and programs were successful, and they outline the compelling reasons why more such evaluations are not undertaken.

The chapter ends with a proposal for a three-phase approach to economic development linking community, firm, and community impact evaluations. This type of analysis, they maintain, can give us a comprehensive picture of a policy's performance and, when finally coupled with greater federal financing of evaluations, can result in greater professionalization of the economic development profession; it can also have the demonstration effect of good evaluations, letting development professionals know what works and what does not work.

My disappointment with the chapter lies not in what it accomplishes (a systematic review of current evaluation methods) but in the issues it fails to raise in answering the opening question. First, at the core of most evaluations is the implicit question "Do the benefits outweigh the costs of a program?" This question, as Bartik and Bingham carefully point out, is what often prevents politicians from embarking on any sort of evalu-

AUTHOR'S NOTE: These comments reflect research by Laura Reese and me on the nature of market paradigms for assessing local economic development and the need for a shift to a social economy perspective in setting evaluation criteria for development policies.

ation in the first place. But what the authors fail to address (either explicitly or implicitly) is that the notions of benefits and costs are themselves problematic. The method is a powerful tool for evaluating economic outcomes, but it is not without its faults (see Fasenfest & Ciancanelli, 1988, for an example of its misuse in community development decision making). Central to any analysis of costs and benefits is the calculation of an opportunity cost structure associated with pursuing any given course of action or policy and the determination of which benefits are to be counted. To derive such a cost and benefit regime, both a calculus with which to measure activities and a conceptual framework through which to determine what activities warrant measurement must be available. Implicit in Bartik and Bingham's chapter are traditional market criteria measuring aggregate or firm-level data, and there is little apparent concern about whether these criteria are appropriate or whether the results are generally applicable to all segments of the community. Too often cost-benefit analyses are not seen as social constructs but somehow as neutral techniques to arrive at some objective truth that, in the hands of planning technocrats, can result in mistaken or misleading decision rules.

This leads to my second concern with the chapter: How are the time frame, constituencies, and venues for economic development to be determined? Development is often a long-term process, and evaluations may have to be made before final costs or benefits materialize. Evaluations are called for by politicians or planners who are themselves often not answerable to those most immediately affected by any negative effects of a policy. As Cox (1993) pointed out, local development is undertaken with little agreement about how the locality is to be defined, what the concepts of *economic* and *development* really mean, and which of a community's competing interests are to be served. Economic development seems to be motivated by a simple notion that "more jobs are good" (Beaumont & Hovey, 1985), and the piecemeal approach to development in general reduces programmatic impacts and limits the attainment of stated development goals (Bingham & Blair, 1984). The purposes and goals of programs, specifically who in the community is to be helped and why, all seem to go unquestioned in the process of merely evaluating the outcome of these programs by some market-based set of costs and benefits.

Market-driven economic development policy is a response to social problems and a reflection of the ability of particular social actors to define these social problems. There is no reason to assume that (market-centered) profit-maximizing and (community-centered) social-maximizing behavior will necessarily coincide. The degree to which social goals are realized is a function more of serendipity than of design, yet most program evaluations measure outcomes in terms of the market frames within which the programs were initially conceived. Markets, it would seem, dominate over communities, but, to paraphrase Polanyi (1944/1957), we need to begin to understand the community as something in which the economy is embedded as part of the overall social relations; markets do not emerge out of a vacuum but out of the social circumstances that surround them. We must see community interests as distinct from and encompassing more than allocation efficiency criteria.

Consequently (my third concern), as Eisenschitz (1993) pointed out, we have to separate our understanding of community development from economic growth. Most of the methods and approaches discussed in this chapter presume that the two are the same. That a policy or program succeeds at increasing aggregate (or even firm-level) wealth says little or nothing about its impact on resource reallocation or redistribution. Displacing development with growth means that the outcome, whether more jobs or higher incomes, shifts attention from more essential problems such as the maldistribution of jobs and resources, creating even larger problems such as income inequities, poverty, crime, and chronic unemployment. It means that in the end we measure the wrong outcomes: Evaluating programs by using job growth or per capita income changes will not address whether these community problems have improved.

Clarke and Gaile (1992) pointed out that often economic development is a quagmire of good intentions and bad measures. Perhaps the problem is rooted in a failure to understand that we must measure more than basic economic (market) outcomes. As Feagin (1988) argued in his discussion of Houston, communities and cities reflect both the social requirements of a given set of economic relationships (in this case, a market economy) and the struggle within those communities that shapes and selects from among a number of possible outcomes. We measure market outcomes and confuse them with community outcomes. Because the overriding goal of local economic development efforts has been to attract jobs and support lagging tax bases, it is not surprising that the most frequently used outcome measures revolve around economic health and performance indicators. More narrowly, many evaluative studies have focused almost exclusively on measures of change in local per capita income and job growth (Clarke & Gaile, 1992). What such measures have in common is a tendency to reflect only certain types of economic change, to overestimate positive outcomes, to be inflated due to their stimulation by negative social externalities, and to devalue or ignore entirely nonmarket impacts on the community, the household, or the neighborhood economy.

Measures such as local per capita income, income growth, and employment or job growth are misleading. Often research has confused economic health as measured by the number of jobs per resident with the economic health of city residents measured by their per capita income (Ladd &

Yinger, 1989). Measures of growth in jobs may be distorted by income inequalities and increased service costs at high levels of in-bound commuting workers (Clarke & Gaile, 1992). Thus, at the root, commonly used economic development outcome measures fail to make a connection between changes in individual wealth and local fiscal health. In other words, "A city can be healthy in the sense of generating many jobs per resident at the same time that its residents remain impoverished" (Ladd & Yinger, 1989, p. 17).

Bartik and Bingham are correct in asserting that evaluations of economic development programs and policies can and should be done. And they are correct when they discuss the methodological issues of how such an evaluation might be undertaken. The issue this chapter begs, however, is whether there is more to an evaluation than simply the *form* of the evaluation. The title of the chapter leads us to expect that the question "Can a program be evaluated?" should encompass a discussion of what such an evaluation will uncover. As their last two sections are written, evaluation criteria center on aggregate impacts of changes in firm behavior and how to assess whether and to what degree changes in firm behavior can be attributable to particular programs. There is no discussion, at the very least, of any attempt to examine whether social goals and objectives (such as, perhaps, greater equity within the community) were met by a particular policy even if the firm-based indicators produce positive results.

During the urban renewal programs of the 1950s and 1960s, "community development" often meant displacing low-income families and replacing them with higher-income families. Average per capita income of community residents rose, the average quality of housing in the neighborhood improved, and it would have appeared that these programs for given communities were a success. On reflection, however, the national stock of adequate low-income housing during that period fell, and the displaced poor people were driven into ever-worsening living situations in their search for any housing.

Employing better, though more sophisticated, evaluation techniques at that time would not have produced a change in policy as long as the measure of success was aggregate change in a community. What was needed then, and what is needed now, is a critical evaluation of the *purpose* as well as the *outcome* of public policy: How well did it serve the social goals of community development writ large rather than the narrow sense of improved economic indicators or economic growth, regardless of how that growth may be distributed? Community development is more than economic growth. Yet as this chapter attests, the distinction in evaluation is usually blurred, to the likely detriment of community development overall.

REFERENCES

Beaumont, E. F., & Hovey, H. A. (1985). State, local and federal economic development policies: New federal patterns, chaos or what? *Public Administration Review, 45,* 327-332.

Bingham, R. D., & Blair, J. P. (1984). *Urban economic development.* Beverly Hills, CA: Sage.

Clarke, S. E., & Gaile, G. L. (1992). The next wave: Postfederal local economic development strategies. *Economic Development Quarterly, 6,* 187-198.

Cox, K. (1993). The concept of local economic development policy: Some fundamental questions. In D. Fasenfest (Ed.), *Community economic development: Policy formation in the U.S. and U.K.* (pp. 45-64). New York: St. Martin's.

Eisenschitz, A. (1993). Business involvement in community: Counting the spoons or economic renewal? In D. Fasenfest (Ed.), *Community economic development: Policy formation in the U.S. and U.K.* (pp. 141-156). New York: St. Martin's.

Fasenfest, D., & Ciancanelli, P. (1988). Public costs and private benefits: The pitfalls of capital budgeting for reindustrialization. *Journal of Urban Affairs, 10,* 291-307.

Feagin, J. (1988). *Free enterprise city: Houston in political-economic perspective.* New Brunswick, NJ: Rutgers University Press.

Ladd, H. F., & Yinger, J. (1989). *America's ailing cities: Fiscal health and the design of urban policy.* Baltimore: Johns Hopkins University Press.

Polanyi, K. (1957). *The great transformation.* Boston: Beacon. (Original work published 1944)

REJOINDER

Timothy J. Bartik
Richard D. Bingham

We agree with David Fasenfest's position that the perfect economic development evaluation would measure precisely the economic and social impact of the project on all individuals, including and especially the poor. But such perfect evaluation is rarely feasible. We strongly disagree with Fasenfest's position that more feasible evaluations, which attempt to measure how an economic development program has affected an area's job growth or some other business-related variable, are misleading.

Fasenfest's position is that whether "a policy or program succeeds at increasing aggregate (or even firm-level) wealth says little or nothing about its impact on resource reallocation or redistribution. . . . Evaluating programs by using job growth . . . will not address whether . . . community problems [such as income inequities, poverty, crime, and chronic unemployment] have improved." Our judgment is that the empirical research literature says quite the opposite. In general, tightening local labor markets through faster job growth will have progressive effects on the distribution of income, providing greater-than-average percentage increases in income for the poor, minorities, and the less educated (see Bartik, 1991, 1993, 1994, for reviews of the research literature).

Each economic development program or project is, of course, unique. A particular program may increase area job growth yet not help much to reduce poverty, for example, if the jobs provided all go to upper-income in-migrants and if these in-migrants spend the bulk of their money on nonlocal goods. But increased overall local job growth is so likely to provide progressive social and economic benefits that increasing such growth is not a bad intermediate goal for policy. Better evaluations that get at ultimate outcomes are always desirable. Feasible studies that focus on intermediate outcomes with a strong connection to such ultimate outcomes are also desirable.

Fasenfest compares our approach with urban renewal evaluations that regarded a project as successful if it increased a neighborhood's property values and per capita income. In the urban renewal case, increasing neighborhood property values and per capita income are likely to have a *negative* correlation with the well-being of the original residents. Increasing neighborhood property values and per capita income would indeed be a problematic intermediate goal for policy. But in the case of economic development programs, increased metropolitan-area job growth on average will have a strong *positive* correlation with the incomes of the poor.

We strongly agree with Robert Giloth that process evaluations of economic development programs can be quite useful. In some cases, such evaluations are the only game in town because outcome evaluations are politically or technically unfeasible. Even if outcome evaluations are possible, process evaluations provide needed additional information about why a program did or did not work. Such information helps program administrators improve program design. As we said in the chapter, process evaluations and outcome evaluations should be seen as complementary, providing different information for different purposes.

But we would still defend our position in the chapter that the greatest need right now in economic development evaluation is for more rigorous outcome evaluations. Although more evaluations of all types are needed, there have been a fair number of process evaluations of economic development programs. There have been very few outcome evaluations of economic development programs, and even fewer that could be regarded as at all rigorous.

Furthermore, because of American political traditions, the legitimacy of economic development programs is frequently challenged. Only rigorous outcome evaluations can show whether economic development programs actually are effective in achieving public goals at an affordable price. Learning more about the design of economic development programs will not help much if the programs are abolished because of lack of proven cost-effectiveness.

REFERENCES

Bartik, T. J. (1991). *Who benefits from state and local economic development policies?* Kalamazoo, MI: W. E. Upjohn Institute for Employment Research.
Bartik, T. J. (1993). Who benefits from local job growth, migrants or the original residents? *Regional Studies, 27,* 297-311.
Bartik, T. J. (1994). The effects of metropolitan job growth on the size distribution of family income. *Journal of Regional Science, 34,* 483-501.

11

Is There a Point in the Cycle of Cities at Which Economic Development Is No Longer a Viable Strategy? Or, When Is the Neighborhood Too Far Gone?

WILLIAM W. GOLDSMITH

The control and regulation of economic life by progressive, democratic and accountable governments is the first condition for the alleviation of the ills plaguing the men and women who are not part of the small minority which prospers in conditions where the organizing principle of social life is competition.

—Ralph Miliband and Leo Panitch,
Real Problems, False Solutions (1993)[1]

The past forty years have not been kind to American cities; almost all older industrial cities are losing population and economic investment and are becoming locations of concentrated poverty and unemployment. Much of the population remaining in these cities lives in areas that provide poor educations, have high crime rates, and in every respect offer a sharply lower quality of life than that enjoyed by other Americans.

—Norman Krumholz and Pierre Clavel,
Reinventing Cities (1994)[2]

There is no [neighborhood] situation that is so bad that it warrants public inattention.

—Robert Mier, former Commissioner of
Economic Development, City of Chicago,
personal communication (1994)

When Richard Bingham and Robert Mier invited me to write this chapter, I was struck by the difficulty of dealing simultaneously with three complex notions. First, *which* city? The word *city* means different things in different contexts, from the encompassing notion of the metropolitan area to the limited notion of the legally bounded center of the metropolis, which in the United States on average houses less than 40% of the metropolitan population. Things look very different from perspectives of metropolis and central city: Over the long period from 1950 to 1990, for example, many central cities have lost population—and therefore tax base, municipal resources, and services—especially old industrial centers such as Detroit, Cleveland, and Harrisburg, whose population losses ran from 42% to 45%. In the same 40 years, however, metropolitan areas grew. Even in the areas including those three cities, the populations grew 35%, 19%, and 42%, respectively. The boundary lines we set for analysis make a great deal of difference.

Second, *what* cycle? Regions and metropolitan areas surely have good times and bad, depending on the timing of business cycles; product cycles; patterns of resource discovery, use, and depletion; and major industrial shifts, such as from manufacturing to services. Areas also follow physical trajectories constrained by the dominant construction and transportation technologies of their boom growth years: Railroads, streetcars, and elevators make for dense cities; highways, tract houses, and automobiles for sprawl. City fortunes probably do follow broad political cycles of the sort discussed by social theorists such as Raymond Williams, E. P. Thompson, and Albert Hirschman, in which first progressive reformers and then selfish reactionaries may take their turns (see, e.g., Hirschman, 1991). That is, there are *times* of urban rise and decline, but any predictable cyclical *pattern* seems highly unlikely.

Third, *whose* economic development? Shall we be concerned with the economic development of the city's business firms or the economic well-being of its citizens? To which citizens do we refer, those with stable, well-paying jobs and therefore with good schools and services, even if they have to purchase them privately, or those without such benefits? Are there times when most citizens benefit because business is doing well? When not?

To proceed with the analysis, I must arrive at tentative answers to, or at least formulations of, these questions. Bingham and Mier, I believe, meant for me to pay attention to *central* cities and their neighborhoods; to look at times of stress, whatever the origin and regardless of the absence of cyclicity; and to think of the needs of poor and dependent citizens. In this context, we come to a fourth question: How are we to think about the words *viable strategy?*

AUTHOR'S NOTE: I thank my Cornell colleagues Matthew Drennan and Robert Abrams and referees Peter Meyer and Carol Waldrop for challenging comments.

In 1995, as I write this, the outlook is decidedly pessimistic. Almost no strategy of economic development for American cities—and now I mean central cities—seems viable. The combination of aging buildings and utility systems, declining blue-collar jobs, impoverished populations, and intensified international competition seems overwhelming. No one contending for national leadership will admit that there might be an urban strategy.

Why is it, in any case, that the United States needs an *urban* policy? Why in the United States do social and economic problems so afflict *cities*, rather than the country as a whole? European cities, even in the 1990s with extraordinarily high national unemployment rates, do not seem to suffer disproportionately. In the case of the struggling cities of Africa, Asia, and Latin America, whose citizens certainly suffer deep indignities, scholars write persistently of "urban bias," that is, of national policy that *favors* city dwellers over those who live in the more heavily populated countryside. The American city is peculiar, indeed. Planners, scholars, activists, and politicians who would repair and revive the city must reach for new ideas, and they must put attempts at strategic thinking high on the agenda.

■ Market Forces and Poor Neighborhoods

In this chapter, I aim to support the following sorts of contentions in thinking about troubled places: Public authorities should rarely give up on economic development in neighborhoods, cities, and regions that have been damaged by past decline and decay. They should not abandon places whose resource bases have shifted for the worse. They should adapt and redevelop places whose most skilled and energetic people have fled for better opportunities, or whose remaining residents are impoverished and seem to be without initiative. To put it differently, public authorities should not ignore the needs of those people made invisible by history; instead, they should work to make them visible and help them to meet their needs.

There are three reasons to promote such equitable policy. It is the decent thing to do, as it provides help to those most in need and is, in an immediate sense, fair. It protects those also who are better off, especially in the case of city neighborhoods and suburbs, by stemming the growth and spread of problems as diverse as street violence and tuberculoses, which have in recent years begun to force themselves outward from the central cities. In the long run, equitable urban policy would be efficient and beneficial economically for the whole society—save perhaps the very rich in some narrow, short-run, and undemocratic sense—through reduction in wasted human and material resources and through a more inclusive and reasonable politics.[3]

In normal urban practice and in the design of urban policy in the United States, however, those who are better off typically abandon the old and the weak. When cities wear down, as their infrastructures wear out and their cultural appeal wears thin, privileged people move up and move out, sometimes to newer cities, most often to ever-newer suburbs. When regions wear out, then investors move to greener fields, to untapped resources, to active markets. What is left behind—the public works, the buildings, the trapped political systems, the stranded businesses, and many of the poorest people—pays a steep price. So, I will argue below, does nearly everyone else.

Although private-sector real estate markets and labor markets serve well to distribute and allocate space and jobs, they also reinforce inequality through their normal operation, and in U.S. cities they are notoriously biased against people of color. Because these inherently biased markets provide much of the context for urban change, that change therefore often works against the interests of the poor and of minority persons. Particular elements of public policy—taxes, subsidies, and regulations, as well as patchworks of local political jurisdictions that are formidably unequal in their power and influence—frequently buttress these privately biased processes of selective change.

Despite widespread agreement that the politics of market individualism and rapid change have historically been hard on some people, during the long economic boom that followed World War II, progress and expansion gave hope for the future to many even as it left them temporarily behind, often in inner-city neighborhoods. Some observers of that historical process judged it to have been adequate along lines of equity, a trajectory along which they wanted the country continue, eventually incorporating the poor into the middle class. Others—the majority of careful observers, I suspect—were less sanguine, wanting less division and quicker progress. Between these positions lies ample space for debate about the distributive effects of postwar urban development.[4]

In the period since 1980, however, the space for debate has diminished, squeezed down by two forces that have worsened the situation. The first force is a trend resulting from economic and technological changes toward more services and less physical production, on the one hand, and toward cheaper and more effective communications facilities, on the other. These combine to threaten the very existence of many central urban areas, as they decrease the need for many central locations and increase the accessibility of spread-out places, including suburbs. The second force toward more pronounced social inequality is a not unrelated politics of fear and selfishness that has accelerated the tendency toward physical separation of people who are rich and poor, lighter skinned and darker skinned.

A rightward shift in U.S. politics has plagued the country, exacerbating division and inequality while inflicting deepened poverty on ever-larger numbers of people. The politics of the 1980s and early 1990s has fostered divisions in U.S. society unparalleled since the 1920s and unequaled in any other wealthy, modern industrial society. These divisions are reflected in book titles such as *Caught in the Crisis* (Amott, 1993), *The Great U-Turn* (Harrison & Bluestone, 1988), *Separate Societies* (Goldsmith & Blakely, 1992), *Savage Inequalities* (Kozol, 1991), and *Faces at the Bottom of the Well: The Permanence of Racism* (Bell, 1992). Even highly conservative election strategists acknowledge how bad the policies of the 1980s were. For example, from Kevin Phillips we learn that "the top two million income earners in this country earn more than the next one hundred million" (quoted in Bell, 1992, p. 9). As Reaganomics prevailed, poverty rates rose, and city neighborhoods across the country fell into disrepair.

Fortress suburbs are now being erected by the hundreds to protect the well-off, who hide inside; the cities are battle zones of the poor. Among the many social forces prompting the shift to the suburbs and neglect of the cities, the changing focus of federal attention is significant. In the Reagan-Bush years, for example, the federal government slashed urban and regional programs so that transfer payments to localities fell nearly 60% in real value from 1981 to 1989 (Judd & Swanstrom, 1994, p. 297). The anecdotal evidence is everywhere: walls, gates, and guards in the suburbs; homelessness; drive-by shootings, muggings, and dangerous schools in the cities. The statistical evidence is mounting rapidly as well. Here are just two indicators: Violent teen deaths in the United States, already alarmingly high, have been rising astronomically—they rose by 13% in only 6 years, from 1985 to 1991—and California's prison population increased fivefold from 1980 to 1990.[5]

With the new and fiercely conservative Congress elected in 1994, even the most hardened political observers should expect central-city neighborhoods, and the poor and minority people who live in them, to suffer further. Congress is likely to expand privileges for the rich (justifying this perverse generosity as a stimulus for economic growth that they claim will trickle down to the rest of the country), while hoping to maintain the living standard of the majority of suburban residents. But the meager aid now available for the destitute, the working poor, and other city residents of modest means will surely be diminished still further.[6]

It is within this dismal political situation, combined with and partly a result of global economic pressures, that planners in the 1990s must confront questions of local economic development. Although even in better circumstances many planners harbor doubts about efficacy and even viability of redevelopment in neighborhoods that suffer from decay and in regions in decline, such doubts assume new urgency in this era of highly fragile welfare guarantees and virulent neoliberalism. With the passing of what is sometimes called the American century of economic growth that characterized the 30-odd years after World War II, economic declines have threatened marginal populations everywhere. Globalizing competition abetted by public policy has pushed groups apart, especially in countries that have opted for hyperderegulation and all-out privatization, and poverty has exploded in the most extreme cases, including the United States. None other than the president of the Federal Reserve Bank of New York, speaking publicly in November 1994, wrote of

> the growing disparity in wages earned by different segments of our labor force. . . . We are forced to face the question of whether we will be able to go forward together as a unified society with a confident outlook or as a society of diverse economic groups suspicious of both the future and each other. (quoted in Cassidy, 1995)[7]

> In the United States, well-being and poverty are very much matters of ethnicity and especially race—even though most poor people are white. The politics of poverty, particularly urban poverty, is a politics of race. As Supreme Court Justice Ruth Bader Ginsburg wrote in 1995, the "system of racial caste only recently ended [is] evident in our workplaces, markets and neighborhoods" (*Nation*, July 1995, p. 39). Racial segregation and geographic isolation are the hallmarks of urban poverty. Minority populations are disproportionately poor; African Americans are nearly four times as likely to be poor as are whites. In addition, gender matters. Three fifths of all households headed by women with children are poor. Women of color and their children are therefore forced into a debility that threatens to be permanent, and many are fenced, along with single men and other women of color, into well-defined ghettos. These zones of widespread poverty and decay are highly isolated, but swelling out. "The geographic spread of the ghetto . . . exploded in the 1980s." (Jargowsky, 1994, p. 194)[8]

Difference and isolation yield "invisibility," and invisibility yields ignorance and hostility.[9] Together, these yield the violence that now spills beyond the boundaries of deprived zones of U.S. central cities.[10]

These zones are the most "underdeveloped" parts of U.S. cities, and they are the ones that present the most serious dilemmas to many development practitioners. As I write this, the country is preoccupied with the anger of the rural right-wing fringe in the aftermath of the Oklahoma City bombing. There is little doubt, however, that the hostilities of conservative politics in the United States will soon be directed once again toward the poor people who inhabit central cities. In fact, the connection is more direct. A good part of the paranoid thinking of the militia movement—

which only rarely breaks out in psychotic episodes such as the Oklahoma City bombing—appears to hinge on simplistic victim blaming.[11] Right-wing politicians encourage this sort of thinking, as they cynically blame the country's economic and social ills on advantages and expenditures (such as welfare) supposedly paid out generously to African Americans, Hispanics, immigrants, and poor residents of central cities.[12] This vicious myopia is encouraged by geographic separation and social isolation.

For some time, those who govern cities have worked to isolate further their cities' worst ills—to conceal the results of superexploitation and concentrate it in what Camilo José Vergara (1995) called the "new ghettos," where the poor, the homeless, and the ill, especially minorities, are warehoused, overloading neighborhoods with needy, institutionalized populations.[13] Now Congress has mounted a matching federal attack that looks even worse than the Reagan administration's enterprise zone idea for its ability to denude such neighborhoods of income and social protections. When House Speaker Newt Gingrich, the Christian Right, and other arch-conservative groups speak of gutting the "welfare system" and penalizing "welfare mothers" either to cut taxes for the rich and business or to reduce the national budgetary deficit, they aim with a sharp focus at poor people of color who live in central-city neighborhoods. Critics have shown that the right-wing attack is economically absurd. There are about 2 million poor black families in the country, but nearly 4.5 million poor white families; less than 1% of the federal budget goes for welfare (Aid to Families with Dependent Children [AFDC]) and only 2% of state budgets, on average. AFDC benefits paid per family *diminished* by 37% in constant dollars between 1975 and 1994; counting the better distributed food stamps on top of AFDC, the benefits still fell 27% between 1972 and 1993.[14] Nevertheless, there can be little doubt: These neighborhoods and the people in them are demonized, are blamed for myriad failures in U.S. society today, and are likely to suffer the bad effects of misguided policy (Corbett, 1994).

This conservative attack, insofar as it shifts the blame on to residents themselves and makes them appear both helpless and undeserving, only obscures the real causes of underdevelopment and thereby holds back progress in these inner-city zones.[15] To get around some of that difficulty, in the next section I shift the discussion temporarily away from the inner city.

■ Invisibility, Profitability, and Decline

The problems confronting U.S. central cities and the people who live in the poorest neighborhoods are particular indeed. The process of impoverishment, however, the obstacles to improvement, and the mechanisms that would serve for development are more general. The generality becomes apparent as we survey other troubled places. To push questions about local economic development to the limit, I first consider small and declining rural towns, then "hopeless" nations on the periphery of the global economy, and finally passed-over mining regions.

For the first case, consider the isolated small towns and scattered rural settlements of the Dakotas. These areas may indeed stretch the idea of local economic development, but for that reason alone they are instructive. In this limiting case, one of the essential constraints against local development appears starkly: the problem of affluent outsiders pushing less powerful locals away, ignoring them, rendering them invisible.[16] According to poet and writer Kathleen Norris, in her 1993 book, *Dakota: A Spiritual Geography,*

> A sense of loss has begun to haunt . . . residents of western Dakota. . . . Less visible than the poor in urban areas, they are afflicted by physical isolation in particularly severe ways. People who are laid off . . . have to travel one hundred miles to a Job Service office in order to register for benefits, and there is no public transportation on the route. . . . Our inability to influence either big business or big government is turning all Dakotans into a kind of underclass. Native Americans are by far the poorest and most invisible among us . . . [but] western Dakotans are increasingly invisible to more affluent eastern Dakotans, and Dakotans as a whole are invisible to the rest of America. (pp. 30-31)

The second case is international. In development circles, some people speak of the poorest countries as "lost causes." They refer to these countries as hopeless "basket cases," eligible for handouts but not expected ever to prosper from their own strength. They speak of Bangladesh, Tanzania, or Guinea-Bissau in these ways.

A telling example involves the Dinka people of Sudan, who suffered severely from a famine from 1985 to 1989. To outsiders, when the Dinka are not invisible, they are guilty of the incompetence that causes their own starvation. Like other famines the world round, this is seen to be a local failure of physical and human resources. In contrast to this invisibility and these ideas of incompetence, however, critics now bring a much more forceful idea into play, the idea of purposeful exploitation.

In his 1994 book *The Benefits of Famine,* David Keen focused attention on local vulnerability to outside forces. He asked how beneficiaries create and perpetuate a famine, who these beneficiaries are, and how they rationalize their gains. In this account, people, businesses, and agencies benefit, often directly, from real suffering, and their victims are falsely made to appear passive and all but invisible, as the sad objects of abstract, disembodied processes.

The famine—with deaths across the south in 1989 put at over 500,000—promised, and to a large extent yielded, significant benefits (many of them economic) for a variety of powerful Sudanese interests, who helped to create the famine. . . . These "beneficiaries"—at the local level and in the central government—were able (and permitted) to manipulate famine relief for their own purposes. For the donors, although the famine itself was not functional, the pursuit of narrowly defined relief agendas served important functions, even as it tended to allow the creation and perpetuation of famine. (p. 1)

Famine can serve important functions. . . . When famine occurs, we may hear a great deal about "market failure" and "policy failure." But who exactly is failing (and who succeeding) when markets behave in particular ways, or when the expressed policy of relieving famine does not work out in practice? (pp. 11-12)

It is one thing to urge that people behave better, it would probably be more helpful to suggest ways in which those who are concerned with improving relief might take into account—and counter—the priorities of those who are not. (p. 13)

For the third case, consider Appalachia. The despair of old mining areas in Appalachia was severe for decades, long before presidential candidate John Kennedy read Michael Harrington's muckraker *The Other America.* The despair continues today. Although there has always been a body of analysis attributing Appalachian underdevelopment to internal weakness and failure, many insiders suggest that outsiders are myopic, if not blind. The first statement below (criticizing the main federal agency of the 1960s) is by one of the most experienced and respected of all Appalachian commentators, Harry M. Caudill, and the second is by a young West Virginia mountaineer:

As a watchdog of the region [the Appalachian Regional Commission is] blind and toothless. It is sorry in conception, sorry in execution. (quoted in Whisnant, 1980, p. 99)

I was at a meeting last week, and we all got to talking, and they had some "observers" there. That's what they called them, an expert on "community action" one of them was, and an expert on "group process" the other one was. We were all saying the same thing, what a rotten shame it is, the way things have gone in the sixties—the hopes raised and the disappointment. "You all seem discouraged, depressed," the "group process" guy said. I didn't say anything. No one did. Then he asked us if we wanted to *talk* about that, our discouragement. I couldn't take it any longer. I said, no, I didn't; but I'd like for him to go over to Washington, D.C., and get the

President and some Congressmen and ask them if *they* wanted to talk about anything. (Coles, 1971, quoted in Whisnant, 1980, p. 126)

In his insider's account, *Modernizing the Mountaineer*, David Whisnant (1980) contrasted two very broad models for viewing underdevelopment. In the first and familiar model, a brief but unusually clear expression of the ideas of the "culture of poverty," he laid out the mind-set that motivates powerful outsiders, such as those who set up federal development programs:

> Poverty
> leads to
> Cultural and environmental obstacles to motivation
> which lead to
> Poor health, inadequate education, and low mobility
> limiting earning potential
> which leads to
> limited income opportunities
> which lead to
> Poverty.
>
> (p. 101)

In his criticism of this mind-set, Whisnant pointed out that even Oscar Lewis, who propounded the "culture of poverty" idea in the 1960s, recognized the outside sources of such destructive cultures. In 1966, Lewis wrote that such pathologies are "likely to be found where imperial conquest has smashed the native social and economic structure and held the natives, perhaps for generations, in servile status" (qtd. in Whisnant, 1980, p. 99; see also Goldsmith & Blakely, 1992, pp. 4-10). Whisnant therefore suggested a different model, one that provides a more comprehensive and also a more accurate view:

> Corporate monopolization of major resources
> leads to
> an inequitable and undemocratic economic and
> political system
> which leads to
> political powerlessness, economic and cultural exploitation
> and environmental destruction
> which leads to
> poor education and social services, minimal income,
> hopelessness, and out-migration
> which facilitate further
> Corporate monopolization of major resources.
>
> (p. 101)

Those who offer these three examples suggest that places can be victimized and that regional (or national, or neighborhood) decline can be provoked and promoted by outsiders (and collaborating insiders) who find benefits to be gained at the expense of others. People benefit, that is, from economic decline; some, whether they choose to know it or not, even profit from hunger and starvation. Privilege and money are the rewards that such oppressors seek. Invisible people cannot sufficiently resist; often they cannot make effective demands, even for the most elementary services for their own defense. These lessons are important to remember as we shift attention to the poorest city neighborhoods, where it has become commonplace to blame the victims.

■ Views From the City

To move from the sparsely inhabited Dakotas, the refugee camps of Sudan, or the hills of Appalachia to the crowded streets of a large U.S. metropolis, we must, of course, travel far in all senses, but when we arrive, we hear surprisingly similar plaints, posed in familiar terms that emphasize invisibility and subjugation to outside power. Here, for example, is what Maynard Jackson said in his inaugural address as mayor of Atlanta in 1974: "We must open our eyes if we are to begin to deal with the systematic eradication of poverty and the diminution of crime. We must render visible the invisible." Twenty years later, in 1993, when David Dinkins was mayor of New York City, he complained against those who would give up on the ghetto, suggesting, it would seem, that they found invisible the faces, names, and tongues of those who live there. "Those who bemoan the death of the old neighborhood," Mayor Dinkins said,

> reject the economics of upward mobility for all New Yorkers. They fail to understand, or refuse to understand, that the old neighborhood is alive and well in our city, that little has changed in our New York neighborhoods except the faces, the names, and the languages spoken. The same decent values of hard work and accomplishment and service to city and nation still exist.

Indeed, outsiders employ a language of destruction and abandonment in addition to invisibility to disguise their ignorance, shield and protect their benefits, or simply distance themselves from neighborhoods of very poor people in the cities. The language is apt because public policy so often destroys, abandons, and makes invisible. Gordon Parks, writing of "The Cycle of Despair" in *Life* magazine in 1968, demanded not only that outsiders see more clearly but that they connect with what they see: "For

I am you staring back from a mirror of poverty and despair, of revolt and freedom. Look at me and know that to destroy me is to destroy yourself." Martin Luther King, in a *Look* magazine article, also in 1968, denounced exploitation as well as abandonment as he insisted on parallels in his protests against the war in Vietnam and against the war on the cities:

> We are on the side of the wealthy and secure while we create a hell for the poor. Somehow this madness must stop. We must stop now. . . . I speak for those whose land is being laid waste, whose homes are being destroyed, whose culture is being subverted. I speak for the poor of America.

Some who fight back speak a language of vision. In his dissent from the Supreme Court's 1974 decision in *Milliken v. Bradley*, which rejected lower-court arguments that Detroit's suburbs should share responsibility with the city for its needy schoolchildren, Justice Thurgood Marshall warned that "in the short run, it may seem to be the easier course to allow our great metropolitan areas to be divided up each into two cities—one white, the other black—but it is a course, I predict, our people will ultimately regret." When Harold Washington announced his candidacy for mayor of Chicago in 1982, he used these words:

> Finally, "the city that works" doesn't work anymore. . . . I see a Chicago in which the neighborhoods are once again the center of our city, in which businesses boom and provide neighborhood jobs, in which neighbors join together to help govern their neighborhood and their city. Some may say this is visionary—I say they lack vision.

■ Efficiency and Equity

From these perspectives—in which abandoned, degraded, and miserable neighborhoods (or regions or even countries) are not just ignored, but also pushed out and exploited—the question becomes not *whether* but *how* to promote local economic development. The old efficiency-equity trade-offs, by which the whether-or-not questions are so often judged, need to be reframed. Although one cannot ignore how sparseness of population and the great distances of the plains confront the Dakotas with special problems, it would be even less sensible and also unfair to ignore how outside forces and decisions brought about the Dakota situation and maintain the subordination. And although it would be unreasonable not to take into account the paucity of resources available to the Dinka in Sudan or the mountain people of Appalachia, it would be irresponsible to ignore how outsiders have profited and still profit from the ownership, extraction, and sale of those resources.

In formal and academic discussions of efficiency-equity trade-offs, efficiency usually wins. Analysts have typically answered questions about the viability of economic development for regions, cities, or neighborhoods with abstractions—after calculation of the net balance of costs and benefits, trade-offs of equity and efficiency. For people trained in economics and planning, and for many others influenced by the last generation's experience of steady material progress, this sort of thinking comes easily: Costs borne by some can after all be compensated with portions of the greater benefits bestowed on others. This is the whole notion of Pareto optimality, a central axiom of economic theory.[17] With the rise of right-wing influence in U.S. politics, efficiency arguments are likely to win more and more in the real world. In any case, in places where the costs of rehabilitation are very high, expending resources locally and thereby losing the opportunity to expend them elsewhere does risk high costs and excessive loss in efficiency.

From this perspective, winning the inning to save the neighborhood may mean losing the game over saving the city; or, on a larger scale, wasting funds on a dying city may weaken the economy of the nation. This is the hard-boiled, analytic answer: Why throw good money after bad? If East St. Louis is in collapse, better to let it go and save the money to put new investments in booming suburbs, where the payoff will be higher. The logic, of course, is that once an area suffers from a circular process of decline, it is too expensive to rescue. Inadequate services drive out small business and residents, the tax base shrinks and reduces municipal revenues, services are cut further, and so on endlessly. East St. Louis is a favorite target. Its population fell 50% in the 30 years up to 1990, its property values collapsed, and its services and infrastructure were ruined. The tax rate quadrupled, the budget collapsed, and a court even awarded City Hall to a creditor.[18]

The 1979 transition team on cities selected by outgoing President Jimmy Carter observed that limits against wasted effort ought to be part of federal urban policy. Some years earlier, policy makers had become enamored of a strategy called "triage," whereby neighborhoods or even cities would be divided into three groups: healthy places likely to develop on their own, without any assistance; transitional places that would multiply aid and investment into self-sustained development; and deteriorated places almost sure to fail, whose residents (and perhaps businesses) ought *not* be encouraged to develop but might be offered services and subsidies to move to places with more opportunities.[19]

Such triage policy in cities was not uncommon in the 1970s. In numerous cities, planners and consultants followed federal guidelines to recommend the reduction of services in blighted neighborhoods and to discourage investment, saving resources for use in salvageable—that is, more efficient—neighborhoods (Judd & Swanstrom, 1994, p. 288). Researchers have observed elsewhere that such policy can be doubly beneficial to some parties, as the resulting swings in real estate value allow big investors to buy at the bottom of the market, cashing in when market forces or public subsidies finally stimulate redevelopment.[20] In the penurious budgetary atmosphere of the 1990s, where any increase in funding is likely to imply another cut, this sort of triage logic becomes more and more difficult to escape.[21]

■ Politics and Poor Neighborhoods

There is an alternative logic—one suggested by the Dakota, Dinka, and Appalachian examples, and one urged by an informed set of progressive urban activists and writers. This logic begins with the assumption that human communities enjoy an inherent capacity to develop and improve. When a community fails, therefore, investigators should look not only at internal problems (evidence of inefficiency?), but even more at external forces (of inequality?) that are holding the community back. Until something is done to stem the damage from those past and present outside forces, the logic argues, local development will be hindered. Opposing groups, with interests in different places, put their power and privilege at stake so that even definitions of equity and efficiency become hard to pin down for agreement. The prospect of this analytical impasse suggests that the whole idea of (local) economic development may need reconsideration. Before one can understand how areas are to be redeveloped, one must acknowledge how they (and their people) come to be underdeveloped, how different places are (economically) connected, and how they profit from one another's gains or losses.

Relief From Unfair Burdens

What kinds of burdens do outside forces impose? In the case of impoverished and especially minority big-city neighborhoods, the burdens are manifold. To begin, residents of these neighborhoods typically bring home little money; if they can find work at all, they are paid low wages. That is, through their employment, they frequently come away with less, often much less, than necessary for maintaining a decent standard of living. Residents are employed, unemployed, or underemployed, typically through labor markets biased by racial segmentation and discrimination. These markets are powerful mechanisms through which more established groups in the society transform their influence over economic affairs into the urban landscape of wealth and poverty. These markets not only operate like a superstructure built on foundations of unequal opportunity in education, unequal skills, and unequal confidence. They also reinforce the underlying inequalities. These inequalities are key, and anyone truly interested in problems of local economic *under*development must grapple with them. People need formal education, job training, and the expectation that work will be rewarded. Even in the absence of the differential advantages of good schools and job training, however, there are many ways in which other groups—from corporate stockholders and managers to middle-class professionals, and from unionized craft workers to petty bureaucrats and shopkeepers—use connections, networks, and influence over jobs to distribute opportunities to kin, friends, neighbors, and other look-alikes. In poor (and minority) neighborhoods, these connections are rare, thereby depriving ghetto residents (and minority persons generally) of (good) jobs.

Problems with neighborhood services are more within the usual ambit of concern for planners involved with local economic development. Residents of less well-off neighborhoods pay taxes but do not get their value returned in benefits conferred; they pay premium rents because their options are limited; they enjoy less than a fair share of public services, from street sweeping to public transit; they get less money, and less for their money, in schools. In addition, poor and minority neighborhoods suffer extra demands for accommodating society's disruptive members (such as people in halfway houses, or those being warehoused for various antisocial behaviors), and they must often accommodate institutions such as hospitals that not only generate traffic and other bothersome activity but also turn an indifferent or even antagonistic face toward the neighborhood and its concerns.[22]

Scholars sometimes undertake global calculations of the high costs or biased advantages imposed by external agencies. One of the best known distortions of public policy is in housing, through tax deductions on mortgage interest payments. The benefits, running in the 1990s between $50 billion and $80 billion per year, accrue to those who own homes and file IRS long forms—that is, almost entirely to well-off suburban homeowners, who are disproportionately white. These enormous federal subsidies to suburbs (called "tax expenditures") are comparable in magnitude to federal *direct* expenditures on urban assistance overall.[23] Transportation policy is biased too. Various hidden subsidies to automobile travel harm those who must use public transportation, either within or between cities. Even *public* mass transit policy is sometimes biased: Where public investments have produced enormously costly new subway systems, taxes are generally broadly based, but benefits are quite narrow, mainly to upper-middle-class riders.[24]

On occasion, scholars calculate not only to estimate the general burdens on cities, such as the multibillion-dollar mortgage deductions, but
also to estimate the specific burdens imposed on a particular city or
neighborhood. For example, consider redistribution between New York
City and its suburbs and within neighborhoods of the city itself. Fiscal
wounds now redistribute benefits upward, from poor neighborhoods to
rich and from the city to better-off suburbs. Robert Fitch (1995) suggested
several ways to stanch the hemorrhaging.[25] The city could tax business
services and impose equal taxes on commuters and residents. It could tax
the richest nonprofit corporations, and it could reduce tax abatements to
luxury residences and office buildings. A great deal of money would be
at stake in such shifts from the false efficiency of low taxes and service
slashing to the equity these taxes would offer. Even if Fitch's estimate of
$6.4 billion in annual new revenue is wildly optimistic, and even if
needs-based tuition and subway fares might be fairer and more efficient
ways to allocate costs, the list of potential benefits makes the proposals
worthy of close study; according to Fitch, "The city could not only close
its budget gap, it could eliminate tuition at CUNY, fund community
development banks, get rid of the subway fare and have billions left over
to rebuild our falling bridges, overcrowded schools and collapsing tene-
ments."

Expanding Cities Into the Suburbs

One of the biggest burdens for U.S. cities is the deficit they suffer
because of the suburbs. The line between city and surrounding suburbs is
a political artifact inappropriate to dense metropolitan patterns of social
and economic interdependence. Almost by definition, this arrangement is
guaranteed to produce city misery. David Rusk, in *Cities Without Suburbs*
(1993), had a big part of the argument right when he attributed city
problems to suburban racial hostility. Rusk wrote that "racial and economic segregation is the heart of America's 'urban problem,' " and he
asserted that until we confronted "deep-rooted fears about race and social
class," we would not be able to reshape urban America. The real city is
the total metropolitan area—city and suburb. "Segregating poor urban
Blacks and Hispanics has spawned physically decaying, revenue-
strapped, poverty-impacted, crime-ridden 'inner cities.' These inner cities
are isolated from their 'outer cities'—wealthier, growing, largely White
suburbs" (pp. xiii-xiv, p. 1).

Rusk, who was mayor of Albuquerque before conducting a study of the
country's 100 largest urban areas, found roughly half of those metropo-
lises working reasonably well, with central cities that tended to be able to
solve their problems. The key, for Rusk, was what he called "elasticity"—

the capacity of the central city either to expand its city limits or to densify as the area's population grows. A city that literally stretches its limits or increases its population by adding density "captures" its suburban growth, whereas cities that keep old boundaries and do not densify "contribute" to suburban growth and themselves decline. "Inelastic" cities are hobbled by bad state laws, are trapped by their surrounding neighbors, and are more highly segregated by race and income.

The argument is deceptively simple: Without the artificial wall that separates well-off suburbs from poor cities, their mutual dependencies can be mutually productive, the burdens of services and development can be fairly shared, and the benefits can be more widespread. European cities serve as examples. There, municipal boundaries are more likely to stretch to incorporate growing needs, cities expand their boundaries (naturally) to incorporate their growing outskirts, and national budgetary allocations to cities tend to multiply in proportion to the size of the expanding population. Indeed, Rusk found that elastic cities in the United States not only deconcentrate their poverty and diminish their racial segregation but cause their metropolitan areas to grow more rapidly. He gave numerous examples: Perhaps the most notable case of elastic city growth is the New York State legislature's creation of New York City in 1898, which merged the five mostly empty boroughs and abolished all other municipalities, including Brooklyn, which was then one of the largest cities in the country. This led to many decades of successful growth for the entire metropolis, *most of which was inside the city limits*, a growth so powerful it included the capacity to absorb and utilize the energy of countless immigrants. It can be argued that the *city's* severe problems began to mount only later, when its physical, fiscal, and political growth was choked off by the suburbs, which from then on were able to capture relative prosperity for themselves.[26]

Fifty-five years later, in 1953, the government of Ontario created Metro Toronto, an overreaching municipal authority that dominates its five constituent parts, including the City of Toronto. Metro Toronto was set up to seize the advantages of internalized metropolitan growth, rationalized services, and fair distribution of burdens and benefits. Nearly all observers agree that the Toronto experience with its "captured" suburbs has been extremely successful.

Unfortunately, the original Metro Toronto was not set up to continue expanding and capturing, so that by the 1990s, more than 40 years later, the population has spilled far beyond the fixed metropolitan boundaries. Only about half the urban area's 4 million inhabitants now live inside the Metro Toronto limits; the other half live in separate suburban jurisdictions. Even though the whole Toronto area works remarkably well compared to most U.S. cities, and even though racial inequalities are much limited by history, by national policies of redistribution, and by national-ized public services, economic downturns and a quest for both efficiency and fairness have led to new efforts to expand the Metro boundaries. But there are problems: Entrenched suburban interests appear eager to resist, wanting to retain their hidden subsidies. The consequent strains between this large "central" city and its surrounding suburbs are pronounced. The provincial government seems to have more pressing concerns, and legis-lators appear to fear the reaction of large suburban electoral groups.[27] In a reflection of the U.S. urban scene, some who advocate an expansion of Metro boundaries fear that without expansion, ethnic and social contrasts between the multiracial city and its overwhelmingly white suburbs will lead to a circular decay—neglect of central neighborhoods, social harm to their residents, and economic harm to the area as a whole.

The Environment

Increasing ecological pressures and rising environmental awareness confront cities and suburbs in very different ways. These differences illustrate once again how unfairly are burdens shifted to city residents and how they weigh especially on those families stuck in the poorest neigh-borhoods—those people without automobile access to suburban park-ways, shopping centers, parks, and beaches.

Suburban needs for extensive mobility constitute principal sources of urban environmental stress, from local air pollution from automobile and truck exhaust and garden maintenance machinery, to local water pollution from lawn fertilizer and weed killer runoff and detergent use, to distant resource use caused by the more extensive public-utility distribution systems. Even air pollution, ozone depletion, and global warming from stationary, point-source fuel oil and coal burning can be construed as largely suburban-generated problems insofar as suburbanites consume more, demand more and more spread-out facilities and parking lots, and cause more construction of highways. These burdens drive down the quality of life in cities even more than in the suburbs. And almost none of the environmental costs of these suburban excesses are charged directly or entirely to suburbanites. Indeed, such problems are frequently viewed as products of dense urbanization, as if the green and apparently clean suburbs somehow lived on their own.[28]

The legitimate desires of modern, middle-class urban residents for almost unrestricted mobility are coming into intense and stressful conflict with their equally legitimate demands for clean air and other environ-mental goods. As of now (the mid-1990s), severely air-polluted cities such as Los Angeles, Detroit, and Houston seem exceptionally unlikely to meet statutory air quality requirements without virtual transportation shut-downs. The conservative-dominated Congress seems only too happy to postpone the crisis, ignoring the dilemma by postponing enforcement of deadlines, in effect by a harsh watering down of the Clean Air Act. It is hard to say, but public opinion polls suggest that citizens do not favor this sort of a trade-off. They want their automobile-based mobility, yes; but they also want clean, and ever cleaner, air. It would seem that a solution will be impossible without changes in urban spatial form: That is, people can have high mobility among multiple job sites, residences, shops, and recreation areas, as well as a clean environment, only if metropolitan areas are reorganized into more densely settled subcenters. For this to happen, at least during the period of transition, suburban residents will have to pay for things they now get for free.[29]

Recent journalism and research on environmental equity and environmental racism have shown that those at the bottom end of the social scale, living in the poorest and especially the darkest skinned neighborhoods and rural zones, share space with dirt and disease at the bottom end of the environment scale. It is poor people who nearly always live nearest town and city dumps and transfer sites, toxic waste storage sites, and nuclear storage areas. The residential areas of poor people, African Americans, and Latinos are most often immediately downwind or downstream from various obnoxious facilities, such as dumps, chemical plants, and dust-producing industries. None of this is surprising: "Lower class" (and its U.S. coordinate, "darker skinned") means, after all, just that—at the bottom of the choosing order, last in line to get the good places to live and thus first in line to get the worst.[30]

Creating Progressive Alternatives

From this analysis, I am able to suggest two approaches to progressive planning for neighborhood revitalization, the first cooperative and the second conflictual. In the best of worlds, these two approaches would be mutually supportive, rewarding the optimists among us.

Cooperation on programs to stimulate the development and well-being of impoverished neighborhoods—between central cities and their suburbs, say—may be stimulated by the joint benefits available or similarly by fear of the joint costs to be borne if poor neighborhoods are left to decay. As I have noted, there is considerable evidence, even in formal epidemiological studies, that central-city "illnesses" are contagious and are spreading to the suburbs. It is hard to say how far out from the center one must live to be sure that the shopping center will be free of drug trades, to avoid drive-by shootings, or to be assured of safety in the public schools. Ever higher walls for protection and ever more prison construction are not only grossly expensive and inefficient but ultimately ineffective. There is an inescapable logic to arguments that the costs of localized underdevelopment are borne globally; unfortunately, given the possibilities of spatial and temporal avoidance and displacement, speculation about this sort of cooperation may be academic.

More generally, one can imagine wide distribution and acceptance of arguments and evidence contending that the healthiest metropolitan areas are those with the healthiest central cities. Were such information to buttress public debate, then the self-interests of suburbanites (decency and fairness aside) might stimulate more cooperation, generosity, and progressive alliances. In this way, rising environmental concerns might fuel a search for "sustainable development," in which case those moving cooperatively would fly alongside the angels. Peter Meyer (personal communication, 1995) argued that "[the sustainability] argument for urban areas is analogous to global-local arguments for the planet." Metropolitan (global) inhabitants are learning that they benefit or suffer consequences from what happens to neighborhood (local) inhabitants in four specific ways: (a) Suburban sprawl is costly in time, effort, and auto-generated pollution; (b) construction of suburban infrastructure combines with decay of city facilities to reduce efficiency of public investment, thereby reducing the metropolitan potential for economic growth; (c) housing abandonment reduces the stock and drives up housing prices, impeding in-migration and thereby similarly reducing growth; and (d) abandonment of neighborhoods is likely to generate contamination of land, contributing to areawide pollution. Combine such cooperative tendencies with conflicts about environmental racism and environmental equity, and perhaps we can find the basis for a broader, more inclusive view. But it is possible to think of the conflictual approach separately.

Some planners and other city leaders have built (and are building) their careers on the idea of redistributive planning; they see efficiency not as a separate category but as something that is promoted by equity, antidiscrimination, and fairness. Beginning from this sort of a logic, Robert Mier, from his experience in Chicago and his studies about other places, took the position that "there is *no* situation that is so bad that it warrants public inattention," and he speculated that other development directors and city planners, such as Cleveland's Norman Krumholz, though recognizing "the need to target resources," would feel likewise. Planner Kenneth Reardon and his team of students from Urbana, responding to pressures on the University of Illinois from the state legislature, have found remarkable opportunities for development even where many see the limiting bad case, in much-maligned East St. Louis. Outside aid added to organized community effort has produced improvements in housing, markets, and street conditions.[31]

Such "equity planners," whether or not they so conceive of themselves, work as *advocates* of common citizens, poor and working-class people, men and women of color, the elderly, children, the disabled, and, ultimately, city dwellers. In the tradition of advocacy, these planners choose sides when necessary, which is often, to work *against* unfair privileges of elites, the upper and middle classes, whites, and suburbanites.[32] Norman Krumholz and Pierre Clavel (1994) have documented efforts by numerous progressive planners in a dozen U.S. cities, and they suspect that they could find "hundreds of other planners like them" (p. ix).[33] These advocates are ineluctably drawn into conflict. They confront those not part of the solution with being part of the problem. They fight for the rights of the powerless against the powerful and self-interested in staff offices, in city halls, in neighborhood meetings, in corporate offices, and, not infrequently, in courts of law.

To make scarce funds available for poor, elderly, or physically disabled riders who are transit dependent, planners must deny funds for suburban highways or trains. To the extent that they can hold down payments through rent control, or exact "linkages" that tax new developments to finance public services or investments in poor neighborhoods, or prevent condominium conversions that expel renters, or prevent the destruction of single-room occupancy housing for the poor, planners must limit incomes of landlords and real estate investors.[34] To establish minority or low-income residences in suburban neighborhoods, planners must do battle with those who wish to live exclusively.

The more they push for locally controlled economic development, for "neighborhood capital accumulation," the more they come into conflict with outside-headquartered corporations. As Wiewel, Teitz, and Giloth (1993) wrote, "Neighborhood ownership and control over land, business, investment, and financial capital [is necessary] because neighborhood resources are being drained, disinvested, or neglected" (p. 85).[35] When the city of Chicago sued Playskool, the pretext was misuse of city funds, but the deeper motive was to stem the loss of jobs, as against corporate efforts to consolidate, streamline, and profit (Mier & Giloth, 1993).

Progressives making these arguments turn up in the most surprising places, and their efforts and innovations have remarkably lasting effects. Even when they are removed from office, their influence persists in systems of broader participation, in new patterns of democracy, and in an ethos of reasonable regulation and redistributive fairness. Thus, there exists an enthusiastic and optimistic reading of the future of a locally led renaissance of cities based on reason and on the efficient consequences of fairness.

In the end, however, no matter how courageous and convincing these city progressives have been, they have been stymied by their surroundings and thus limited in their effect. As Krumholz and Clavel (1994), two of the most enthusiastic interpreters of progressive planners, wrote, "They have not . . . reduced poverty, unemployment, or dependency. . . . They have also left unimproved the neighborhoods of the working class and poor. And they have left virtually untouched the widening economic disparities between central cities and suburbs."

▪ Conclusion

How, then, are we to approach the problems of impoverished, declining, neglected, and abandoned neighborhoods? How can citizens and governments of central cities set their sails as they are buffeted by the strong and shifting winds of the global economy and national politics? How can others help—those who hold power in other positions or at higher levels of government?

At the outset, I argued that programs to support poor neighborhoods, regions, or nations in their quest for economic development are fair, likely to mitigate the spread of problems to better-off places, and, ultimately, efficient. But many barriers are in the way, not the least of which come from the mundane but recently much-celebrated competition of the marketplace. Thinking of such barriers, in 1992 Edward Blakely and I opened the concluding chapter of *Separate Societies* with great optimism:

> There is a potential cycle for change. It begins with the local problem of urban poverty and central-city decay, then moves to local public recognition, which generates a local response. That response is severely constrained . . . by lack of resources and power. In the best of circumstances . . . the conflict between attempts to deal locally with the problems of poverty, on the one hand, and lack of resources, on the other, will lead to coalitions and pressures on Congress, the federal judiciary, the White House, and federal agencies. In the face of these pressures, Congress will pass better federal laws and offer more generous budgets, and industry will develop a more progressive response to competition in the global economy. These changes, in turn, will lead not only to better conditions, such as stronger labor demand, more attention to education, and broad health care coverage, but will also provide the funds municipalities need to become better places in which to work and live.

However, few coalitions have formed; little effective pressure has worried Congress, the courts, or executive agencies; and industry has turned vengefully against even the most established regulatory practices. Instead, "conservative" policy is ascendant, in favor of radical, right-wing change. There are many local and regional examples of damages done, but best known and appearing in clearer view are cases of international market liberalization, in which national interests are taken as paramount. As all now know, from the Zapatista opposition in Chiapas, rampant inflation and currency devaluation, and near economic collapse, the Mexican free-market experiment failed, pushed to the limit by NAFTA. The rush by eastern Europe and the former Soviet Union to drop all redistributive policy and substitute competitive markets has been catastrophic, but conservative U.S. spokespersons still applaud the changes as they posit a simpleminded equation of competitive markets with democracy. The United States pressures Japan fiercely not only to open its markets but to deregulate its traditional system of extensive protection of small retailers. Such changes not only would threaten to undermine Japanese prosperity but also would damage or perhaps destroy the extensive and nationwide social networks that have yielded very broad measures of equality and social peace.

In the case of actions that might help struggling neighborhoods or central cities in the United States, the record of recent years is at least equally dismal. The Republican "Contract With America" set a legislative agenda that is actively hostile to redistributive programs and to city demands, and there is scant evidence that Democrats are prepared to respond. Needs for health care reform are unattended. Court decisions, reflecting broad public opinion, have turned against affirmative action, and there is broad support for anti-immigrant (and closely associated anti-nonwhite) legislation and biased enforcement of existing laws.

In the 1995 debate over affirmative action, it was suggested by conservatives that color-coded preferences should be eliminated, replaced by advantages offered to *all* who are poor. Imagine! Any such policies in kind and quantity sufficient to expand democracy to the economic sphere and to provide truly equal opportunity to the poor would indeed provide assistance to cities and to their residents. "A preference for the poor," as the maxim of liberation priests in the Third World puts it, would serve as a foundation for public policy conflict with the very basis of class distinctions, and thus many of the operating principles of market capitalism. Such reflections suggest that when conservatives (who after all aim to liberate capitalist markets as much as possible) propose to substitute low incomes for dark skins as the basis for preferential treatment due those who start from handicapped positions, they are being either simpleminded, cynical, or covertly racist.

In this generally oppressive economic and political situation, advocates for central-city improvements, residents of poor neighborhoods, and constructive, progressive city governments must be tenacious and creative. Not only must neighborhoods organize to protect their territories against depredations and to demand a fair share of public services, but they must join citywide groups to find ways to attack the subsidies and other privileges offered to suburbs. Community development corporations must continue to seek local investors and purchases, but they must also demand subsidies matching those offered to suburban developers. Cities and poor neighborhoods must demand that environmental movements recognize how unfairly burdens are spread, and suburbanites must be challenged to pay their fair share.

NOTES

1. Miliband and Panitch (pp. vii-viii) argued that neoliberalism, far from the solution to our problems, generates many of them.

2. Krumholz and Clavel (1994) have published numerous studies documenting how "progressive" planners may succeed even under the most adverse conditions.

3. The first and third reasons—fairness and long-run inclusion—may not count for much under the assumption that a majority derive psychic or even material benefits from knowing that others are kept in subordinate positions.

4. For a brief survey of these policies, see Goldsmith and Jacobs (1982), and for an early argument that urban change almost inevitably imposes high costs on those least able to bear them and least responsible for them, see Cottrell (1951).

5. Not all the prison increase reflects rising crime, for drug use has been increasingly criminalized, but even the application of drug laws reflects the city/suburban, poor/rich, black/white divides. "While drug use is spread relatively evenly throughout the entire U.S. population, the arrest and imprisonment rates for Blacks and Latinos are significantly higher than for whites" (Kohl, 1994a; also see Kohl, 1995, and Davis, 1995).

6. This Congress has enabled outspoken racists Jesse Helms and Strom Thurmond to chair important committees, and it eagerly accepts bill-writing initiatives from antitax and antiregulation corporations and antigovernment libertarians.

7. Others, looking worldwide, refer to capitalism's golden age. See, for example, Singh (1989, 1992) and Esping-Anderson (1994). Observers have noted the harm inflicted by uncontrolled ethnic rivalries elsewhere in the world. See Young (1994). Bell (1992) offered a chilling interpretation of interethnic and cross-class civility in the United States by emphasizing lack of severe conflict among *white* groups. He speculates that the advantages to whites of oppression of blacks leads *all* whites to recognize the advantages of racist solidarity. Also see Wilkins (1995).

8. For detailed information on "ghettos" (defined as census tracts 40% of whose black population is officially poor), see Jargowsky (1994). Adding the one half of rural blacks who live in high-poverty tracts, and some more who are "black" Latinos, probably more than one third of all African Americans live in "ghetto" isolation. For comprehensive evidence that vast majorities of African American city residents, poor and nonpoor, are highly segregated, see Massey and Denton (1993).

9. The phrase is, of course, associated with Ralph Ellison's *Invisible Man* (1952).

10. Statistics on racial and gender disparities in the United States are well known. The poverty rate for black families was 28% versus about 7.8% for white families in 1989; see Goldsmith and Blakely (1992, Figure 2.12, p. 40). Statistics for the mid-1980s show female-headed households facing extraordinarily high poverty risks. Esping-Anderson (1994, note 23, p. 30), who defined poverty as 50% of (adjusted) median income, found that 60% of female-headed households are poor in the United States. Comparable international figures are: Canada—57%, Germany—27%, France and Italy—19%, and Sweden—4.5%. A vivid account of the fencing-in process by which minority neighborhoods and their residents are literally squeezed—by economics, legislation, and physical hostility—appears in Davis (1990). The psychological aspects of difference and isolation were noted years ago by Sennett (1970) in *The Uses of Disorder*. Wallace, in several papers, has used epidemiological methods to show the spilling from central cities to suburbs of such things as disease, crime, and even housing deterioration.

11. For analysis of militia and militialike movements, see Beerlet and Lyons (1995), Ridgeway (1990), and the Southern Poverty Law Center's *Klanwatch Intelligence Report* newsletter.

12. For a thorough study of militant right-wing movements in the United States, see Beerlet and Lyons (1995).

13. For dramatic portrayals of such exploitation, see, for example, Kozol (1989), Goldsmith (1982), and Vergara (1995).

14. Some of these statistics appear in "The Welfare Quiz" (1995).

15. For a discussion of the corrosive effects of rhetoric, see Hirschman (1991).

16. Norris (1993) ridiculed a proposal to create the Buffalo Plains. In the 1980s, the Poppers offered a promise of broad efficiency to compensate for narrow and short-run inequity; the project would depopulate America's Great Plains and return them to the buffalo, subsidizing the departure of those who now live on the plains. The change would promote efficiency and sustain the environment. They argued that the gains could easily compensate those who would have to move: economic gains mainly from reduced need for transportation and ecological gains in saved topsoils, reduced flooding, and other such improvements (see Popper & Popper, 1994).

17. Other people, too, influenced by their imagined experience competing for opportunities against invisible or despised opponents, discover that efficiency arguments work well to justify lack of equity. I think this mode of thinking undergirds recent attacks on affirmative action by white men who believe they have suffered from discrimination.

18. The city at first leased the building back, but on appeal, the court voided the original sale (Judd & Swanstrom, 1994, p. 334).

19. The word *triage* comes from the French, who developed a system to assign wounded soldiers to field hospitals in the carnage of World War I: The "healthy" would survive on their own until doctors and nurses had spare time; the "transitional," who were expected to die without treatment but live with it, got maximum treatment; and the "deteriorated," those too badly wounded to save, got painkillers. Chicago's Anthony Downs, who is a consultant to the Department of Housing and Urban Development (HUD) and president of the Real Estate Research Corporation, suggested a mild form of triage for urban policy that would distinguish among healthy, transitional, and deteriorated areas. New York's Roger Starr "advocated the idea of reducing public services and expenditures in blighted urban areas" (Judd & Swanstrom, 1994, pp. 287-288).

20. Private investors not only benefit from such shifts in markets but sometimes are thought to *cause* the shifts.

21. In this vein, in an early and hotly argued debate on central-city economic development, Kain and Persky (1988) proposed that federal subsidies ought not "gild the ghetto," for that would be throwing good money after bad. Instead, they argued that subsidies should be allocated more efficiently, to programs aimed at needy persons. Persons, not places. They estimated that if public policy would help people to escape the ghetto, it would increase the net, overall benefits. Opponents then argued, and opponents of policies of spatial abandonment generally point out, that gains are often not shared, mainly because losers are too weak politically to demand compensation. No matter how many benefits the system offers to middle-class blacks who live in suburbs, it offers few benefits to the poor who remain stuck in central cities. It is at this sort of margin (which to neighborhood people is not marginal at all) that questions of viability must be answered.

22. Vergara's (1995) "New Ghetto" photographs provide stunning evidence of the many ways that problems are piled on those neighborhoods least able to handle them but also least able to resist their imposition.

23. In 1980, when total expenditures for urban programs (for cities *and* suburbs) by all federal agencies—including HUD, the Economic Development Administration, the Environmental Protection Agency, the Department of Transportation, the Department of Labor, the Small Business Administration, the Community Services Administration, and the Treasury—

ran to $28 billion, the mortgage tax deduction alone cost $12 billion (Goldsmith & Jacobs, 1982, p. 61).

24. In Webber's (1976) words, "The poor are paying and the rich are riding" (p. 93). On the broad and general benefits of mass transit investments, see Black (1995). Goldsmith and Jacobs (1982) traced the historical biases in public policies for transportation, housing, urban redevelopment, and industrial development.

25. Fitch (1995) also recommended reimposition of the tax on stock transfers, but my colleague Matthew Drennan warns that the location effect could push the stock market outside New York to elsewhere in cyberspace.

26. In the other direction, Frances Piven and Richard Cloward (1979) long ago warned that southern whites would expand central-city boundaries to dilute African American voting power. Furthermore, Drennan, Tobier, and Lewis (1996) observed that in cities specialized in producer services, African American incomes rose dramatically in the 1980s, compared to manufacturing cities, where African American incomes fell markedly. Notably, among the 10 cities with the highest income rise, only San Jose and San Diego have high elasticities.

27. For observers from the United States, the Canadian political system seems relatively transparent and benign. Interested parties—from industry, real estate, and government at all levels—meet openly to discuss the options, and there is an appearance, at least, of yielding parochial interests both to common needs and to the needs of those most dependent (Municipality of Metropolitan Toronto, 1994).

28. It is abundantly unclear how rapid technological advances in telecommunication will affect the economics of suburbanization. On the one hand, they would seem likely to loosen the bonds by which suburbs are tied to the central city, as people employed in administration, services, and other efforts of symbolic manipulation are able more and more to work at home or in small, dispersed offices. On the other hand, these electronic advances may increase the viability of very dense and walkable central cities, as least for those cities that serve as the administrative centers and subcenters of global capitalism. This is one of the inferences I make from the analysis by Drennan et al. (1996). Also see Schuler (1993).

29. In a proposal for such changes, Victorisz and Goldsmith (1995) suggested a focus on a triad of changes—in clean fuels, green vehicles, and urban spatial form. Also see Freund and Martin (1993) and Bonham, Burnor, and Marzotto (1995).

30. For more detailed discussion of these and related urban environmental issues, see Olpadwala and Goldsmith (1992), Bullard (1993), and Bryant and Mohai (1992).

31. The 1995 conference of the Planners Network took place in and about East St. Louis, where participants could see for themselves the bad conditions created by external burdens and hostility but also the opportunities for progress that new investments of time and money can yield.

32. Many of these planners do not work consciously with a model of society in deep or fundamental conflict, though, of course, some do. One planner who did recognize such conflict was Robert Mier (1993), who reflected on the racism (potentially) involved in nearly every "nonracial" planning decision. Progressives in places where bias, inequality, and unfairness are too evident ever to be avoided often can work as planners only when they are employed by governments of formal opposition. For two cases in which the lineup is presented by social class, see Kowarick's (1994) collection on Sao Paulo, Brazil, and Schneider's (1995) history of protest in Santiago, Chile.

33. See also Krumholz and Forester (1990), Clavel (1986), and Clavel and Wiewel (1991) for documentation on progressives in the following cities: Burlington, Cleveland, Berkeley, Jersey City, Dayton's Miami Valley, Chicago, Portland, Boston, Denver, Hartford, Santa Monica, and San Diego.

34: The New York State Commissioner of Housing was unceremoniously fired by a conservative governor in June 1995 for suggesting how deep such conflict may be after he mentioned (at a meeting of landlords) that New York City rent control was just like communism.

35. Wiewel et al. (1993) did not pose this drain as a conflict, but their cited source, Gunn and Gunn (1991) saw the democratization of the economy as a form of conflict and struggle.

REFERENCES

Amott, T. (1993). *Caught in the crisis: Women and the U.S. economy today.* New York: Monthly Review.

Bell, D. (1992). *Faces at the bottom of the well: The permanence of racism.* New York: Basic Books.

Berlet, C., & Lyons, M. L. (1995, June). Militia nation. *Progressive, 59*(6), 22-35.

Bingham, R. D., & Mier, R. (Eds.). (1993). *Theories of local economic development: Perspectives from across the disciplines.* Newbury Park, CA: Sage.

Black, A. (1995). *Urban mass transportation planning.* New York: McGraw-Hill.

Bonham, G. S., Burnor, V. M., & Marzotto, T. (1995). EPA clean air requirements and employee commuting in Baltimore. *Journal of Urban Affairs, 17*, 165-182.

Bryant, B., & Mohai, P. (1992). *Race and the incidence of environmental hazards: A time for discourse.* Boulder, CO: Westview.

Bullard, R. (1993). *Confronting environmental racism: Voices from the grassroots.* Boston: South End.

Cassidy, J. (1995, October 16). Who killed the middle class? *New Yorker,* pp. 113-124.

Clavel, P. (1986). *Progressive cities.* New Brunswick, NJ: Rutgers University Press.

Clavel, P., & Wiewel, W. (Eds.). (1991). *Harold Washington and the neighborhoods: Progressive city government in Chicago, 1983-1987.* New Brunswick, NJ: Rutgers University Press.

Cobb, C., & Halstead, T. (1994). *The Genuine Progress Indicator: Summary of data and methodology.* San Francisco: Redefining Progress.

Corbett, T. (1994, September). The Wisconsin welfare magnet debate: What is an ordinary member of the tribe to do when the witch doctors disagree? *Focus, 13,* 19-28.

Cottrell, J. (1951, June). Death by dieselization: A case study in the reaction to technological change. *American Sociological Review, 16,* 358-365.

Davis, M. (1990). *City of quartz.* New York: Verso.

Davis, M. (1995, February 28). Hell factories in the field: A prison-industrial complex. *Nation,* pp. 229-234.

Drennan, M., Tobier, E., & Lewis, J. (1996, February). The interruption of income convergence and income growth in large cities in the 1980s. *Urban Studies, 33*(1), 63-82,

Ellison, R. (1952). *Invisible man.* New York: Random House.

Esping-Anderson, G. (1994). *After the golden age: The future of the welfare state in the new global order.* Geneva: United Nations Research Institute for Social Development.

Ferman, L. A. (Ed.). (1988). *Poverty in America.* Ann Arbor: University of Michigan Press.

Fitch, R. (1995, May 8). The New York budget war: "Spread the pain"? Tax the gain! *Nation,* pp. 628-631.

Freund, P., & Martin, G. (1993). *The ecology of the automobile.* New York: Black Rose.

Frey, D. (1994). *The last shot: City streets, basketball dreams.* Boston: Houghton Mifflin.

Goldsmith, W. W. (1974). The ghetto as a resource for black America. *Journal of the American Institute of Planners, 40*, 17-30.

Goldsmith, W. W. (1982). Bringing the Third World home: Enterprise zones. *Working Papers Magazine, 9*(2), 24-30.

Goldsmith, W. W., & Blakely, E. J. (1992). *Separate societies: Poverty and inequality in U.S. cities.* Philadelphia: Temple University Press.

Goldsmith, W. W., & Jacobs, H. M. (1982). The improbability of urban policy: The case of the United States. *Journal of the American Planning Association*, pp. 53-66.

Gunn, C., & Gunn, H. D. (1991). *Reclaiming capital: Democratic initiatives and community development.* Ithaca, NY: Cornell University Press.

Harrison, B., & Bluestone, B. (1988). *The great U-turn: Corporate restructuring and the polarizing of America.* New York: Basic Books.

Hirschman, A. O. (1991). *The rhetoric of reaction.* Cambridge, MA: Belknap.

Jargowsky, P. A. (1994). Ghetto poverty among blacks in the 1980s. *Journal of Policy Analysis and Management, 13*, 288-310.

Judd, D. R., & Swanstrom, T. (1994). *City politics: Private power and public policy.* New York: HarperCollins.

Kain, J. F., & Persky, J. (1988). *Alternatives to the gilded ghetto.* Cambridge, MA: Harvard University, Program on Regional and Urban Economics.

Keen, D. (1994). *The benefits of famine: A political economy of famine and relief in southwestern Sudan, 1983-1989.* Princeton, NJ: Princeton University Press.

Kohl, B. (1994). *Analysis of coca/cocaine control policies in Bolivia and the United States.* Unpublished manuscript, Cornell University, Department of City and Regional Planning.

Kohl, B. (1995, April). War on Drugs, Inc. *Bookpress*, pp. 13-14.

Kowarick, L. (Ed.). (1994). *Social struggles and the city: The case of São Paulo.* New York: Monthly Review.

Kozol, J. (1989). *Rachel and her children: Homeless families in America.* New York: Ballantine.

Kozol, J. (1991). *Savage inequalities.* New York: Crown.

Krumholz, N., & Clavel, P. (1994). *Reinventing cities: Equity planners tell their stories.* Philadelphia: Temple University Press.

Krumholz, N., & Forester, J. (1990). *Making equity planning work: Leadership in the public sector.* Philadelphia: Temple University Press.

Lawson, B. E. (Ed.). (1992). *The underclass question.* Philadelphia: Temple University Press.

Massey, D. S., & Denton, N. A. (1993). *American apartheid: Segregation and the making of the underclass.* Cambridge, MA: Harvard University Press.

Mier, R. (1993). Community development and diversity. In R. Mier (Ed.), *Social justice and local development policy* (pp. 182-199). Newbury Park, CA: Sage.

Mier, R., & Giloth, R. P. (1993). Democratic populism in the U.S.: The case of Playskool and Chicago. In R. Mier (Ed.), *Social justice and local development policy* (pp. 135-143). Newbury Park, CA: Sage.

Miliband, R., & Panitch, L. (Eds.). (1993). *Real problems, false solutions: The Socialist Register 1993.* London: Merlin.

Mullane, D. (Ed.). (1995). *Words to make my dream children live: A book of African American quotations.* New York: Anchor.

Municipality of Metropolitan Toronto. (1994, November). *The Economic Forum on the Future of the Greater Toronto Area.*

Norris, K. (1993). *Dakota: A spiritual geography.* Boston: Houghton Mifflin.

Olpadwala, P., & Goldsmith, W. W. (1992). The sustainability of privilege: Reflections on the environment, the Third World city, and poverty. *World Development, 20,* 627-640.

Piven, F. F. (1995, February). Poorhouse politics. *Progressive, 59,* 22-24.

Piven, F. F., & Cloward, N. A. (1979). *Poor people's movements: Why they succeed, how they fail.* New York: Random House.

Pizzigati, S. (1992). *The maximum wage: A common-sense prescription for revitalizing America—by taxing the very rich.* New York: Apex.

Popper, F., & Popper, D. (1994, Winter). Great Plains: Checkered past, hopeful future. *Forum for Applied Research and Public Policy,* pp. 89-100.

Ridgeway, J. (1990). *Blood in the face: The Ku Klux Klan, Aryan nations, skinheads, and the rise of a new white culture.* New York: Thunder's Mouth.

Riley, D. W. (Ed.). (1993). *My soul looks back, 'less I forget: A collection of quotations by people of color.* New York: HarperCollins.

Rusk, D. (1993). *Cities without suburbs.* Washington, DC: Woodrow Wilson Center Press.

Schneider, C. L. (1995). *Shantytown protest in Pinochet's Chile.* Philadelphia: Temple University Press.

Schuler, R. E. (1993). Transportation and telecommunications networks: Planning urban infrastructure for the 21st century. *Research in Urban Economics, 9,* 43-58.

Sennett, R. (1970). *The uses of disorder: Personal identity and city life.* New York: Knopf.

Singh, A. (1989). Third World competition and deindustrialization in advanced countries. *Cambridge Journal of Economics, 13,* 103-120.

Singh, A. (1992). The lost decade: The economic crisis of the Third World in the 1980s. *Contention, 1*(3).

Stein, J. M. (Ed.). (1995). *Classic readings in urban planning: An introduction.* New York: McGraw-Hill.

Talking 'bout a devolution. (1995, February 4). *Economist, 334,* 23-24.

Vergara, J. C. (1995). *The new ghetto.* New Brunswick, NJ: Rutgers University Press.

Vietorisz, T., & Goldsmith, W. W. (1995). *Autocity project: Long-term urban environmental policy for automotive transportation.* Ithaca, NY: Cornell University, Department of City and Regional Planning.

Wallace, R., & Andrews, R. F. H. (in press). The spatiotemporal dynamics of AIDS and TB in the New York metropolitan region from a sociological perspective: Understanding the linkages of central city and suburb. *Environment and Planning.*

Wallace, R., & Wallace, H. A. D. (in press). *AIDS, TB, and violent crime and low birthweight in eight U.S. metropolitan areas: Public policy and the regional diffusion of inner city markets.*

Webber, M. (1976, September). The BART experience: What have we learned? *Public Interest, 45,* 79-108.

The Welfare Quiz (courtesy of World Hunger Year). (1995, July). *Poverty and Race, 4*(4).

West, C. (1993). *Race matters.* Boston: Beacon.

Whisnant, D. E. (1980). *Modernizing the mountaineer: People, power, and planning in Appalachia.* Boone, NC: Appalachian Consortium.

Wiewel, W., Teitz, M., & Giloth, R. (1993). The economic development of neighborhoods and localities. In R. D. Bingham & R. Mier, *Theories of local development: Perspectives across the disciplines.* Newbury Park, CA: Sage.

Wilkins, R. (1995, March 27). Racism has its privileges: The case for affirmative action. *Nation,* pp. 409-416.

Young, C. (1994). *Ethnic diversity and public policy: An overview.* Geneva: United Nations Research Institute for Social Development.

COMMENTARY ON
THE CYCLE OF CITIES

Peter B. Meyer

It is impossible to overemphasize the importance of William Goldsmith's comments on the city-suburb interaction and the need to consider "viability" or the efficiency-equity interaction from a metropolitan-area or broad-conurbation, not specifically municipal, perspective. His three organizing questions—*which* city, *what* cycle, and *whose* development—are central to defining the scope of inquiry and the answer to the question posed in the chapter. They effectively define *current political* viability, so I cannot understand why Goldsmith treats the viability issue as a separate matter.

Accurate as he is in describing the policy environment that tends toward abandonment when investment is preferable from a broad social stability (not to mention equity) perspective, Goldsmith never makes explicit the link between his three questions and his answer regarding viability. The reality is that those with the power to decide on viability also dictate the answers to the questions about the city, the cycles, and the relevant economic development measures. It follows that no objective response to the question posed in his chapter title is possible.[1]

Goldsmith's clearly identified personal perspective, that of a progressive planner, leads directly to his contention that when a community fails, "investigators should look . . . at external forces" as key causal factors to be addressed. Externally generated spatial inequalities (whether or not such differences are characterized as "injustices") provide the underpinning for his assertion that "the whole idea of (local) economic development may need reconsideration."[2] I only wish he had given more prominence to this point, given its centrality, importance, and policy significance: Strictly *local* economic development, whether neighborhood specific or narrowly and parochially municipal, has rarely been efficient. The returns on most area-specific regeneration efforts have been limited by the *nonlocal* institutional factors that such policies fail to address or challenge.[3]

The chapter proposes an explicit response to nonlocal pressures, calling for "relief from unfair burdens." Much as I am similarly inclined to raise equity and justice concerns above those of efficiency, I find this approach troublesome: To the extent that our values are *not* accepted by decision makers outside the impoverished areas, our recommendations and calls for actions will fall on deaf ears. How, then, can an ethical basis for action be made politically salient?

Goldsmith's answer is the evidence compiled by David Rusk on the gains attainable from urban "elasticity," which he uses to support his claim that abandonment may never be appropriate, even from efficiency criteria. But this approach relies on the transformation of a question of "local" economic development into one of "metropolitan" development, a redefinition that remains difficult to attain and sustain in a hostile political environment.

The very prominence of Robert Mier's work on transformative neighborhood redevelopment in Chicago and the few other cases that we can document attest to the paucity of successes on the record, and to their transitory nature. Goldsmith never asks whether there might be other grounds on which we could argue for higher levels of effort to minimize inequality and differential deterioration in urban areas. The ideal would be to identify bases for policy reorientations that *serve the self-interest of the nonpoor* and thus do not rely on any change in social mores or attitudes toward the unequal impacts of market processes on individuals or neighborhoods. If new approaches address the relative efficiency of metropolitan areas writ large, they may serve the self-interest of suburbanites as well as inner-city residents.

One set of bases for such policy shifts lies in the concepts of "sustainable development," which are getting increasing attention on the world stage. Sustainability is replacing the focus on economic development in even the rhetoric of the World Bank and other institutions shaping international capital flows.[4] The argument for urban areas is analogous to that for the planet: The ecological web is such that inhabitants of the totality (globe or metropolis) are affected by what happens in individual subunits (nation-states and regions, or neighborhoods and communities). The idea of the "ecological city" or "sustainable urbanism" has broad appeal, and the negative impacts of neighborhood or community abandonment on suburbanites and others geographically distant from the affected area become obvious in such an approach.[5] Consider some of the potential consequences of abandonment:

- Accelerated suburban sprawl increases commuting time and effort, and fossil-fuel-dependent transportation systems contribute to areawide environmental problems, affecting all residents.

- New infrastructure construction to cater to suburban growth, combined with underutilization and/or inadequate maintenance of existing inner-city infrastructure, reduces the overall efficiency of public-sector capital, adversely affecting metropolis-wide economic growth potential.

- Competition for residential facilities outside abandoned areas drives up housing costs, a trend that, though beneficial for existing owners, impedes in-migrations, further retarding economic growth.

- Abandoned areas are more likely to generate new instances of severely contaminated lands as the result of neglect, thus contributing to water, land, and air pollution for the entire ecological region (watershed or air basin), affecting all nearby residents or workers.

Some of these effects are clearly more concentrated than others, and there is ample evidence of conscious and systemic environmental racism at work in promoting inequality. Nonetheless, broader recognition of the effects of abandonment on households across a conurbation will increase the political support for the pursuit of neighborhood regeneration throughout the metropolis.[6]

Thus, I conclude that however one may define the "cycle of cities," there exists a political arena, that of sustainable development and environmental concern, within which the economic "development"—or regeneration—of even the most severely depressed and impoverished neighborhoods or urban areas may always be "viable" for a majority of the socioeconomic interests in a metropolitan area. This finding does not, however, address a question that Goldsmith and I have both avoided: whether the political will and economic power of even an entire metropolis can regenerate previously abandoned areas in the face of the indifference or active opposition of global capital interests. Much as I would like to believe that the ethos and political expediencies underlying the world environmental summit in Rio in 1992 carry sufficient weight to support sustainable urban economic redevelopment, I have little evidence to support that hope. Electoral developments in the United States in the mid-1990s suggest that unfortunately, the politically expedient answer to the title question in Goldsmith's chapter may be "yes."

NOTES

1. I have argued this point elsewhere (Meyer, 1995).

2. Anne Shlay (1993) has provided a stimulating discussion of the roots of such inequalities, underscoring how little we really know about the causal processes.

3. The point need not be belabored here. Two decades of conflicting assessments of programs run by the U.S. Economic Development Administration and the Department of Housing and Urban Development (HUD) have highlighted the relative "inefficiency" of the federal programs relative to other public investments and the desperation of the cities' pursuit of the external subsidies.

4. This new view of development arguably has its roots in the work of the World Commission on Environment and Development (1987). The concepts of sustainability and

their global consequences moved onto center stage after the Rio conference on the environment in 1992. Notwithstanding conflicting definitions of sustainability and argument over appropriate responses, the discussions all recognize that environmental consequences are shared by all, not merely those experiencing the most immediate local impacts. Consider, for example, the divergent approaches of *The Ecologist* (1993), Makhijani (1992), the National Commission on the Environment (1993), and economists Pearce and Warford (1993).

5. Meyer, Williams, and Yount (1995) addressed these consequences and the associated political imperatives in some detail, although the discussion is centered on policies toward previously polluted urban sites. The increasing national concern over the costs of abandonment is evidenced by the recent joint funding by HUD and the Environmental Protection Agency for a national study of "The Impact of Environmental Hazards and Regulations" on prospects for, and rates of, urban regeneration.

6. The evidence on substantial environmental inequality is stark and suggests strongly that the experienced patterns constitute consistent racism. See, for example, Bryant and Mohai (1992). The experience of efforts to challenge such environmental injustices, however, suggests that a broad environmental health focus may facilitate coalition building, even across groups normally not allied in pursuit of specific forms of local *economic* development. Suggestive examples abound in Bullard (1993) and Hofrichter (1993).

REFERENCES

Bryant, B., & Mohai, P. (Eds.). (1992). *Race and the incidence of environmental hazards: A time for discourse*. Boulder, CO: Westview.

Bullard, R. D. (1993). (Ed.). *Confronting environmental racism: Voices from the grassroots*. Boston: South End.

The Ecologist. (1993). *Whose common future? Reclaiming the commons*. Philadelphia: New Society.

Hofrichter, R. (Ed.). (1993). *Toxic struggles: The theory and practice of environmental justice*. Philadelphia: New Society.

Makhijani, A. (1992). *From global capitalism to economic justice*. New York: Apex.

Meyer, P. B. (1995). Measurement, Understanding and Perception: The Conflicting "Realities." In R. Hambleton & H. Thomas (Eds.), *Urban policy evaluation* (pp. 89-109). London: Paul Chapman.

Meyer, P. B., Williams, R. H., & Yount, K. R. (1995). *Contaminated land: Reclamation, redevelopment and re-use in the United States and the European Union*. Cheltenham, UK: Edward Elgar.

National Commission on the Environment. (1993). *Choosing a sustainable future*. Washington, DC: Island.

Pearce, D. W., & Warford, J. J. (1993). *World without end: Economics, environment and sustainable development*. New York: Oxford University Press.

Shlay, A. (1993). Shaping place: Institutions and metropolitan development patterns. *Journal of Urban Affairs, 15*, 395-412.

World Commission on Environment and Development. (1987). *Our common future*. New York: Oxford University Press.

COMMENTARY ON
THE CYCLE OF CITIES

Carol Waldrop

There is a point in the cycle of cities and their neighborhoods when economic development is no longer a viable strategy for revitalization. Even when the traditional concept of economic development as business attraction is broadened to mean job creation, the question becomes what kind of jobs, for whom, and where. As Goldsmith demonstrates, the "trickle-down" effect of economic development efforts does not automatically or necessarily address the problems of decay and isolation we find in our cities and their neighborhoods, nor does it change the pattern.

At the policy level, Goldsmith argues that redistribution of revenue and resources through such measures as densification in urban areas and elasticity of urban/suburban boundaries are fundamental to an equitable urban policy and strategy. However, enacting measures for redistribution of revenue and resources requires political will and may well be beyond the planner or the community to affect. Such efforts are likely to be overwhelmed by well-organized interest groups.

At the level of individual practice, Goldsmith cites examples of the work of "equity planners," noting that they are inevitably drawn into conflict, fighting for the rights of the powerless against the powerful and self-interested. Goldsmith concedes that little substantive change has occurred as a result of their efforts. His conclusion is that "in this generally oppressive economic and political situation, advocates for central city improvements, residents of poor neighborhoods, and constructive, progressive city governments must be tenacious and creative."

Goldsmith's analysis stops short of exploring what such creative activity and approaches might be and leaves the reader with the impression that adversarial activity is the only correct position for a progressive "equity planner" to take.

What would be a promising approach to urban revitalization, which stimulates capacity building at the most basic, individual level in neighborhoods and communities; which stimulates economic vitality at the city and regional levels (translating to jobs); and strategically creates linkages between people and jobs (or, more broadly, economic opportunity) through transportation, education, cooperative action, capital availability?

How can each of us in our respective capacities as practitioners and community members be more effective, particularly in an era of shrinking resources, in achieving the vision of a city, articulated by Harold Washington, in which the "neighborhoods are once again the center of our city, in which businesses boom and provide neighborhood jobs, in which neighbors join together to help govern their neighborhood and their city"?

First, it is important to understand that the role of government action in revitalization and community building should be to stimulate expansion of opportunity and increase access to opportunity for those who have been historically underserved.

Second, in order to link opportunity to the "inherent capacity to develop and improve," as a community we must approach revitalization in a "hands-on," persistent, incremental manner within the context of a comprehensive vision or strategy, forming partnerships to develop the tools for redistribution and to develop opportunities to foster neighborhood capital accumulation.

We must deliberately and strategically target, layer, and combine existing resources; coordinate their distribution; provide services directly rather than via the middleman of federal bureaucracy or state or local service provider; eliminate barriers; capture opportunity. We need to look more closely at the small successes to replicate what made them work and understand that small successes add up over time to substantive change.

An example of a promising approach at the federal level is the U.S. Department of Housing and Urban Development's challenge to major urban areas to develop community-based strategic plans for revitalization which would form the basis for competing for designation as an Empowerment Zone or Enterprise Community. Such designation would mean eligibility for wage tax credits as a lure to business investment, and $100 million in social service funds. The intention of the program is to use the tax system to encourage private investment and use public resources to leverage additional funds from businesses, foundations, and state and local governments. The approach has its foundation in the belief that the residents themselves are the most important element of revitalization and the belief in the power of creating partnerships for real change. The strategic vision for change created by the community and its partners would demonstrate how economic opportunities would be created, how the community would achieve streets that are safe to walk, housing that is secure, human services that are accessible, and how a vital civic spirit is nurtured.

The challenge of the application process was for city governments and their communities to work together and form partnerships that draw on federal, state, regional, local, private sector, and community commitments to create and carry out a strategic vision for change. For example, in support of their strategic vision for change, the City of Detroit secured over $2 billion in private-sector commitments to match federal dollars. In Baltimore, $1 billion has been committed by the private sector.

It will be instructive for those of us committed to the process of community building to follow the progress of cities that were awarded the federal designations and the accompanying dollars based on their strategic plans as well as to study the progress of other efforts that, although not selected, were jump started into a strategy by the application process.

For example, even though Los Angeles's application was not one of the six awarded an Empowerment Zone designation, the process forced some issues and provided a needed stimulus to create partnerships for community building in the disadvantaged neighborhoods of Los Angeles. Perhaps most important, the process made visible the numerous quietly effective grassroots efforts already under way, promoting the opportunity for linkages and for leveraging their effectiveness. Some examples of promising partnership efforts are the following.

The business community, the City of Los Angeles, the County of Los Angeles, the federal government, private lenders, and community organizations have come together to establish the largest community lending institution in history, the Los Angeles Community Development Bank, creating the opportunity for Los Angeles to invest more than $1 billion of working capital, equity, and grants over the next 10 years into organizations in those areas needing an economic stimulus, utilizing funds from HUD, private lending institutions, and proceeds of the sale of bonds. The Los Angeles Community Development Bank model allows for a range of lending activity from small micro loans to large venture-capital findings. Making funds and technical assistance available primarily through investment in existing community-based lending organizations allows most decision making to take place at the local level. By capturing the opportunity to leverage loan loss reserves and loan guarantee funds, $25 million has the potential to create $100 million to $150 million in lending activity in the marketplace, leading to significant leverage for the community. The intention is for the bank to become a self-perpetuating institution, and to the extent that income is created, those dollars could be used for social service programs. Not unexpectedly, the most difficult issues to be resolved among the parties involved in structuring the bank are governance and independence.

The City of Los Angeles' Redevelopment Agency is in the process of redefining its mission and being dramatically restructured to sharpen its focus as a tool for redistribution and stimulus for opportunity, particularly coordinating its activities with the mission of the Los Angeles Community

Development Bank to strengthen and leverage investments in neglected areas.

Based on the premise that job growth and inner-city revitalization come from neighborhood-based small businesses that can get the capital to grow, Rebuild Los Angeles (RLA), now under new leadership, has refocused its mission from attempting to lure large corporations to locate in impoverished areas. RLA's efforts now focus on creating jobs by assisting small businesses, facilitating business networks, and redeveloping lots left vacant in the historically neglected neighborhoods, particularly those most heavily hit by the fires and vandalism of the 1992 civil disorders. RLA has organized networks of companies in the furniture, toy manufacturing, and metalwork industries to address common obstacles to growth such as access to investment capital and new markets, finding skilled workers and negotiating environmental and other regulations. They have developed a computer online system to link companies with purchasers at major corporations.

An example of a partnership to provide quality educational opportunities and support services to at-risk and low-income children and youth is LEARN (Los Angeles Educational Alliance for Restructuring Now). LEARN is a broad-based coalition of civic leaders and representatives of Los Angeles' diverse education, ethnic, business, labor, academic, religious and social advocacy constituencies who came together in 1991 to draft and implement a community-generated education reform and restructuring agenda adopted by the Los Angeles Unified School District (LAUSD) Board of Education in 1993 for implementation by the LAUSD. Serving in the dual roles of ombudsman and community consensus builder, LEARN has suc- ceeded in bringing together all of the disparate factions within LAUSD, as well as over 200 community based organizations in it. The LEARN plan reflects a community's strategic vision for change, focusing on and leveraging the power of the school site as the single defining and stabilizing institution in each community and a center for social economic uplift in the neighborhoods.

An example of persistent effort and individual leadership is found in City Councilman Mark Ridley-Thomas' district, which includes part of South-Central Los Angeles, Leimert Park, and the Crenshaw district. The councilman's aggressive prodding of government and businesses to rebuild in South Los Angeles has resulted in a 12-screen theater at the Baldwin Hills Crenshaw Mall, three new grocery stores put into operation, and development of commercial and residential projects in the Manchester-Vermont Avenues area.

One of the numerous examples of community-based initiatives that became visible to the larger community as a outcome of the process to create a strategic vision for change is the Dolores Mission Day Care

Center and the Homeboy Industries Bakery. In 1992, 15 ex-gang members built the Dolores Mission Day Care Center, which is staffed by local mothers and provides low-cost day care and a job referral services. The Homeboy Industries Bakery, formerly known as "Homeboy Tortillas," was created as an enterprise under the Dolores Mission's "Jobs for a Future" program in 1992. Ex-gang members started this enterprise at the Los Angeles Grand Central Market, making tortillas. Currently, a building at the Dolores Mission site is being renovated into a bakery which will employ 20 youth, making rye and pumpernickel bread for one of the biggest institutions in LA's Eastside, the USC Los Angeles County Medical Center.

In cities across the country, there are examples of promising approaches, productive partnerships, and individual and neighborhood initiatives that should be studied to understand what individual and cumulative ingredients succeed in creating opportunity, maximize effectiveness of resources, and leverage the "inherent capacity to develop and improve." Leadership and political will are significant ingredients for success and for resolving inherent conflicts, but even in the absence of political will, each person who has any role in the process must seek out the appropriate level of intervention and arena of effectiveness in order to resolve conflicts, open opportunity, or to help people organize themselves, and, in so doing, incrementally move a system toward equity.

REPLY TO PETER MEYER
AND CAROL WALDROP

William W. Goldsmith

Thanks very much to Peter Meyer and Carol Waldrop for their critical and constructive comments. I have incorporated in the text some minor changes in acknowledgment of their suggestions; here I will respond more directly, in the hopes that the conversation may be continued. Meyer and Waldrop both explicitly question whether the "political will" exists to enact many of the changes we all seem to desire. I assume not, at least not now; our job, then, it seems to me, is to describe and understand the neighborhood and city situation, including the political constraints, but always to try to find ways in which political pressure from below can work with collaboration from above, to look for and work on "nonreformist reforms," in Andre Gorz's good phrase, that engender further reforms—those "small successes," says Waldrop, that "add up over time to substantive change." I remain pessimistic, not hopeful, but still hoping—and trying. No fun to be a naysayer, but it is important to be honest. The situation in this country's poor neighborhoods is simply horrible, inhuman. Kids in the seventh grade, lots of them, write papers about selling drugs, murder on the block, and mothers who are sick without care or, worse yet, can't care for them. *This* in the middle of the materially richest civilization in history? We have big problems.

Waldrop suggests that planners look for areas of cooperation rather than conflict. She is joined here by Meyer, who looks to sustainability as a nontargeted good toward which all might join effort. She urges that government's role is to help "those who have been historically underserved." I agree, but tell it to the suburbs, or to Newt Gingrich! With them, cities *are* in conflict. To some considerable extent, their gain *is* the city's loss.

Waldrop correctly urges deliberate, strategic, coordinated action (such as LEARN), and she looks with hope at empowerment zones (EZs) or enterprise communities (ECs). From the perspective of a particular city, the opportunity of gain from an EZ or EC must be seized. But from a national perspective, I must disagree. The EZ idea is, as Waldrop says, "to use the tax system to encourage private investment." That is not the same as believing, in her words, "that the residents themselves are the most important element of revitalization," though they may be. I am convinced that Henry Cisneros's HUD wants to help cities and their poorest residents by shifting resources, but I do not think HUD has the resources. The EZ/EC idea is, in the end—as were its enterprise zone predecessors under antiurban, antipoor conservatives—a sop, a bribe, a pretense (Goldsmith, 1987). (I admit that $100 million in social service funds is a considerable sum, but it should be measured in the context of federal changes generally *reducing* social service allocations.) This is not the place for technical arguments about an EZ helping Detroit or Baltimore leverage private funds—I will just say I am very skeptical. Community development banks can be extremely useful, but even the (optimistic?) $1 billion in Los Angeles must be compared to the $9 billion in venture capital alone raised in Silicon Valley in the 1980s, or $3 billion for Route 128 companies (Saxenian, 1995, Figure 3, p. 106). If we really want to use federal tax policy to help cities, let's go after some of the big items: military spending, or even the auto and home mortgage subsidies. But I admit my response is weak; as Willie Sutton found out, it takes more than knowing where the money is.

Waldrop points to the importance of small business, much as Harold Washington's city of Chicago stressed assistance to existing, and normally very small, businesses. Point taken, and the examples from Los Angeles are heartening. But again, the context must be used as the base for measurement. Here we should reflect on the fact that big firms really do matter, not just in airframe decline, but in informatics growth. In 1987, for example, although it is true that 85% of U.S. computer firms had fewer than 100 employees, the 5% of firms with more than 500 employees "accounted for fully 91 percent of all employment" (Harrison, 1994, p. 5).

Peter Meyer is right that analysts of community economic failure should prominently identify external causal factors, and he is right to point out that "strictly *local* economic development" is a rare phenomenon. One can extend that argument: Most localities—most mainly residential areas, most suburbs, and many city neighborhoods, that is—do not *need* to "develop" economically. Their residents bring home wages and salaries from work typically undertaken elsewhere, from jobs located in some place that *is* developed economically. No economic development problem for Scarsdale. The ghetto, the central city, the poor inner-city neighborhood is where we say the problem is lack of development. Well, the main problem really is lack of good jobs, wherever they may be, from which the residents can earn good incomes. So, first, jobs and income. Not available? Then the residents need transfer payments—and social services. Second, residential neighborhoods need services—garbage collection, schools, police, playgrounds and parks, and so on. "Good" neighborhoods get better services. Finally, good neighborhoods are the beneficiaries of massive income transfers through (usually hidden) public subsidies, the main ones being the mortgage tax deduction and subsidies for highways and automobile drivers. Something needs to be done to right this balance.

Peter Meyer is also right to point out that it will be difficult to force suburbs to do anything constructive about these imbalances unless we can identify policies that serve the interests of the nonpoor. From a very broad vantage point, his "sustainable urbanism" might be one of those nontargeted approaches, but unfortunately, for both of us the scene is bleak.

REFERENCES

Goldsmith, W. (1987). Bringing the Third World home: Enterprise zones. In R. Peet (Ed.), *International capitalism and industrial restructuring*. New York: Routledge.

Harrison, B. (1994). *Lean and mean: The changing landscape of corporate power in the age of flexibility*. New York: Basic Books.

Saxenian, A. (1995). *Regional advantage: Culture and competition in Silicon Valley and Route 128*. Cambridge, MA: Harvard University Press.

Index

Alabama, 144, 145
Amenities. *See* Quality of life
Appalachia, 299-300
Assessments, 181-185, 247. *See also*
    Program evaluations
Attraction factors, 62-66. *See also*
    Location decisions; Quality of life

Benchmarking, 58-59, 70, 84-85, 188,
    198-199, 202
Bureaucracy, professional, 229-231,
    235-236
Business cycles, 250-251
Business displacement, 30, 32-33, 37-38,
    50-51

Capital, 111-112, 123-124. *See also*
    Human capital; Infrastructure
    investment; Venture capital
Central cities, 292-298, 301-314, 320-330
Chicago, IL, 36-43, 46-48, 54-55, 207-208,
    266, 279-280
Cities:
    burdens on, 304-306
    cycle of, 291-314, 320-330
    elasticity and, 306-307, 321
    environment of, 308-309
    equity-efficiency and, 293-297,
        302-304, 320-321
    innovation in, 205-212
    invisibility and, 297-302
    outsiders and, 299-302, 304-312
    revitalization of, 304-312, 321-330
    suburbs and, 295, 306-314, 330
    sustainable development for, 321-322,
        330
Cost-benefit analyses:
    evaluations by, 249, 255-256, 285

of evaluations, 267, 274-275
of infrastructure investment, 95-96
of university impact, 138
spreadsheet for, 36-43, 46-48, 50-55
See also Game theory; Program
    evaluations
Culture of poverty, 300
Cycle of cities, 291-314, 320-330

Dakotas, small towns in, 298
Deadweight spending, 30, 31, 36-37
Decision making, 227-231, 235.
    See also Location decisions
Deterrence, modeling of, 31, 34, 39
Detroit, MI, 236, 237
Dinka people, 298-299

Economic development:
    bureaucracy of, 229-231, 235-236
    goals of, 69
    organizational arrangements of, 229
    professionalization of, xiii-xiv, 275
    viability of, 291-314, 320-330
Economic Development Administration,
    275
Education:
    for economic development, xiii, xiv
    idea generation and, 212-216
    infrastructure of, 113
    investment in, 216, 222-223
    neighborhood, 215-216
    rate of return from, 108, 109, 118
    urban revitalization and, 327
    See also Private universities; Public
        universities
Elasticity:
    of city limits, 306-307, 321
    of output, 91

Employment:
  business displacement and, 30, 32-33
  new versus retained, 30, 31-32, 37
  opportunity costs and, 30, 33-34, 39-41
  universities and, 118
  valuing of, 11
  See also Human capital; Jobs;
    Unemployment
Empowerment zones, 16, 325, 326, 329
Enterprise communities, 325, 329
Enterprise zones, 16-17, 251, 264, 270,
  297
Entitlements, 16-17, 214. *See also* Welfare
  system
Environmental burdens, 308-309
Equity-efficiency interaction, 293-297,
  302-304, 320-321
Equity planners, 310-311
Evaluations. *See* Incentives, ex ante
  evaluations of; Incentives, ex post
  evaluations of; Program evaluations
Exports, 33, 118, 273. *See also* Migration

France, 209
Free-market policies, 12-13

Game theory:
  limitations of, 24-25
  modeling by, 2-11, 24-27
  negative-sum games, 6-8, 12-17
  positive-sum games, 3-6, 12-18
  zero-sum games, 8-9, 12-17, 32
  See also Cost-benefit analyses;
    Program evaluations
Government, versus governing, 225

Hierarchy of needs, 66
Higher education. *See* Private universities;
  Public universities
Housing and Urban Development, U.S.
  Department of (HUD), 46-48, 325,
  329
Human capital, 108-109, 118, 120, 136,
  138, 203-205, 219-223

Ideas, incubation of, 203-216
Illinois, 260-262. *See also* Chicago, IL
Incentives:
  equity of, 23, 26-27
  game theory and, 2-18
  job bidding and, 144-147, 203
  strategies for, 13-18, 21-23
  tax breaks as, 144-145
  versus idea incubation, 203, 215-216
Incentives, ex ante evaluations of, 28-29
  business displacement and, 30, 32-33,
    37-38, 50-51
  employment and, 30, 31-32, 37
  limitations of, 29, 48, 49-53
  modeling issues for, 29-36
  opportunity costs and, 30, 31, 33-36,
    39-42
  spreadsheet for, 36-43, 46-48, 50-55
  tax revenues and, 42, 48, 54-55
  See also Cost-benefit analyses;
    Program evaluations
Incentives, ex post evaluations of, 28,
  29-35, 49. *See also* Cost-benefit
  analyses; Program evaluations
Income, 108, 118, 214, 221, 222
Indiana, 264
Inducements. *See* Incentives
Industrial clusters, targeting of, 17-18, 23
Industry performance assessments,
  181-182
Industry resource assessment, 184-185
Industry targeting:
  as more art than science, 190
  background on, 172-179
  benchmarking for, 188, 198-199, 202
  benefits of, 171-172, 174, 176-177
  competition and, 173, 174, 183,
    198-199
  criticisms of, 177-178
  data collection for, 179-185, 198
  definition of, 173-174
  economic profile for, 180-181
  fit analyses for, 186-187
  game theory and, 17-18, 23
  goals and, 174, 186-187, 195-196
  growth and, 181-182, 197
  impact of, 190-191, 196-197

input-output analysis for, 189-190
policies and, 174, 176-177, 191, 197,
  199-202
politics of, 178, 199, 200
resource assessments for, 182-185
steps in, 179-187, 198-199
types of, 176, 178-179, 187-190
undersupply analysis for, 187-189
Inequality, burdens of, 304-306
Information asymmetries, 15-16, 25
Infrastructure:
  core, 83
  definition of, 83
  growth and, 95, 99-103, 209-210
  innovation and, 209-210
  knowledge-based, 113, 124, 126, 136
  measurement difficulties and, 84-86
  taxes and, 87, 88, 159, 165
  universities and, 111-112, 113, 123-124,
    126
Infrastructure investment:
  benchmarking of, 84-85
  benefit-cost analysis of, 95-96
  causation and, 86, 91, 99-101
  competition and, 86-87
  contingent capacity and, 88, 99-103
  econometrics and, 84-93
  location decisions and, 86-87
  policy suggestions on, 93-96
  productivity and, 87-88
  quality of life and, 94-95
  services and, 83-85, 87
  taxonomies for, 99-103
Innovation, 110-111, 122-123, 204-212
Input-output analyses, 30, 33, 51-52,
  189-190
Invention, 205-206
IQ theory, 213

Jobs:
  bidding for, 144-147, 151, 163-164,
    203-216
  creation of, 203-216, 219-223
  See also Employment; Unemployment

Know-how transfer, 109-110, 121, 128,
  129
Knowledge-based infrastructure, 113, 124,
  126, 136
Knowledge creation, 107-108, 117-118

Leadership:
  big events and, 230-231, 241
  cooperative, 232, 236-238
  education for, 215
  in individuals, 227-237
  in public universities, 112-113, 124
  meaning of, 241, 242, 244
  politics of, 224-238, 242-245
  professional bureaucracy and, 229-231,
    235-236
  socioeconomic factors and, 242,
    243-244
  structure and, 224-225, 227-235,
    241-244
Location decisions:
  assessment of, 173, 183, 184-185,
    186
  game theory and, 4-5, 7
  infrastructure investment and,
    86-87
  innovation and, 208
  quality of life and, 56-66, 68, 76-79
  universities and, 120
  See also Exports; Industry targeting
Los Angeles, CA, 209, 326-328,
  329-330

Massachusetts, 126, 265
Mexico, 312-313
Michigan, 262, 264-265. *See also* Detroit,
  MI
Migration, 34, 47, 77, 78, 150-151, 273
  universities and, 109, 114, 118, 120,
    136
Mobility, 150-151, 220-222. *See also*
  Migration
Morrill Act, 104

Nantes, France, 209
Negative-sum outcomes. *See* Game theory
Neighborhood education, 215-216
Neighborhood impact index, 42-43
Neighborhoods, poor, 304-312.
    See also Cities
New York City, 307
New York State, 163-164, 257, 258-259

Ohio, 259-260
Opportunity costs, 30-31, 33-36, 39-42,
    115, 285
Oregon, 264
Outcomes, 231-232, 241, 254-255.
    See also Program evaluations
Output elasticity, 91
Outsiders, and exploited cities, 298-302,
    304-312
Oversubsidization, 8-9, 14, 16-17

Patents, 205
Perpetual inventory method, 84
Political economy, features of, 56
Politics:
    conservative, 295-297
    of industry targeting, 178, 199, 200
    of leadership, 224-238, 242-245
    of program evaluations, 269, 274
    of revitalization, 304-312, 322, 324,
        328
    of taxes, 146-154, 159, 166, 168-169
    pressures of, 15, 49, 53, 55
    progressive, 304, 309-312
    symbolic, 236-237
Positive-sum outcomes. *See* Game theory
Private universities, 136-137. *See also*
    Public universities
Process evaluations, 247, 256-257, 267,
    278, 289
Profits, expected, and game theory, 2-11
Program evaluations:
    causation and, 252, 254
    continuum of, 247-250
    cost-benefit analysis of, 267, 274-275

definition of, 246-247
designs for, 251-273, 285
growth and, 286-287, 288-289
inadequate, reasons for, 267-269, 274
of big projects, 281-282
of communities, 270
of community impact, 273-274
of firms, 270-273
of incentive impact, 28-55
of outcomes, 247, 249-267, 270-274,
    280-283, 289-290
of process, 247, 256-257, 267, 278, 289
of purpose, 287
politics of, 269, 274
social goals and, 285-286, 287, 288
standards for, 274-275
types of, 246-250, 270-274
See also Incentives, ex ante evaluations
    of; Incentives, ex post evalu-
    ations of
Public universities:
    as multiproduct organizations, 106-115
    as supportive organizations, 137, 139
    capital investment by, 111-112, 123-124
    changes in, 104-106, 126-130
    comparative advantages of, 127-128
    credibility of, 127
    human capital formation by, 108-109,
        118, 120, 136, 138
    impact of, 106-126, 136-139
    infrastructure of, knowledge, 113, 124,
        126, 136
    infrastructure of, physical, 111-112,
        123-124
    knowledge creation by, 107-108,
        117-118
    leadership by, 112-113, 124
    legitimacy of, 130
    migration and, 109, 114, 118, 120, 136
    outputs of, 106-126, 136-139
    quality of life and, 112-113
    rate of return from, 108, 109, 118
    taxes and, 112
    technology and, 109-111, 122-123,
        128, 129
    See also Education; Private universities

Quality of life, 56-71, 74-79, 81
    definition of, 77, 80
    infrastructure investment and, 94-95
    taxes and, 150
    universities and, 112-113

Racial bias, 294, 295, 304-305, 306, 309, 313
Regime theory, 227
Regional Economics Models Inc (REMI) model, 30, 37-39, 47, 52, 266
Regional purchase coefficient (RPC), 33, 51-52, 54
Rent, and growth, 208-209
Resource assessments, for industry targeting, 182-185
Revitalization, of cities, 304-314, 321-330. *See also* Cities

Science, spending on, 205-206
Social goals, and program evaluations, 285-286, 287, 288
Social targeting, 174
Socioeconomic factors, and leadership, 242, 243-244
Standard Industrial Classification (SIC) codes, 182, 185, 198
St. Louis, MO, 207, 208
Suburb-city interaction, 295, 306-314, 330
Sudan, 298-299
Sustainable development, 321-322, 330
Symbolic politics, 236-237

Targeting, types of, 174. *See also* Industry targeting
Taxes:
    burdens of inequality in, 305-306
    ex ante evaluations and, 42, 48, 54-55
    infrastructure and, 87, 88, 159, 165
    program evaluations and, 262-263
    sales, 151-152
    universities and, 112
Taxes and the states:
    ambiguity and, 147-154, 166, 169

competition and, 144-147, 151, 163-164
    differential effects and, 152-153, 156-157, 162
    economic changes and, 140-144
    economic performance and, 155-159
    growth and, 148-150, 152-153, 163, 166, 169
    local factors and, 162-165, 167-168
    measurement issues and, 150-153, 155, 156-159
    modeling for, 154-159, 162-163, 166
    period effects and, 153-154, 158
    policy choices and, 147-159, 165-167, 168-170
    politics of, 146-154, 159, 166, 168-169
    quality of life and, 150
Technology, 58, 110-111, 122-123, 204-209
Technology transfer, 109-110, 121, 128, 129
Toronto, Canada, 307-308
Total quality management (TQM), 64
Training programs, evaluations of, 264-265, 267, 278
Transportation, 207-208, 305
Triage policy, 303-304

UICUED Cost-Benefit Spreadsheet, 36-43, 46-48, 50-55
Unemployment, 12. *See also* Employment; Jobs
Universities, 136-137, 212-213. *See also* Public universities

Venture capital, 210-211
Virginia, 256-257

Washington State, 265
Water services, 85
Welfare system, 297

Zero-sum outcomes. *See* Game theory

About the Editors

Richard D. Bingham is Professor of Public Administration and Urban Studies and a Senior Research Scholar of the Urban Center at Cleveland State University. His latest books include *Theories of Local Economic Development*, edited with Robert Mier (Sage, 1993); *Managing Local Government*, authored with other Urban College faculty (Sage, 1991); and *State and Local Government in a Changing Society* (1991) authored with David Hedge. He is Founding Editor of the journal *Economic Development Quarterly*.

Robert Mier was Professor of Urban Planning and Public Administration at the University of Illinois at Chicago from 1975 until his death in 1996. He founded the UIC Center for Urban Economic Development in 1978. The Center is acclaimed for its technical assistance to community-based organizations and its policy research on local economic development.

He left the University in 1983 to serve as Director of Development for the City of Chicago under Mayor Harold Washington and, later, Mayor Eugene Sawyer. He was the architect and chief implement of Chicago's highly regarded 1984 development plan, a national model for equity-oriented local municipal development planning. After returning to UIC in 1989, he became a nationally know consultant on urban economic development and equity planning. His 1993 book, *Social Justice and Local Economic Development*, was the 1995 recipient of the prestigious Paul Davidoff award from the Association of Collegiate Schools of Planning.

About the Contributors

David Arsen is Associate Professor of Political Economy in the James Madison College at Michigan State University. His research focuses on the impacts of international trade and investment on the regional structure of industry and employment, as well as state and local government economic development policies.

Timothy J. Bartik is Senior Economist for the W. E. Upjohn Institute for Employment Research, a nonprofit research organization in Kalamazoo, Michigan. He served as Legislative Assistant for Housing and Urban Policy to U.S. Senator Donald Riegle, Jr., from 1975 to 1978; Assistant Professor of Economics at Vanderbilt University from 1982 to 1989; and Associate Director of the Center for Regional Growth and Governance for the Vanderbilt Institute for Public Policy Studies from 1987 to 1989. His research focuses on regional economic development and local labor markets. His book *Who Benefits from State and Local Economic Development Policies?* was published in 1991.

Robert A. Beauregard is Professor in the Robert J. Milano Graduate School of Management and Urban Policy of the New School for Social Research. He writes on economic development and cities and recently published *Voices of Decline: The Postwar Fate of U.S. Cities* (1993).

John P. Blair is Belinda A. Burns Scholar and Professor of Economics at Wright State University. He has published widely in the field of economic development in such journals as *Urban Affairs Quarterly, Review of Regional Studies,* and *Economic Development Quarterly.* His latest book is *Local Economic Development: Analysis and Practice* (Sage, 1995). He has served as a consultant to businesses and governments and is a member of the Montgomery County Planning Commission. Prior to joining the faculty of Wright State, he taught at the University of Wisconsin-Milwaukee and served for 2 years as a Policy Analyst in the Department of Housing and Urban Development.

William M. Bowen is Associate Professor of Urban Studies and Public Administration at the Levin College of Urban Affairs, Cleveland State University. His research and teaching interests are in decision science applications in economic development, environmental affairs, and energy policy.

Paul Brace is Clarence L. Carter Professor of Political Science at Rice University, with research interests in state, judicial, legislative, and presidential politics. His research has appeared in the *American Political Science Review, Journal of Politics, Western Political Quarterly/Political Research Quarterly, Polity, Social Science Quarterly, American Politics Quarterly, Legislative Studies Quarterly*, and *Spectrum*, among other journals. He is the author of *State Government and Economic Performance* (1993) and coauthor of *Follow the Leader: Opinion Polls and the Modern Presidents* (1992), winner of the Neustadt Award for the best book published on the American presidency in 1992. His research on state taxation won the Best Paper Award from the State Politics and Policy Section of the American Political Science Association in 1995. He has served, or is serving, on the editorial boards of the *American Political Science Review, Journal of Politics, Political Research Quarterly*, and *American Politics Quarterly* and is an Associate Editor for *Spectrum: The Journal of State Politics*.

Brian J. Cushing is Associate Professor of Economics and Research Associate Professor of the Regional Research Institute at West Virginia University, where he has been a faculty member since 1981. He served as Acting Director of the Regional Research Institute in 1991. He has published numerous articles in his three primary research areas—poverty, population migration, and urban structure—including two articles directly related to the effect of amenities on migration (*Annals of Regional Science*, 1987; *Journal of Regional Science*, 1987) and one article related to intraurban residential and employment location (forthcoming in *Urban Studies*). He currently serves on the Executive Council of the Southern Regional Science Association.

Sabina E. Deitrick is Assistant Professor at the Graduate School of Public and International Affairs, University of Pittsburgh, where she teaches urban and regional planning. Her research focuses on economic development and regional planning both in the United States and abroad. She is coauthor of *The Rise of the Gunbelt* (1991) with Ann Markusen, Peter Hall, and Scott Campbell.

Steven C. Deller is Associate Professor and Extension Community Development Economist with the Department of Agricultural and Applied Economics, University of Wisconsin-Madison. His research and extension education programming focuses on regional economic structure and modeling, the efficiency and effectiveness of local governments in rural areas, and rural economic development policy. His work has appeared in the *Review of Economics and Statistics, Economic Development Quarterly, Regional Science and Urban Economics, Public Choice, Land Economics*, and the *American Journal of Agricultural Economics*.

David Fasenfest is an economist and sociologist, currently Research Associate Professor at the Great Cities Institute, University of Illinois at Chicago. His research focuses on the impact of economic restructuring on the spatial and social distribution of income. He received his graduate training at the University of Michigan and has authored several articles appearing in *Urban Affairs Quarterly, Journal of Urban Affairs,* and *Economic Development Quarterly.* He is the editor of *Community Economic Development: Policy Formation in the U.S. and U.K.* (1993).

Daniel Felsenstein is Lecturer in the Geography Department and Institute of Urban and Regional Studies at the Hebrew University of Jerusalem. His current research interests are in impact analyses of local economic development programs and university-industry relations. His recent publications have dealt with estimating the employment effectiveness of small-firm assistance schemes, examining labor market extension and network formation among high-technology firms, and analyzing the impacts of "induced migration" associated with the presence of a major university. His published research on these and related issues has appeared in journals such as *Urban Studies, Regional Studies, Small Business Economics, Entrepreneurship,* and *Research in Higher Education.*

Robert Giloth is Senior Associate at the Annie E. Casey Foundation in Baltimore and manages its new employment initiative. A former Deputy Commissioner of Chicago's Department of Economic Development, he directed the Research and Development Division, which was responsible for program evaluations. He has also directed Community Development Corporations in Baltimore and Chicago. His research and writing are concerned with community development, economic development, and planning theory.

Robert E. Gleeson is Associate Professor of Business Administration at the A. J. Palumbo School of Business Administration, Duquesne University.

William W. Goldsmith is Professor of City and Regional Planning at Cornell University. With Edward Blakely, he won the 1994 Davidoff Prize for *Separate Societies: Poverty and Inequality in U.S. Cities.* His first paper on ghetto development was published in 1974; his most recent appeared in the *Review of Black Political Economy* in 1996. He teaches and studies urbanization and regional development, problems of poverty and racism, and the urban environment. He has worked extensively in Latin America.

Harvey A. Goldstein is Professor of City and Regional Planning and Director of the PhD program in planning at University of North Carolina-Chapel Hill. He is a consultant on labor market issues to the Bureau of Labor Statistics and is a member of the Editorial Board of *Cityscape.* With Michael Luger, he is coauthor of *Technology in the Garden: Research Parks and Regional*

Economic Development (1991) and of a book in progress on the changing role of universities in regional economic development.

Brett W. Hawkins is Professor of Political Science at the University of Wisconsin-Milwaukee. He has also taught at Washington and Lee University and the University of Georgia. He was educated at the University of Rochester and at Vanderbilt University. He is author of more than 45 publications on urban political phenomena.

Donald T. Iannone is a nationally recognized leader in the economic development field with 20 years' experience in various professional, academic, and consulting capacities. He has headed the Economic Development Program in Cleveland State University's Urban Center for the past 10 years. As a former board member of the National Council for Urban Economic Development and the American Economic Development Council, he has contributed to the national advancement of economic development education and policy development. He has participated in and directed more than 125 major technical assistance, training, and policy research projects for public- and private-sector clients during his career, including nearly 20 target industry studies. As a private consultant, he has worked on industry site location and economic development strategies in 13 countries and 24 U.S. states. He has written and published numerous technical reports, journal articles, and book chapters on economic development policy and corporate site selection issues.

Valerie B. Jarrett has been the Commissioner of the Department of Planning and Development for the City of Chicago since 1991. Prior to this appointment, she was the Deputy Chief of Staff to Mayor Richard M. Daley.

Judith Kossy is Associate Director of San Diego Dialogue, a community-university public policy group at the University of California-San Diego, focusing on economic development issues in the San Diego/Northern Baja, California, region. She recently was president of the Western New York Economic Development Corporation in Buffalo. She has held various positions in neighborhood revitalization and economic development at the U.S. Department of Housing and Urban Development in Washington, D.C.

Rishi Kumar is the Dean of the College of Business and Administration at Wright State University. He earned his BA with Honors from University of Delhi and studied in the United States with the support of a Fulbright-Smith-Mundt Fellowship. He has written numerous articles in such journals as the *Southern Economic Journal*, the *Journal of Regional Science*, and *Public Choice*. He is active in various community programs and has been the recipient of numerous awards, including the Trustees' Award for Faculty Excellence and the Presidential Award for Outstanding Faculty.

Larry Ledebur is Director of the Urban Center, Maxine Goodman Levin College of Urban Affairs, Cleveland State University. Prior to coming to Cleveland State, he directed the Center for Urban Studies at Wayne State University. He has also served as Director of the Urban Institute's Economic Development Program and as Principal Researcher for the White House Conference on Balanced National Growth and Economic Development. His books include *Economic Disparities: Problems and Strategy for Black America; Urban Economics: Processes and Problems;* and *The U.S. Common Market* (forthcoming) with William Barnes of the National League of Cities.

Nancey Green Leigh is Associate Professor in the Graduate City Planning Program at the Georgia Institute of Technology. She teaches and conducts research in urban and regional development, local economic development planning, industrial restructuring, and changing standards of living. She is the author of *Stemming Middle Class Decline: Insights for Economic Development Planning* (1994).

Michael I. Luger is Carl H. Pegg Professor of City and Regional Planning and Chairman of the Curriculum in Public Policy Analysis at the University of North Carolina at Chapel Hill. He has served on several gubernatorial advisory commissions and task forces related to economic development and is a member of the Editorial Board of *Economic Development Quarterly.*

Peter B. Meyer is Professor of Urban Policy and Economics and Director of the Center for Environmental Management at the University of Louisville. Prior to moving to Kentucky in 1988, he taught at the Pennsylvania State University, where he directed the Local Economic Development Assistance Center from 1979 to 1987. He has written extensively on community and local economic development and on comparative public policy, recently editing *Comparative Studies in Local Economic Development: Problems in Policy Implementation* (1993) and coauthoring *Contaminated Land: Reclamation, Redevelopment and Re-Use in the United States and the European Union* (1995).

Joseph Persky is Professor of Economics at the University of Illinois at Chicago. His primary research interests include urban economic development, regional income distributions, and the history of economic thought. He has worked extensively with the Center for Urban Economic Development and the Great Cities Program at the University of Illinois at Chicago. He is an Affiliate of the Institute of Government and Public Affairs of the University of Illinois and a Research Associate of the Economic Policy Institute in Washington, D.C.

Laura A. Reese is Associate Professor in the Department of Geography and Urban Planning at Wayne State University. Although she has published articles in the areas of urban politics and urban service delivery, her work focuses on local economic development. She is currently studying economic development policy making in U.S. and Canadian cities, focusing on the role of development professionals.

Signe M. Rich is Associate Director of Economic Development for the City of Albuquerque. The office oversees financial incentives, such as tax increment financing, tax abatement, and industrial revenue bonds. From 1979 through 1981, she received a Fellowship in Intergovernmental Management and was assigned as Development Officer in the Urban Development Action Grant (UDAG) program and Office of Federal Urban Policy at the Department of Housing and Urban Development (HUD), Washington, D.C. She manages the Shared Vision project for the city of Albuquerque and manages a regional economic development strategy through a grant from the U.S. Department of Commerce, Economic Development Administration. She is on the Board of Directors for the National Council of Urban Economic Development (CUED).

James A. Segedy is Associate Professor of Urban Planning at Ball State University. He is also Director of the Community-Based Projects Program of the College of Architecture and Planning and Ball State University. He is a member of the American Institute of Certified Planners. He is codirector of the Indiana Total Quality of Life Initiative, serves as President of the Indiana Planning Association, and is Chair of the Small Town and Rural Planning Division of the American Planning Association. He is also Past-President of the Indiana Community Development Society.

Ron Shaffer is Professor of Agricultural and Applied Economics and Community Development Economist, University of Wisconsin-Madison/Extension, where he currently directs the National Rural Economic Development Institute, which is part of the National Rural Development Partnership. He has served as Director of the University of Wisconsin-Madison/Extension Center for Community Economic Development since 1990. His extension efforts have emphasized working with Wisconsin communities to create comprehensive development strategies. He also teaches a graduate course and does research in the area of community economics. He has worked in several U.S. states and internationally on questions of local development strategies.

Donald F. Smith, Jr., is Visiting Assistant Professor and Executive Director, Center for Economic Development, H. John Heinz III School of Public Policy and Management, Carnegie Mellon University.

Leon Taylor is Assistant Professor of Economics at Tulane University in New Orleans. He has written about economic competition between jurisdictions for *Economic Development Quarterly*, the *Review of Regional Studies*, and the *Journal of Public Economics*. His research interests include public and environmental economics.

Bronwen J. Turner is Deputy Director for the Mayor's Office of Economic Development, City and County of Denver, Colorado. She has been with this office for 6 years. Her areas of responsibility include managing the business development staff in the office; redevelopment planning for Stapleton International Airport, an inner-city redevelopment site; managing target-industry marketing activities and public relations strategies; economic and fiscal impact modeling of major projects; job training strategies; and economic development policy formulation for the city. Most recently, she was a member of the team that negotiated the sale of two large parcels of land at Stapleton to two major Denver employers.

Kenneth Voytek is Chief Economist at the National Alliance of Business (NAB). Before coming to NAB, he was on the staff of the Office of Policy Research and Development of the Department of Housing and Urban Development. He was previously with the Michigan Department of Commerce and, before that, the Urban Institute. He is the author of *Local Economic Analysis: Using Analysis to Guide Local Strategic Planning* (with Mary McLean).

Carol Waldrop is Principal of Carol Waldrop & Associates and specializes in providing project management and public polity interface services for complex public- and private-sector development projects and joint ventures. For the last 4 years, she has served as Project Director in a consulting capacity to the city of Los Angeles in the preparation and implementation of the City of Los Angeles Citywide General Plan Framework, a comprehensive plan designed to stimulate development in targeted urban growth areas appropriate for increased density and concentration of uses, particularly related to transportation nodes, and to direct public expenditures and other regulatory incentives to encourage economic revitalization. She also served in a project management capacity to the City of Los Angeles Community Development Department and the Coalition for Neighborhood Developers, a coalition of nonprofit community-based organizations, in their efforts to secure a federal empowerment zone designation.

Wim Wiewel is Dean of the College of Urban Planning and Public Affairs and Professor of Urban Planning and Policy at the University of Illinois at Chicago (UIC). He formerly served as the Special Assistant to the Chancellor, in charge of planning and implementing the Great Cities Initiative, UIC's commitment to metropolitan Chicago. From 1983 through 1993, he directed UIC's Center for Urban Economic Development, which provides technical assistance and research on economic development issues to community organizations and local governments. He has published many articles and reports on community, urban, and regional economic development and is co-editor of *Harold Washington and the Neighborhoods: Progressive City Government in Chicago, 1983-1987* and *Challenging Uneven Development: An Urban Agenda for the 1990s*, both published in 1991.